RADIOECOLOGY
OF AQUATIC ORGANISMS

The accumulation and biological effect of
radioactive substances

G. G. Polikarpov

Radioecology of Aquatic Organisms

The accumulation and biological effect of radioactive substances

Translated from the Russian by
SCRIPTA TECHNICA, LTD.

English translation edited by
VINCENT SCHULTZ AND ALFRED W. KLEMENT, JR.

1966
REINHOLD BOOK DIVISION—NEW YORK

Publishers:

NORTH-HOLLAND PUBLISHING CO. — AMSTERDAM

Sole distributors for U.S.A. and Canada:

REINHOLD BOOK DIVISION — NEW YORK

Printed in The Netherlands

TABLE OF CONTENTS

V

Part III. The effect of nuclear radiation on marine organisms

Part IV. Conclusions

Appendix I

Appendix II

LIST OF FIGURES AND TABLES

Figure

EDITORS' PREFACE TO ENGLISH EDITION

In *Radioecology of Aquatic Organisms* G.G. Polikarpov has produced the first monographic account of marine radioecology with emphasis on the interaction of marine organisms with environmental radioactivity and has formulated concepts and problems of this relatively new and important field of knowledge. In addition, considerable reference is made to studies on freshwater organisms, Russian and non-Russian.

The book contains a summary of information on the concentration factors of artificial radioisotopes of the eight groups of the periodic system of chemical elements in marine and freshwater plants and animals, and also gives information on the radiosensitivity of aquatic organisms. Upon learning of plans to translate this book, Dr. Polikarpov with considerable effort kindly undertook to update references and text material. On the basis of current knowledge on the concentration of radionuclides in the hydrosphere, on the maximum permissible concentrations of radionuclides in sea water for man, and on the radiosensitivity of marine organisms, the author concludes that further contamination of the seas and oceans with radioactivity is inadmissible.

It is of interest to note that for the most part radioecologists have not taken advantage of modern statistical procedures. Thus, many results are merely point estimates derived under specific conditions and thus have questionable value. Answers to questions posed in Dr. Polikarpov's excellent monograph will undoubtedly be derived only by use of this powerful tool.

The book will be of considerable value to persons concerned with radiation biology, hydrobiology, oceanography, and sanitary engineering. It is of direct concern to ecologists, health physicists, and administrators responsible for or concerned with radioactive fallout and the disposal of radioactive wastes.

Further, this English translation focuses attention on Russian activities in this field of endeavor and serves for ecologists as a companion reference to two other publications: 1) *Radioecology* (V. Schultz and A. W. Klement, Jr., editors), published in 1963 by the Reinhold Publishing Corporation and The

American Institute of Biological Sciences, and 2) Bibliographical Series No. 5: *Disposal of radioactive waste into marine and fresh waters*, published in 1962 by the International Atomic Energy Agency.

Even though quite time consuming, editing the English translation has been a distinct pleasure, as we have had the assistance of associates who contributed to the task even though our requests for assistance were accompanied by a note of urgency.

We are especially indebted to Dr. G. G. Polikarpov who, despite a distance barrier, cooperated in all our idiosyncrasies. His independent editing of the page proofs was quite intense and contributed much to the quality of the English edition of this book. Although his transliteration of some Russian names differed from ours, we have chosen to retain our initial spellings. These differences are as follows, where Dr. Polikarpov's spelling is enclosed by parentheses: Gus'yev (Gusev), Parchevskiy (Parchevsky), Skul'skiy (Skulsciy), Timofeyev-Resovskiy (Timofeeff-Ressovskiy), Yemel'-yanov (Emel'yanov), and Zaytsev (Zaitsev). We only hope that our attempt to bring his monograph to the attention of the English speaking peoples will please him.

To an unknown translator employed by Scripta Technica Ltd., London, we give thanks for his excellent translation of the Russian edition into English.

Believing that this monograph will be an invaluable reference for many years, we have expended considerable effort expanding source material on references listed by the author. Dr. Charles F. Lytle, Pennsylvania State University, in spite of heavy academic commitments during the editing period, made a major contribution by editing the references with Russian titles.

We have listed articles by Russian scientists appearing in non-Russian periodicals under the reference section *d. Non-Russian titles*, e.g., Fedorov, 1964. Similarly in a very few cases articles by non-Russians were translated into Russian. Where we were unable to locate the original source of these translations, we listed the reference under section *c. Russian titles*, e.g., Matsuda and Hayashi, 1959.

The assistance of John K. Hartsock in translating specific Russian words is gratefully acknowledged.

Vincent Schultz   Alfred W. Klement, Jr.
Department of Zoology   *Nevada Operations Office*
Washington State University   *U.S. Atomic Energy Commission*

March 1, 1966

EDITOR'S PREFACE TO FIRST RUSSIAN EDITION

There has been and continues to be contamination of the environment by the fission products of heavy elements as a result of the nuclear weapon tests carried out before signing of the Moscow agreement banning such tests in space, the atmosphere, and under the water, and also as a result of the operations of the nuclear industry. The seas and oceans which occupy the greater part of the surface of our planet have been significantly affected by contamination. The oceans are an important source of human food supply and of raw materials, in addition to their role in transportation; towns and other centers of habitation are situated along coasts.

Although short-lived fission products are gradually excluded from the natural cycle, the long-lived products (including strontium-90, cesium-137 and cerium-144) remain and will continue to participate in the cycle of matter in the atmosphere, the soil and waters for some long time, which will be mainly dependent on the half-life of each nuclide.

Fission products may enter the oceans in a variety of ways.

Even now that nuclear weapon tests have ceased, fallout will continue for many years owing to the existence of a stratospheric reservoir of strontium-90 and cesium-137. Moreover, some countries (the United States and the United Kingdom) use the seas and oceans to dump some of the radioactive wastes of the nuclear industry.

Shipping hazards to nuclear-powered vessels (tankers, submarines, ice-breakers) may lead to significant radiocontamination of the water, especially in small and closed basins.

Radioactive substances may have an adverse effect on biocoenoses. The life cycle is of longer duration in the sea than on the land. The food chain incorporates phytoplankton, zooplankton (small crustaceans and other animal organisms) and, finally, fishes. If the distribution of radioactive substances in all links of the chain is established it will be possible to study the features of migration. It is known that many chemical elements, including radionuclides, have concentration factors of several thousand or more in marine organisms. Migration of radioactive substances may, moreover, be promoted by the vertical and horizontal migrations of the organisms, and some radionuclides may be transported from the depths into the upper layer in greater amounts than are transported as a result of physical circulation

Data on the migration of fission products from the nuclear weapon tests in the area of the Marshall Islands in 1956 were presented to the Ninth Pacific Science Congress at Bangkok (MOISEYEV [1958]). This information revealed that radioactive contamination was not confined to such wide-ranging fishes as sharks, tunny, and swordfish, but also affected the fishes of coastal areas, including reef bass and pollack. This shows that there is a risk of the radio-contamination of man by consumption of sea fishes containing radioactive substances.

Different organisms do not respond identically to the effect of radiation. Plant organisms and invertebrates are, for example, highly radioresistant forms, whereas it has been established that the developing eggs of fishes and amphibians are very radiosensitive.

The radioecology of marine organisms is that branch of radiobiology concerned with the effect of radioactive substances on oceanic and marine biocoenoses.

There is not enough information in the literature on the significance of artificial radioactive substances as an ecological factor in the life of marine organisms, and especially their young. There is need for a detailed study of the role of somatic and genetic effects in the vital processes of fishes and other marine organisms, and for data on the concentration factors of radio-nuclides in them.

All the factual material is systematically set out and scientifically evaluated in this book. Its author, G. G. Polikarpov, has played a large part in the significant contribution of Soviet scientists to the development of this new branch of science – the radioecology of seas and oceans.

The book will certainly be a valuable reference source for radiobiologists, oceanologists, ichthyologists, geophysicists, radiochemists and other specialists concerned with the radioactivity of the environment, and will have a great influence on the further development of radioecology.

Professor V. P. Shvedov

AUTHOR'S PREFACE TO REVISED EDITION

It was not long after the discovery of radioactivity had heralded the birth of nuclear physics and radiochemistry that radiobiology began to develop. Further progress in the science of the atomic nucleus has led to the creation of artificially radioactive substances; their escape from human control and appearance in the biosphere have led to the development of radioecology (KUZIN and PEREDEL'SKIY [1956], ODUM [1956]), which is a new facit of radiobiology having connections with ecology and biogeochemistry.

Radioecology, which is concerned with the interaction between a radioactive medium and living organisms, has been faced from the beginning with the difficult task of providing a theoretical basis for prediction of the biological consequences of radioactive contamination of the natural environment, and thus for all necessary safety precautions and for recommendations and measures to reduce and eliminate radiation hazard in the biosphere.

It is understandable that the first studies of ecological aspects of radiobiology were undertaken in the USSR and the USA – the principal nuclear powers. The major centers for such research in the Soviet Union, centers whose research staffs are known for their many publications, are headed by V. M. Klechkovskiy (KLECHKOVSKIY [1956], KLECHKOVSKIY and GULYAKIN [1958], KLECHKOVSKIY et al. [1959], GULYAKIN and SELETKOVA [1954], GULYAKIN and YUDINTSEVA [1956, 1958, 1962]); V. P. Shvedov (SHVEDOV [1959], SHVEDOV and SHIROKOV [1962], GEDEONOV [1957], SHVEDOV et al. [1958], SHVEDOV et al. [1959], SHVEDOV et al. [1960]); I. N. Verkhovskaya (VAVILOV et al. [1963], POPOVA and KODANEVA [1965]); Ye. M. Kreps (KREPS [1959], BOGOROV and KREPS [1958], BUROVINA et al. [1962]); N. V. Timofeyev-Resovskiy (TIMOFEYEV-RESOVSKIY [1957, 1962], TIMOFEYEVA-RESOVSKAYA [1956], TIMOFEYEV-RESOVSKIY and TIMOFEYEVA-RESOVSKAYA [1959], TIMOFEYEV-RESOVSKIY, TIMOFEYEVA-RESOVSKAYA et al. [1960], KULIKOV [1957], GILEVA [1960], ZHADIN et al. [1959], GETSOVA [1960]).

Work of considerable fundamental and methodological importance has also been carried out at the Institute of Biophysics and the Institute of Biochemistry of the USSR Academy of Sciences under the direction of

A.M.Kuzin, I.N.Verkhovskaya and A.A.Peredel'skiy (VERKHOVSKAYA *et al.* [1955], KUZIN [1959, 1960, 1963, 1964], KUZIN and PEREDEL'SKIY [1956], PEREDEL'SKIY [1957a,b, 1958, 1964], PEREDEL'SKIY and BOGATYREV [1959]); and under leading scholars at the Institute of Geochemistry and Analytical Chemistry (USSR Academy of Sciences), the Institute of Oceanology (USSR Academy of Sciences), Moscow University and other scientific establishments: BARANOV [1939, 1955], VINOGRADOV [1953], YEMEL'YANOV [1958, 1962], ZENKEVICH [1960, 1963], LEBEDINSKIY [1957], LEYPUNSKIY [1957, 1958], POPOV *et al.* [1962, 1963, 1964].

The growing need to develop the ecological aspects of radiobiological research in seas and oceans was demonstrated by large groups of Japanese scientists in 1956 and American scientists in 1957, by VODYANITSKIY [1958], ZENKEVICH and SHCHERBAKOV [1960], MOISEYEV [1957, 1958], SHVEDOV and SHIROKOV [1962], FONTAINE [1956, 1959], and others.

The development of marine radioecology has become a particularly pressing matter as a result of the extremely intensive radioactive contamination of sea and ocean waters by past underwater nuclear weapon tests, the continuing global fallout of radioactive aerosols, the systematic dumping of liquid and solid radioactive wastes on the sea bed and their discharge in surface waters, and the 'normal' functioning of nuclear-powered ships and submarines.

There is now a body of information on radioecological research at the marine laboratories of the USSR, the USA, the UK, Japan, France, Sweden, Monaco, Italy and some other countries. It is hoped that publication of the present book will remedy the lack of a general account of marine radioecology and bring together the rather widely scattered information.

The author's aims in writing the book were: 1) to set out his views on the position of radioecology in relation to the other sciences, 2) to give information on artificial radioactive substances as an environmental factor in the hydrosphere, 3) to outline the principles and scope of marine radioecology, and 4) to define the initial concepts and relations underlying study of the two aspects of interaction of a radioactive environment with marine organisms – accumulation and effect. Much attention has been paid to an analysis of summary information on the concentration factors of important radionuclides of the eight groups of the periodic system in marine organisms and to the comparative radiosensitivity of aquatic organisms. This is the first attempt to provide a general account of the interaction between marine organisms and radioactive substances. Some, admittedly incomplete information on the radioecology of freshwater organisms is also given for purposes of comparison.

The information given in the book is an argument against the dumping of nuclear wastes in seas and oceans, and against nuclear and thermonuclear weapon tests, because any increase of the radioactive contamination of seawater is a threat to the resources of the sea fisheries. The 1963 Moscow Treaty on the prohibition of nuclear explosions in space, in the atmosphere, and under water, which should have put a stop to further entry of radioactive substances into the environment by nuclear detonations, was a great triumph for human reason. The cumulative contamination of the earth's surface will increase to a maximum by about 1970 (in terms of strontium-90 and cesium-137), because of the existence of a vast stratospheric reservoir of long-lived radioactive substances.

According to the information given in a report of the United Nations Scientific Committee on the Effects of Atomic Radiation (United Nations [1962]) cumulative deposits of strontium-90 would have reached their maximum in 1964 if tests had been discontinued at the end of 1958. If nuclear explosions had ceased in 1962 strontium contamination of the earth's surface should have reached its maximum level by approximately 1970 (fig. 1). After this the strontium-90 content in the biosphere should decline gradually to a half by the year 2000 and to one thousandth within approximately three centuries.

Unfortunately, not all the powers that possess nuclear weapons or that propose to manufacture them in the near future have become signatories to the Moscow Treaty. Moreover, other channels for the radiocontamination of the hydrosphere remain open; these channels are the disposal of nuclear waste from power stations and ships in the sea, and accidents to the nuclear reactors of nuclear-powered ships and installations on the coast. It is common knowledge that the Soviet Union does not dump radioactive waste in this manner, and has always insisted at international conferences on the need to prohibit radioactive contamination of the seas (YEMEL'YANOV [1962, 1964]).

The existence of a radioecological factor, whether natural or artificial in origin, in waters is the basis of the radioecology of aquatic organisms, the scope of which is to study the nature of the interaction between aquatic organisms and the radioactive aquatic environment. The isotopic composition of the environment and the concentrations of radioactive substances in it are modified by accumulation of radionuclides by plant and animal populations, by migrations, and by release of radionuclides into the water in the normal course of metabolism. Radionuclides, in their turn, affect the activity and reproduction of aquatic organisms as external emitters and especially when incorporated into tissues; this effect arises from emission of ionizing radiation and from nuclear transformation of atoms of elements incorporated in key structures of the biological substrate into atoms of other elements.

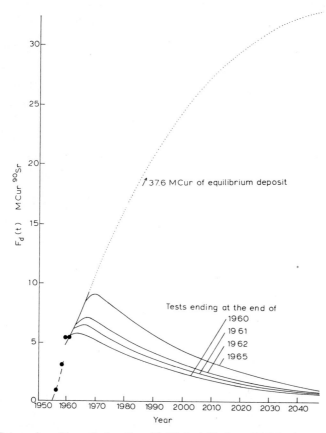

Fig. 1. Future deposition of strontium-90 (United Nations [1962]). (●) data for the soil; dotted line denotes 6 MCur ^{90}Sr/5 yr.

Marine and freshwater radioecology, which are both at a stage of rapid development, are based on the same general principles but may differ in methods and in the solutions of specific problems by virtue of the fact that they are concerned with waters that are not the same in salinity or in constancy of composition. Moreover, freshwaters (streams, lakes and ponds) are small by comparison with seas and oceans, and their populations are different.

The radioecology of aquatic organisms is related to that of terrestrial organisms by the interaction between the terrestrial and aquatic biota and by temporary or permanent transfer from one environment to the other, as well as by certain features in common in the accumulation, distribution and exchange of radionuclides, and in the biological effect of radiation. The specific features of the environment are, however, responsible for the great differences that exist between many radioecological processes in water and on land. Whereas the migration of radionuclides along food chains is, apparently, of great significance on land, direct uptake of radionuclides from the water is perhaps the dominant factor in waters, especially in the sea.

At the present stage of development of the radioecology of aquatic organisms attention is centered on the following topics.

1) Kinetic and physico-chemical aspects. Although the importance of this area of investigation is obvious, developments are recent, and very little has as yet been done. The rate at which information is accumulated must therefore be increased.

2) The capacity of aquatic organisms to concentrate radionuclides, expressed by Soviet and non-Soviet radioecologists alike as the concentration factor, which is the ratio of the concentration of a radionuclide in the organism to that in the water.

When estimating the role played by aquatic organisms in the migration of radionuclides and when calculating absorbed doses from incorporated emitters, concentration factors must be expressed in terms of wet (live) weight. At a symposium on the radioecology of aquatic organisms (see *Problems of hydrobiology*, All-Union Hydrobiological Society [1965]) radioecologists reached complete agreement on the expression of concentration factors and decided that they should be based on wet (live) weight. It is only occasionally that it may be more appropriate to employ concentration factors related to ash weight to convert from radionuclide concentration *in vivo* to concentration in the water (for example, in calculations relating to biological concentrators). A concentration factor of, for example, 1000 units becomes one of 100000 units on an ash weight basis, and 100000 times less ash than water are needed for determination of the radionuclide. This greatly simplifies time-consuming analyses of large quantities of water (especially sea-water).

The concentration factor is the central concept and the quantitative basis of radioecology as applied to aquatic organisms. If reliably established concentration factors are available for radionuclides (elements), it is possible to convert from the concentration in aquatic organisms (especially biological concentrators or biotic indicators) to concentration in the water, to assess the part played by these organisms in extracting radioactive substances from the environment, and to use existing mathematical formulae to calculate many parameters. More needs to be done to establish the concentration factors of the most important radionuclides and their stable carriers in abundant species of aquatic organisms and in those commercially exploited. The production of tables of concentration factors is already overdue. When compiling these tables particular attention should be paid to the biological concentrators (the major accumulators) of each radionuclide. The concept of a biological concentrator may include organs (tissues), species, and even biocoenoses. Radioecological biotic indicators may serve as 'compasses' in the location of naturally radioactive waters, ore deposits and radioactive contamination. Stress should, however, also be laid on the important point made by some authors that attention must also be paid to those aquatic organisms that are poor concentrators of radionuclides.

There is, therefore, much important work to be done in establishing reliable maximum concentration factors of radionuclides for as many aquatic organisms (organs and tissues), populations and biocoenoses as possible. Once reliable concentration factors have been established, time spent on analysis in radioecology and radiation hygiene may be greatly reduced and standard procedures may be programmed for computer solution.

3) The role of aquatic organisms in the migration and distribution of natural and artificial radionuclides. In contrast to terrestrial organisms, aquatic organisms for the most part concentrate more radionuclides by direct absorption from the water than by feeding. In other words it is, apparently, possible to disregard the part played by food chains in the migration of radionuclides in a radioactive aquatic environment. Food relations are of importance in this respect mostly in 'pure' water.

If consideration is given to the biomass of aquatic organisms it is a simple matter to calculate that, at any given moment, aquatic organisms will contain only an insignificant proportion of the total amount of radionuclides present in the environment and these organisms. This proportion is considerably less than 1 % for marine organisms, but may reach a few per cent and, exceptionally, even higher for freshwater organisms.

As generation succeeds generation there is continual displacement (biocirculation) of radionuclides from layer to layer in the water, from the water to the bottom, and from the benthos into the surface layers of the water. The rate of this displacement is related to the rate of reproduction.

One of the major tasks facing radioecologists is to investigate the role of the hyponeuston, plankton, nekton and benthos in the migration and distribution of radioactive substances in sea-water and freshwater. It is of the greatest importance to trace the main pathways of biocirculation and to establish the major depots for the various groups of radionuclides. Information of this nature may find practical application in forecasting the course and rate of deactivation of contaminated waters and in hypotheses concerning the formation of deposits of natural radioactive elements of organic origin.

4) The radioactivity of aquatic organisms under natural conditions. Far too little attention has been paid to this aspect. Studies of the radioecology of natural radioactive elements under natural conditions are of great importance. In the future, attention must be paid to determination of radionuclides and their carriers in aquatic organisms and in the environment with allowance for a number of variables. There is a great need for simultaneous determinations of the concentration factors of radioactive and stable isotopes of elements to establish the size of the exchange fund for each radionuclide.

It is the duty of radioecologists to ensure a continuous flow of information on the actual radioecological situation in different waters and to establish the underlying patterns. This is essential for the proper protection of nature from the harmful consequences of radioactive contamination of the environment.

5) The radiosensitivity of aquatic organisms. This is the youngest branch of radioecology, although it ought, by virtue of its extreme importance, to be one of the long-established and well-developed branches of aquatic radioecology.

The major research tasks include: the search for those links in the biological structure of the hydrosphere that are readily damaged by radiation, and for the most sensitive species (stages of development) and biocoenoses, clarification of the part played by natural radioactivity (especially in radioactive zones) in the life of aquatic organisms and their biocoenoses, study of the stimulating and harmful effects of small doses of ionizing radiation and the features of radiation injury (genetic and somatic damage).

Although these are major problems they must be solved rapidly and efficiently.

6) The effect of radionuclides on populations and biocoenoses. This is, perhaps, the most complicated and vitally important aspect of radioecology and, at the same time, the least investigated. In the radioecology of the hyponeuston the first step has been taken in prediction of the possible consequences of chronic damage to the pelagic eggs of sea fishes. Thus, if 10% of the eggs are damaged in a number of generations of the mullet, the anchovy and the horse-mackerel, the numbers of these fishes may be halved in the ensuing decades. The possible effect of a radioecological factor on the struc-

ture of aquatic biocoenoses is a major and involved problem intimately related to the protection of nature and the conservation of natural resources. Study and prediction of the threat which radiocontamination of inland waters and seas represents to their biological and commercial productivity is the task of applied radiation biocoenology.

The fastest possible development of all the nuclear sciences, including radioecology, is essential in this nuclear age. A general theory of radioecological processes is a vital necessity. The practical applications of science are impossible without adequate theoretical groundwork. To provide this groundwork all radioecologists must coordinate their efforts and ensure the ready exchange of experience and information. The evidence of progress in this relation are the International Symposium on radioecological concentration processes (Stockholm, April 1966) and the Symposium on radioecology and radioactivity of the ocean at the Second International Oceanographic Congress (Moscow, June 1966).

Because of the support of the All-Union and the Ukrainian Academies of Sciences and the Institute of the Biology of the Southern Seas (formerly the Kovalevskiy Biological Station at Sevastopol), the author was able to set up a radiobiological laboratory – the first marine laboratory of its type in the Soviet Union and in the Mediterranean basin – and to carry out research on marine radioecology. The work of the Radiobiological Laboratory (now the Radiobiological Department) has been and continues to be greatly assisted by the scientific committees on radiobiology by the Academies of Science of the USSR and the Ukrainian Soviet Republic, the All-Union Hydrobiological Society of the USSR Academy of Sciences, the USSR State Committee for the Use of Atomic Energy, and the Oceanographic Commission of the USSR Academy of Sciences.

The author wishes to thank Dr. I. N. Verkhovskaya, the head of the Central Isotope Laboratory of the Institute of Biochemistry (Acad. Sci. USSR), Dr. V. I. Korogodin, the head of the Laboratory of General Radiobiology at the Institute of Medical Radiology (Acad. Med. Sci. USSR) and Candidate of Chemistry I. A. Skul'skiy, the head of the Laboratory of Natural Radioactivity at the Institute of Evolutionary Physiology (Acad. Sci. USSR) for careful analysis of the manuscript, and for valuable advice and comments and to Professor A. A. Strelkov (Zoological Institute Acad. Sci. USSR) for the careful examination of species names.

The author would also like to take this opportunity of heartily thanking his teachers Professors Drs. I. N. Verkhovskaya, V. S. Yelpat'yevskiy, B. N. Tarusov and N. V. Timofeyev-Resovskiy.

The author is indebted to many specialists from the Departments of Benthos, Plankton and Necton at the Institute for undertaking the identification of marine organisms, for giving advice where necessary, and taking part in a number of projects, and also to colleagues in the Department of Radiobiology who gave great assistance in planning the arrangement of the book. Dr. Yu. P. Zaitsev, Head of the Department of Hyponeuston at the Institute (Odessa Branch) made a decisive contribution to the creation of the new important field – radioecology of hyponeuston. It should also be mentioned that the collaboration of scientific colleagues abroad has made it possible to assemble fairly complete material on the results of studies on the radioecology of marine organisms and on related matters.

The author is deeply grateful to Professor V. P. Shvedov, the scientific editor of the first edition of this book, for his great labors on the manuscript and to Dr. V. Schultz and Dr. A. W. Klement, Jr., the scientific editors of the English edition of this book, for their admirable work on the text and the references, and their very kind attention to the author's requests.

In the English edition attention was given to the published critical remarks by V. G. Bogorov and N. I. Popov (BOGOROV and POPOV [1965]), as well as to remarks by Dr. Glaser in 1964 and D. G. Fleishman in 1965. The author expresses his gratitude to them for their concern in improving the book and in future development of aquatic radioecology.

The author is fully aware of the extreme complexity and considerable novelty of much of his subject matter, and would welcome readers' comments, which should be addressed to:

G. G. Polikarpov,
Department of Radiobiology,
The A. O. Kovalevskiy Institute of the
    Biology of the Southern Seas,
Academy of Sciences of the Ukrainian
    Soviet Socialist Republic,
Sevastopol.

PART I

Introduction

CHAPTER 1

THE SUBJECT MATTER
AND SCOPE OF MARINE RADIOECOLOGY

Until the discovery of artificial radioactivity, or to be more precise, until the development of the nuclear power industry and the production of nuclear weapons, radiobiologists concentrated on explanation of the mechanisms and features of the biological effect of ionizing radiation from natural radioactive substances and from X-ray tubes, beginning with the macromolecules of the biological substrate and ending with multicellular animals and plants. Only a few papers were concerned with ecological aspects of the effect of natural radioactive elements, and especially radium, in mine workings (STOKLASA and PĚNKAVA [1932]).

The production of vast quantities of radioactive wastes by the nuclear power industry and the radioactive fallout that has followed nuclear weapon explosions has introduced into the biosphere the most effective and harmful of all the ecological factors created by man – the ionizing radiation of artificial radioactive substances.

The specific feature that distinguishes the biological effect of ionizing radiation from that of chemical substances is the absence of a threshold, since the latter become effective above a fairly high 'threshold dose', which differs little from the 100% effective dose. In other words, living systems are affected by any radiation dose, however small (fig.2). The effect of incorporated radionuclides* on the organism will, of course, be related to their chemical nature, and to the type and energy of the radiation.

It has been shown by measurement and calculation that, prior to November 1958, 40% of the total content of strontium-90 and cesium-137 in the biosphere had reached the earth's surface as global fallout, and an equal quantity of these radionuclides had reached the surface as local fallout formed as a result of all the nuclear tests before November 1958, and that the remaining 20% of the total amount of strontium-90 and cesium-137 was then still in the stratosphere (GUS'YEV [1961], Joint Committee on Atomic Energy, Congress of the U.S. [1959b]).

---

* The term 'radionuclides' denotes the radioisotopes of various chemical elements.

3

According to Pritchard the land received approximately 1.8×10^6 Cur of strontium-90 and the seas and oceans received 2×10^6 Cur in addition to the 3×10^6 Cur formed in the Pacific Ocean as a result of underwater and above-water explosions, making a total of 5×10^6 Cur of strontium-90 (PRITCHARD [1961]). SHVEDOV, ZHILKINA *et al.* [1959] established, however, that radio-active fallout per unit of surface of the sea was at least one and a half times to twice as high as on an equal area of land. The different effectiveness of this type of fallout was taken into account by SEREDA [1962], who calculated that

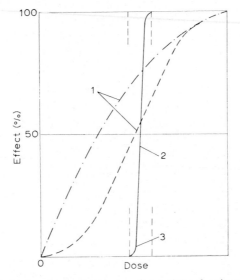

Fig. 2. Diagram illustrating dose-effect curves for action of poisons and of radiation (ZIMMER [1961]). (1) radiation, (2) chemical agent, (3) threshold.

the oceans received 7.6×10^6 Cur of strontium-90 and 13×10^6 Cur of cesium-137 as a result of the tests between 1954 and 1958, i.e., an incomparably higher proportion than that received by the land. Information concerning the probable total strontium-90 and cesium-137 content in the hydrosphere is summarized in table 1.

The data in table 2 show that strontium-90 concentration in sea-water is higher in the vicinity of nuclear weapon tests and where industrial radioactive wastes are discharged. Following the hydrogen bomb explosions at Bikini Atoll in 1954 strontium-90 levels in the northern Pacific reached nearly 10^{-11} Cur/l. The concentration was approximately the same in the surface

Table 1

Total cumulative strontium-90 and cesium-137 fallout in 1963 in the hydrosphere and for the globe as a whole (POLIKARPOV [1963])

Radionuclides	Cumulative fallout in megacuries					
	For the whole globe*	With allowance for differences in density of fallout				
		For the whole globe**	In seas and oceans**	With allowance for local fallout		
				For the whole globe	In seas and oceans	
					A	B
^{90}Sr	7	9	7	12	9	10
^{137}Cs	12	16	12	21	16	17
^{90}Sr + ^{137}Cs	19	25	19	33	25	27

(A) Taking 30% local fallout as a basis.
(B) PRITCHARD [1961].
* Report of UN Scientific Committee on the Effects of Atomic Radiation (United Nations [1962]).
** SEREDA [1962].

waters of the Central Pacific in 1956. In 1960 and 1961 a liter of water from the Pacific Ocean and the Irish Sea contained approximately 10^{-12} and even 10^{-11} Cur of strontium-90. Strontium-90 levels were one order of magnitude lower (10^{-13} Cur/l) in the surface layer of the Atlantic and the Black Sea in these years.

Cesium-137 concentration varied between 10^{-14} and 10^{-12} Cur/l in the Pacific Ocean between 1956 and 1960, and between 10^{-12} and 10^{-11} Cur/l in the Irish Sea in 1959 and 1960. In 1961 the content of this radionuclide in the Atlantic Ocean was 10^{-14} to 10^{-13} Cur/l.

In the Irish Sea in 1959 cerium-144 activity reached 10^{-10} Cur/l zirconium-95-niobium-95 activity reached 10^{-10} Cur/l and ruthenium-106 activity reached 10^{-9} Cur/l.

There is a slowly but steadily increasing concentration of long-lived fission products in sea-water (MAUCHLINE and TEMPLETON [1964]).

Let us take one example of the negative effects of nuclear tests on fisheries. After the explosion of the hydrogen bomb at Bikini Atoll in March 1956 the Japanese authorities rejected and destroyed between March and December of the same year about 4500 net registered tonnage of tunny on 856 of the 1200 vessels in the Japanese tunny fleet (fig. 3). Total losses to the Japanese fishing industry from this source up to 28 September 1954 were 2040000 yen.

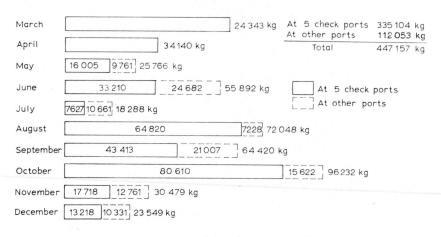

(a)

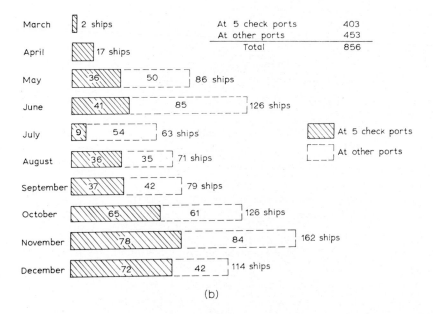

(b)

Fig. 3. Losses of the Japanese fishing industry as a result of the thermonuclear test at Bikini Atoll in 1954 (MATSUDA and HAYASHI [1959]). (a) catches (kg) destroyed because of extensive radiocontamination; (b) number of fishing vessels whose owners suffered losses as a result of radiocontamination of fish.

Consumption of sea fish contaminated by radioactive substances was followed by a number of the symptoms of radiation sickness (MATSUDA and HAYASHI [1959]).

Accumulation of radionuclides by marine organisms is intensively studied in regions near nuclear explosions (BERNER et al. [1962], CHAKRAVARTI and HELD [1960], FUKAI and MEINKE [1962], NAKAI et al. [1960/1961], YAMAGATA [1959], YAMAGATA and MATSUDA [1959a, b]) and in seas far-away from the places of nuclear detonations (CERRAI et al. [1962], POLIKARPOV and PARCHEVSKIY [1961], PARCHEVSKIY [1964a, b, 1965a], CHESSELET and LALOU [1964], COHEN and GAILLEDREAU [1960]).

Table 2

Levels of artificial radioactivity in seas and oceans

Radionuclide and time of analysis	Water	Activity (Cur/l)	Source
Strontium-90			
1954 ⎧ III–IV	North Pacific	4×10^{-13}	a
V–VI	North Pacific	0.7×10^{-11}	a
⎩ VIII–IX	North Pacific	3×10^{-14}–2×10^{-13}	a
1956	East China Sea, depth 50 m	1×10^{-13}	b
1956	Sea of Japan off Shimane, depth 5 m	0.6×10^{-12}	b
1956	North Pacific off Hokkaido:		
	depth 15 m	2×10^{-13}	b
	depth 50 m	3×10^{-14}	b
1956	Central Pacific, at surface	0.8×10^{-11}	b
1957	Baltic Sea	1×10^{-13}	c
1957 ⎧ III–XII	Atlantic Ocean, near Brazil, at surface	$(1.8–4.9) \times 10^{-14}$	d
VII	Same locality	$(3.5–4.1) \times 10^{-14}$	d
⎩ VII	Same locality, depth 100 m	3.7×10^{-14}–0.6×10^{-13}	d
1958	Pacific Ocean along Japanese seaboard	1.3×10^{-12}	e
1959	Pacific Ocean along Japanese seaboard	$(0.6–1.9) \times 10^{-12}$	e
1959	Atlantic Ocean, at surface	1×10^{-13}	f
Before 1959	Pacific Ocean, equatorial area	2.3×10^{-12}*	g
1959	Irish Sea	0.6×10^{-12}–2×10^{-11}	h
1959 ⎧ VIII–IX	Black Sea, Cape Kikkeneiz, surface layer	1.2×10^{-13}–4×10^{-14}	i, j
IX–X	North Atlantic, surface layer	4.10^{-14}–1×10^{-13}	k, l
⎩ X, XII	Indian Ocean, surface layer of entire ocean	5.8×10^{-14}–2.2×10^{-13}	k, l

* Mixture of strontium-89 and strontium-90-yttrium-90

Table 2 (continued)

Radionuclide and time of analysis	Water	Activity (Cur/l)	Source
1960 {III–IX	Black Sea, Cape Kikkeneiz, surface layer	2.9×10^{-13}	i
IX	Same locality, northern area, surface waters	2.9×10^{-13}	i
X	Same locality, surface layer	$(1.2-3.3) \times 10^{-13}$	i
	Same locality, depth 35 m	2.1×10^{-13}	i
	Same locality, depth 100 m	1.8×10^{-13}	i
	Same locality, depth 300 m	4×10^{-14}	i
1960	Irish Sea	$0.8 \times 10^{-12}-1.1 \times 10^{-11}$	h
1960	Winfrith, English coast	1.5×10^{-13}	m
1960	Atlantic Ocean	1.1×10^{-13}	n
1960 {I, II, IV	North Atlantic, surface layer	$3.10^{-14}-2.7 \times 10^{-13}$	k, l
I–IV	Indian Ocean, surface layer of entire ocean		
		$2.3 \times 10^{-14}-2 \times 10^{-13}$	l, o
X–XII	Surface layer	$4 \times 10^{-14}-2.2 \times 10^{-13}$	l, o
1960	Northwestern Pacific	3.1×10^{-12}	p
1961 {I	Black Sea, Cape Kikkeneiz, surface layer	2.8×10^{-13}	i
III	Central and western Mediterranean, surface layer		
		$4.5 \times 10^{-14}-1 \times 10^{-13}$	q
I–III	Indian Ocean, surface layer	$3 \times 10^{-14}-2.6 \times 10^{-13}$	o
III	West Pacific, surface layer	$1.6 \times 10^{-13}-5.6 \times 10^{-13}$	r
III–VI	Central Atlantic, surface layer	$1.8 \times 10^{-14}-0.8 \times 10^{-13}$	q
IV	Gulf of Guinea, surface layer	4×10^{-14}	r
V	Southern part, surface layer	$1.8 \times 10^{-13}-0.7 \times 10^{-13}$	l
VI	Northern part, surface layer	0.6×10^{-13}	l
IX	Northern part, surface layer	$4.1 \times 10^{-14}-1.8 \times 10^{-13}$	l
VI	Southern North Sea, surface layer	1×10^{-13}	r
VII	Southwestern Baltic Sea, surface layer	2×10^{-13}	r
VI–VIII	Norwegian Sea and adjacent waters, surface layer		
		$3 \times 10^{-15}-5 \times 10^{-14}$	s
Cesium-137			
1956 XI	Northwestern Pacific, at surface	$(1.3-3.3) \times 10^{-12}$	p
1958	Japanese coastal waters	$(0.7-1.5) \times 10^{-13}$	t
1959	Japanese coastal waters	3.8×10^{-13}	t
1959	Californian coastal waters	4.5×10^{-14}	u
Before 1959	Pacific Ocean, equatorial area	1.4×10^{-13}	g
1959	Irish Sea	$1.2 \times 10^{-12}-3.4 \times 10^{-11}$	h
1959 XI–XII	Californian coastal waters, at surface	$(0.8-1.6) \times 10^{-13}$	v
1960 I–IV	Californian coastal waters, at surface	$(0.5-2.3) \times 10^{-13}$	v
1960	Winfrith, English coast	1.7×10^{-13}	m
1960	Irish Sea	$(0.9-2) \times 10^{-12}$	h
1961	Atlantic Ocean, Cape of Good Hope	$2.9 \times 10^{-14}-1.6 \times 10^{-13}$	w

Table 2 (continued)

Radionuclide and time of analysis	Water	Activity (Cur/l)	Source
1961 { III	Black Sea, surface layer	5.2×10^{-13}–1×10^{-12}	x
III	Mediterranean, surface layer	1.7×10^{-13}–3.4×10^{-13}	x
III–VI	Tropical area of Atlantic Ocean, surface layer	3.6×10^{-14}–1.4×10^{-13}	x
Cerium-144			
1958 VII	Sargasso Sea	2.4×10^{-13}	y
1959	Irish Sea	1.9×10^{-12}–0.6×10^{-10}	h
1960	Irish Sea	0.0–1.9×10^{-12}	h
Ruthenium-106			
1959	Irish Sea	0.9×10^{-10}–0.5×10^{-9}	h
1960	Irish Sea	0.0–1.1×10^{-9}	h
Zirconium-95-niobium-95			
1959	Irish Sea	0.9×10^{-11}–0.7×10^{-10}	h
1960	Irish Sea	$(1.4$–$2.2) \times 10^{-11}$	h
Promethium-147			
1958 VII	Sargasso Sea	3.2×10^{-14}	y
Mixture of gamma-emitters			
1961 IV–V	Mediterranean	10^{-12}	z

a MIYAKE and SARUHASHI [1960].
b HIYAMA and ICHIKAWA [1961].
c AGNEDAL et al. [1958].
d BOWEN and SUGIHARA [1960].
e FUKAI and YAMAGATA [1962].
f SUGIHARA et al. [1959].
g HIGANO [1959].
h MAUCHLINE [1963].
i SHVEDOV, IVANOVA et al. [1962].
j SHVEDOV, YUZEFOVICH et al. [1964].
k POPOV et al. [1962].
l BARANOV, VDOVENKO et al. [1964].
m TEMPLETON [1962].
n SINOCHKIN et al. [1962].

o POPOV, ORLOV et al. [1963].
p MIYAKE et al. [1961].
q SHVEDOV, YUZEFOVICH et al. [1963].
r AZHAZHA and CHULKOV [1964].
s FEDOROV, PODYMAKHIN, KILIZHENKO et al. [1964].
t YAMAGATA and MATSUDA [1959a].
u FOLSOM and MOHANRAO [1961].
v FOLSOM et al. [1960].
w SCHROEDER and CHERRY [1962].
x CHULKOV and GORBUNOV [1963].
y BOWEN [1956].
z AUBERT et al. [1962].

Large amuunts of radioactive substances, which may enter the hydrosphere, may also arise from the peaceful exploitation of nuclear energy, from the use of nuclear explosions to produce new harbors, and from nuclear submarines and ships, both from normal functioning and especially from accidents, and from reactors situated on the coast or on major rivers discharging into the seas and oceans. Thus, the high strontium-90 level in the Irish Sea is due to the discharge since early 1952 of liquid radioactive wastes from the nuclear power station at Windscale. It was initially intended that the discharge of radioactive waste products into the Irish Sea should have an activity of several hundred curies a month. Discharges were, however, later increased to 100 Cur of beta-radioactive substances and 0.1 Cur of alpha-emitting nuclides a day, and soon after that to 1000 Cur of beta-activity daily, and several curies of alpha-activity. Ruthenium-106 accounted for 8000 curies and strontium-90 for 2800 curies a month. The permissible standard of waste discharge into the Irish Sea has now been increased to 100000 Cur a month (DUNSTER [1958], Anonymous [1960b]).

The example of the United States and the United Kingdom has been followed by France, which is now systematically discharging the untreated radioactive wastes of the nuclear industry via the Rhône into the Mediterranean in amounts slightly less than those at Windscale and Winfrith. It was also planned to bury 2000 tons of solid radioactive wastes in 6500 metal containers in a depression on the floor of the Mediterranean 80 km from the shore (PICCOTTI [1961]), but energetic protests by neighbouring countries and by French public opinion have prevented this (YEMEL'YANOV [1962]). A very significant part was played by the decisive actions of COUSTEAU [1963] the celebrated marine explorer and director of the Oceanographic Museum in Monaco.

Many other countries, including Norway, Sweden, Holland, Italy and Japan, already discharge or plan to discharge liquid radioactive wastes into the seas and oceans.

Wastes containing artificial radionuclides have been discharged for many years from the factories at Hanford on the Columbia River. The activity of the wastes disposed of was approximately 2000 Cur/day in 1957, of which chromium-51 accounted for 1000 Cur/day, and phosphorus-32 and zinc-65 for 15 Cur/day each (HEALY et al. [1958]). It is instructive to note that the wastes disposed of into the river returned unexpectedly from the ocean to the workers of the Hanford enterprises. On one occasion an increased content of zinc-65 for which there was no local explanation was discovered in an individual at Hanford. On investigation it was established that the source was oysters brought to the market from the Pacific coast (250 miles away), which contained considerable quantities of zinc-65 (table 3). The oysters had built up an accumulation of zinc-65 in their tissues 200000 times greater than the concentration in the surrounding sea-water (PERKINS *et al.* [1960]).

It was lower with zinc-65 levels in oysters in the Thames River (Connecticut) (FITZGERALD *et al.* [1962]).

In many publications great attention is paid to problems of radioactive waste disposal and to biological processes in the regions of their disposal (ROBECK *et al.* [1954], MARTIN [1957], SADDINGTON and TEMPLETON [1958], LEVI [1960], International Atomic Energy Agency [1960, 1962], Anonymous [1960a, c], WILLARD [1960], VICHNEY [1960], SCHAEFER [1961], CROSSLEY and HOWDEN [1961], GLASER [1961b], FEDOROV and PODYMAKHIN [1962], SABO and BEDROSIAN [1963], TEMPLETON [1965]), as well as to problems of contamination of water bodies and aquatic organisms by radioactive fallout (BUZZATI-TRAVERSO [1961], MORGAN and STANBURY [1961], SÁNDI [1962]), decontamination of waters and related experiments with these problems (TAYA [1956], ANGHILERI [1957, 1959, 1960a,b], FONTAINE and AEBERHARDT [1961], GRAHAM [1959]) and problems of maximum permissible concentra-

Table 3

Zinc-65 content in oysters sold in local markets (PERKINS *et al.* [1960])

Oysters	Where gathered	^{65}Zn content (10^{-12} Cur/g)
Fresh	Western seaboard of the USA	63.5
Fresh	Willapa Bay, Washington State	56.3*
Fresh	Willapa Bay, Washington State	37.8**
Fresh	South Bend, Washington State	38.1
Preserved	Western seaboard of the USA	15.1
Preserved (stewed)	Seattle, Washington State	4.13
Preserved	Gulf of Mexico	0.1
Preserved	New Orleans, Louisiana	0.30
Preserved	Biloxi, Mississippi	0.1
Fresh	Port Norris, New Jersey	0.19
Preserved	Japan	0.19
Preserved	Japan	0.2
Preserved	Japan	0.1
Preserved	Japan	0.2

* Gathered on 5 September 1959.
** Gathered on 27 January 1960.

tions of radionuclides in air and water (ADAMS *et al.* [1960], SUMMERS and GASKE [1961], TERESI and NEWCOMBE [1961], WYKER [1961]).

Since 1951 the United States has been implementing an extensive program for the burial of solid radioactive wastes in seas (Joint Committee on Atomic Energy, Congress of the U.S. [1959a]).

Fig. 4 shows the points at which containers of radionuclides and contaminated equipment were buried in coastal areas of the Atlantic Ocean and the Gulf of Mexico between 1951 and 1956. Radioactive wastes with a total activity of approximately 6000 Cur were deposited in these waters between 1951 and 1958. Up to 1960 solid wastes with an activity of 10000 Cur had been dumped into the ocean along the California coast. It has, however, been stated that there may be a ten-fold error in the estimate of this amount (Committee on Oceanography [1959b]).

It is planned to construct 10000–20000 MW power reactors in the United States by 1965, and reactors of 100000 MW or more by 1980. The total

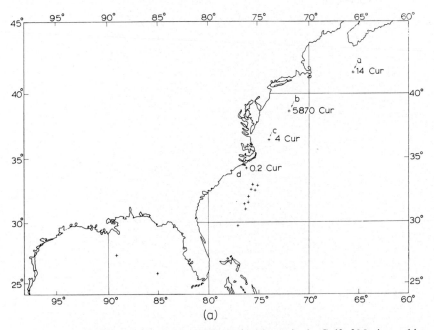

(a)

Fig. 4. Map of the burial of solid wastes by the United States in the Gulf of Mexico and in Atlantic coastal waters (Committee on Oceanography [1959b]).

(a). Areas and approximate amounts of radioactive wastes buried on the ocean floor between 1951 and 1956 (crosses without entries approximately 25 Cur).

activity of the fission products thus formed will be 10^{10} Cur by 1980, including 0.8×10^9 Cur of strontium-90. By 1980 the total volume of highly
active liquid wastes in the US will be 1.4×10^8 l (Anonymous [1960b]).

It is anticipated that there may be as many as 300 nuclear-powered ships
in the world by 1970–1975, which may be the source of 5000 Cur of radioactive wastes discharged into the sea annually from the cooling circuits, of
3400 Cur annually as a result of leakage of liquid wastes, and of 0.9×10^6
Cur annually from ion-exchange resins (Committee on the Effects of Atomic
Radiation on Oceanography and Fisheries [1959], PRITCHARD, [1960, 1961]).

It is clear that the greatest risk arises from damage to nuclear-powered
vessels, especially when in port (DuSHANE [1959]).

After the fuel elements of a 60 MW naval reactor have been functioning
for a year the total activity of the fission products is more than 10 MCur
(Committee on Oceanography [1959b]). It is therefore difficult to over-
estimate the consequences of the tragedy which befell one of the largest US

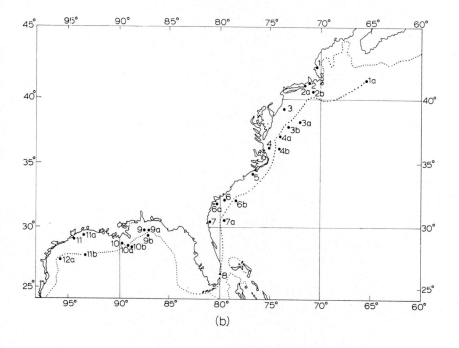

(b)

(b). Map of proposed sites for the burial of slightly active wastes (the letters with the figures
denote reserve (a) and secondary (b) areas).

nuclear submarines in the Atlantic Ocean in April 1963. Crushed by water pressure, the submarine Thresher was a source of radiocontamination whose total activity may possibly have been of the order of a million curies. The volume of sea-water needed to dilute this activity to a concentration of 10^{-12} to 10^{-10} Cur/l is 10^6 to 10^4 km^3, i.e., twice the volume of the Black and Irish Seas, respectively. Uniform mixing does not, however, occur in nature. Radioactive substances are transported in a concentrated form for great distances by streams and powerful currents.

The problem of radioactive waste disposal is not dependent on the type of cooling employed in the ship's reactor. The American regulations for disposal of radioactive wastes in the sea from nuclear-powered vessels allow for replacement of the ion-exchange resins in water demineralisers every two to three months. Contamination of the resin may be approximately 100 Cur by corrosion products, and between 100 and 1000 Cur by fission products (if part of the fuel enters the water). Approximately 0.5 m^3 of water are needed to wash the resin out of the container. These regulations allow nuclear-powered merchant vessels to dispose of up to 1000 Cur of liquid radioactive waste in the inshore zone at distances of one to 12 miles from the shoreline, and of unlimited amounts of radioactive solutions and approximately 500 Cur of contaminated resin in the open sea. The current regulations of the United States Navy (Buships Instruction 9890.5, Department of the Navy, Bureau of Ships, Washington 25, D.C., 12 May, 1958) permit submarines to dispose of resin from the demineraliser at a distance of 12 miles from the coast except in fishing waters. Naval vessels may discharge liquid radioactive wastes without restriction at a distance of more than 12 miles, and with certain restrictions in coastal waters and harbors. It is suggested that radioactive gaseous wastes should be discharged into the wind after storage for one to two days (SMITH [1959]).

The hydrosphere is, therefore, currently being contaminated as a result of mishaps and of the disposal of liquid and solid radioactive wastes by the nuclear industry and by nuclear-powered vessels.

The American ecologist E.P.Odum has stressed the need to develop the branch of knowledge which may be named radiation ecology (or radio-ecology), both in the interests of protecting people from the direct effect of radioactive contamination, and also to establish the prolonged effect of small doses of radiation on aquatic and terrestrial coenoses. Only fragmentary information is available concerning the effect of ionizing radiation on populations, biocoenoses, and the biogeochemical cycles so vitally important to the normal functioning of the ecological systems of the world (ODUM [1956]).

The first expanded definition of radioecology was given by A. A. Peredel'skiy, who suggested: 'Radioecology is the science that deals with the interrelationships in nature between a radioactive environment and organisms and their coenoses, with the migrations and concentrations of radioactive elements as a result of the activity of organisms, with the ecological chains of radioactive nutrition and indicator species pointing to the presence of considerable deposits of radioactive ores or of radiocontamination, and with quantitative and qualitative changes occurring in vegetation and in the animal population under the influence of fluctuations of radioactivity levels, both external and within the organism' (PEREDEL'SKIY [1957b], p.379).

ODUM [1959] notes in chapter 14, that since radioecology is concerned with radioactive substances, radiations, and the environment, it has two fairly different aspects: 1) study of the effect of radiation on the individual, the population, the coenosis, and the ecosystem, 2) study of the fate of radionuclides in a habitat and the role of biocoenoses and populations in the distribution of radioactivity.

The main problems to be considered at the present stage in radioecology have been defined as follows by A. M. Kuzin, the leading Soviet radiobiologist: 'Life has developed on the earth in conditions of continual exposure to low levels of ionizing radiation, referred to as natural radioactivity. The first problem in radiation ecology is to establish the contribution of natural radioactivity to the total irradiation of living organisms, to investigate the migration and concentration of natural radionuclides in the biosphere and to determine the role and significance of natural radioactivity in vital processes and in the general evolution of living matter. From the time that tests of high-yield nuclear weapons were commenced in 1954, the radioactive background on earth began to change, and matters concerned with the spread of the radioactive products of nuclear weapon tests among living organisms and with the effect of various levels of secondary irradiation on the main functions of organisms and on the state of natural biogeocoenoses became of overriding significance [this is the second problem in radioecology: radiocontamination of the environment]. Finally, the peaceful development of the nuclear industry, which involves the production of large amounts of radioactive waste, and the increasing use of radionuclides and ionizing radiation in scientific research, and in industry, medicine and agriculture give rise to a third radioecological problem: determination of the extent of the harmful effect of these factors on man and the elimination of this effect. These three problems are the basis of radiation ecology at the present stage.' (KUZIN [1964], p.360).

Radioecology has now been accorded due recognition as a separate branch of study. Such scientific journals as *Radiobiologiya* and the *Comptes Rendus des Séances de l'Académie des Sciences* devote sections to radioecology. Sections on radioecology are also to be found in various bibliographies and in the catalogues of the major scientific libraries of the United States and the Soviet Union. The All-Union conference on study of the radiocontamination of seas and oceans and their food resources held in Moscow in 1960 was concerned with radioecology (BARANOV and KHITROV [1964], VENDE and PARCHEVSKIY [1964a], PARCHEVSKIY [1964a,b], POLIKARPOV [1964]). The first symposium on radioecology in the United States was held in September 1961 (SCHULTZ and KLEMENT [1963], ANDERSON *et al.* [1963], BONHAM and WELANDER [1963], BOWEN and SUGIHARA [1963], CORCORAN and KIMBALL [1963], DAVIS [1963], FOSTER [1963], GORBMAN and JAMES [1963], LOWMAN [1963a,b], RICE [1963b,c], SEYMOUR [1963], WANGERSKY [1963]). There was a symposium on the radioecology of aquatic organisms at the First All-Union Congress of the USSR Hydrobiological Society in February 1965 (All-Union Hydrobiological Society [1965], BARINOV [1965d], VASIL'YEV and SHERSTNEV [1965], VINOGRADOVA [1965], GILEVA [1965], ZHAROVA [1965], IVANOV [1965b], LEONT'YEV and SKUL'SKIY [1965], LUBYANOV [1965], NESTEROV and SKUL'SKIY [1965], PARCHEVSKIY [1965b], POLIKARPOV [1965], ZAYTSEV [1965], ZAYTSEV and POLIKARPOV [1965], POPOVA [1965], TIMOFEYEVA [1965], FEDOROVA [1965], SHERSTNEV and VASIL'YEV [1965]). In the USSR the ecological aspects of radiobiology are allowed for in the structure of the scientific committee set up by the USSR Academy of Sciences to coordinate all aspects of research on radiobiology. All national nuclear energy committees, commissions, and agencies and international associations have radioecological departments.

The following list of titles containing the word radioecology (or radiation ecology) will suffice to show that the term is already widely employed: *Problems of radioecology* (PEREDEL'SKIY [1958]), *The radioecology of crop plants and farm animals* (YEMEL'YANOV [1958]), *Species response to radiation*; *radioecology* (BUCHSBAUM [1958]), *Radiation ecology of the front range* (HESS [1959]), *Studies in radiation ecology* (PLATT [1959]), *Terrestrial and freshwater radioecology: a selected bibliography* (KLEMENT and SCHULTZ [1962, 1963, 1964, 1965]) and *Radioecology* (SCHULTZ and KLEMENT [1963]).

What is the position of this new branch of knowledge, and to which other branches of science is it related?

A whole new superstructure of nuclear sciences has arisen on the basis of nuclear physics (table 4). Mention may be made of the following pairs of

Table 4

The place of radioecology among the sciences

Biology	Radiobiology
Ecology	*Radioecology*
Physiology, embryology,	*Radiophysiology, radioembryology,*
genetics etc.	*radiation genetics etc.*
Biochemistry	*Radiation biochemistry*
Biophysics	*Radiation biophysics*
Chemistry	Radiochemistry and radiation chemistry
Physics	Nuclear physics

nuclear and non-nuclear disciplines: geology – nuclear geology, hydrogeology – radiohydrogeology, selection – radioselection, phytopathology – radiophytopathology, toxicology – radiotoxicology, hygiene – radiation hygiene, therapy – radiotherapy, medicine – medical radiology.

The need to group the nuclear sciences together and the desirability of so doing have found expression in the publication of special periodicals covering all aspects of the nuclear sciences, and in the setting up of major nuclear centers or the creation of new nuclear institutes from existing institutes concerned with a non-nuclear discipline. The feature in common between all the nuclear sciences and their branches concerned with living or non-living matter is that they study radioactivity and its interactions with substances (including living matter at various levels of organisation, populations and biocoenoses).

Arising from the definition of radiobiology as the science of the reactions of living systems to the effect of ionizing radiation and of ways of controlling these effects, radiobiologists are specialists who study the effect of ionizing radiation on macromolecules of the biological substrate, cells, tissues and organs, entire higher organisms, and, finally, living communities. In addition to radiobiological methods and concepts, this involves in each case aspects of biophysics, biochemistry, cytology, histology, morphology, physiology and ecology. In other words, all levels of organization of living matter are represented in radiobiology. Despite the specific nature of this situation, radiobiology remains separate from the various branches of biology, first because it employs the various biological parameters (biochemical processes, cell division, death etc.) only as indicators of the effect of ionizing radiation, i.e., has as its scope the features of radiobiological reactions, and second because the methods of other disciplines are secondary to its own concepts and methods. This also applies to the use of radiobiological methods in other biological disciplines. In this case nevertheless the research is not radio-biological as in principle the work remains biochemical, cytological, embryological, etc.

Many examples could be given of the contribution of radiobiology to progress in biology as a whole, but the matter has been specially dealt with in a book by PETROV et al. [1962].

In principle it is also possible to use radiation effects on populations and biocoenoses in ecology to study 'purely' ecological features, and some steps have been taken in this direction (O'BRIEN and WOLFE [1964]).

When seen in this light radioecology is a branch not of ecology but of radiobiology. When an attempt is made to classify radioecology purely as a field of ecology it is placed with applied ecology (ODUM [1959], GROMADSKA [1961]), because it 'will not fit' into existing frameworks and categories. The very term 'radioecology' could, however, be quite justifiable rejected on the basis that there are no similar terms such as thermoecology and photoecology. Moreover, the radiological factor as an ecological factor has no equivalent in ecology.

The adaptation of living creatures and their communities to various conditions, which is one of the basic problems of ecology, is solved in a completely different manner when one is considering the biological effect of ionizing radiation. Ecology is concerned with more or less formed ecological systems, which react fairly rapidly to any change in the intensity of an operative factor; harmful factors result in extinction or reconstruction of the community, accompanied by impoverishment of the biocoenoses, whereas improvement of conditions results in an upsurge of life, increased productivity, and the addition of new species and forms. These phenomena are affected by the existence of a threshold of reactions to various environmental factors.

The action of radioactive substances has no threshold (see fig. 2), i.e., even very small doses continue to exert a biological effect. The appearance of radiomutants may lead to the gradual reconstruction and depletion of an entire community owing to the retrogressive evolution and inhibition of species.

All possible objections to the term radioecology are removed if it is treated as a branch of radiobiology at the level of populations and communities. In this sense radioecology and ecological radiobiology are synonyms.

As a term radioecology does not clash with the names of the other radiobiological sciences (see table 4). In many respects the connection between radioecology and radiobiology is similar to that between geochemical ecology and geochemistry, in which geochemical ecology is a part of geochemistry (biogeochemistry) at the ecological level (KOVAL'SKIY [1958]). KUZIN [1963] has convincingly demonstrated that radioecology is a radiobiological discipline.

In nature, ionizing radiation and artificially radioactive substances are not treated as ordinary ecological factors, but as an extremely specific radioecological factor produced and imposed on nature by man. This factor may be extended to include contamination by man of a number of areas of the earth with naturally radioactive elements in the working of ore deposits. Contamination of the environment by artificial radioactive substances is, however, of far greater significance both in terms of concentration and in the abundance of these substances in the biosphere.

Since the interaction between a radioactive environment and living organisms under natural conditions comprises, on the one hand, the accumulation and migration of radioactive substances in the biosphere, and, on the other hand, the effect on the structure and biomass of biocoenoses, it is clear that recourse will have to be had to the methods and concepts of geochemistry and biogeochemistry in explanation of a number of radioecological features. In fact, in order to estimate radiation effect it is not sufficient to know the concentration of radioactive substances in the surrounding environment. It is also important to be in possession of information concerning the capacity of the organisms to concentrate those elements whose radioisotopes are of interest in radioecology. This is the scope of biogeochemistry and biogeocoenology (VERNADSKIY [1929, 1940], SUKACHEV [1947], TIMOFEYEV-RESOVSKIY [1957], TIMOFEYEVA-RESOVSKAYA [1963], VINOGRADOV [1957b], KOVAL'SKIY [1958]). Only by close collaboration between the three sciences of radiobiology, ecology and biogeochemistry will it be possible to solve radioecological problems. Although these problems have many aspects, the main object of solution is to explain, evaluate and foresee the effect which man will exert on nature and, in the last analysis, on himself by his creation of this new ecological factor of the edaphic environment – the artificially radioactive substances in the biosphere, or to be more precise in the zoosphere, and to develop the principles upon which to base recommendations and measures to prevent radioactive substances adversely affecting the organic world.

The specific tasks of marine radioecology (POLIKARPOV [1960a, 1963, 1964, 1965]) are related to the specific features of a habitat in which the exchange of chemical substances between living creatures and the surrounding solution does not take place in the same way as on land, and the effect of radiation from without is extremely limited. The seas and oceans of the world are a single interrelated system (ZENKEVICH [1963]). Currents and mass migrations of aquatic organisms may be responsible for the movement of large quantities of radionuclides, and the contamination of further areas and layers of water. Despite this, and as already noted, a number of countries (the United Kingdom, France, the USA and others) solve the acute problem of disposing of large quantities of radioactive wastes by contaminating the seas and oceans. Moreover, nuclear reactors are being increasingly used to power ships and submarines which, like nuclear reactors on the coast, produce local contamination of the water by the disposal of radioactive wastes and especially by shipwreck.

Consideration of the radioecological problem of interaction between a radioactive environment and marine organisms must not be reduced to estimation of the direct degree of danger of the radioactive substances accumulated in foodstuffs to people. It is of the utmost importance to establish the effect of radionuclides on marine organisms and on their populations and communities, since this is, in the last analysis, bound to affect man.

Three main approaches have been followed in marine radioecology (POLIKARPOV [1960a]). The first is to determine the content of radioactive substances and their carriers in aquatic organisms and in the environment under natural conditions by chemical and radiochemical separation and concentration, or by radiometric measurement and analysis with highly efficient (gas and scintillation) low-background counters, pulse discriminators and analysers, and even optical instruments. These methods establish the absolute amounts and nuclide composition of radioactive substances in aquatic organisms and in the environment. This can be done with particular ease from the gamma spectra of gamma-radiating nuclides, by using various scintillators (VENDE and PARCHEVSKIY [1964a,b], GLAZUNOV et al. [1963], PARCHEVSKIY [1964a,b, 1965a,b]). The information thus obtained can be used, first, to explain the ways in which radionuclides migrate in the sea and, second, to calculate the doses of ionizing radiation received by organisms, and thus to estimate the criticality of the situation both for man and for aquatic organisms. Analysis of radioactive substances in sea-water is so laborious and time-consuming that there have been few determinations under natural conditions of the capacity of marine organisms to accumulate various radionuclides as measured by the concentration factor.*

---

* The concentration factor shows how many times the concentration of a chemical element or a radionuclide in an aquatic organism exceeds its concentration in the water.

The second approach is one of experimental establishment of the concentration factors of various radionuclides in abundant marine organisms and in those of food value.

The capacity of aquatic organisms to concentrate radionuclides is determined in the following manner. The marine organisms under investigation are transferred to aquariums containing sea-water in which the concentration of a radionuclide (in terms of activity) has been accurately measured. It is an extremely important condition, and one which is readily checked and implemented, that the concentrations of chemical elements (by weight) in the sea-water should remain constant, i.e., that the carrier-free isotope should, as a rule, be weightless. Particle counters are used to determine the activity of preparations from samples of the water and aquatic organisms taken at various intervals, and the concentration factors are then calculated (Spooner [1949], Chipman [1960], Polikarpov [1960a-e, 1961a-d]). It is frequently possible to trace accumulation until saturation is reached. When metabolism is slowed down or when there is constant deposition of some elements in plant or animal tissues (e.g., deposition of strontium in the shells of molluscs and in the bones of fishes and whales) it may sometimes not be possible to obtain constant values of the concentration factor. The accumulation of radionuclides under experimental conditions and that of the corresponding chemical elements under natural conditions are compared by invoking biogeochemical data (Vinogradov [1953], Burovina et al. [1962], Mauchline [1963], Ichikawa [1961]), or by special determination of the concentration of the stable elements in aquatic organisms and in the surrounding water (Templeton [1959, 1962], Parchevskiy et al. [1965]) by various methods of chemical separation and concentration and by spectroscopy. This comparison can be used to establish for subsequent explanation whether the concentration factor of a chemical element and the corresponding radioisotopes are the same or different, and by varying the conditions it is possible to establish the role of the various modes of accumulation, including the significance of nutrition. Once in possession of verified concentration factors, it is a simple matter to proceed from the concentration of a radionuclide in the organism to its concentration in sea-water. It is also clearly of importance to establish the rate and nature of accumulation of radionuclides and the rate of their elimination from marine organisms (Polikarpov and Ten [1962], Barinov [1964, 1965a]). One matter of major interest is the distribution pattern of radionuclides in individual aquatic organisms and in their communities, and the search for mathematical means of expressing the relations between the various parameters of accumulation (Polikarpov

[1963], ZESENKO and POLIKARPOV [1965], ZESENKO [1965]).

Finally, the third approach is to study the biological effect of incorporated radionuclides at various levels of radioactivity in sea-water. Information on radiosensitivity is of particular importance in forecasting the radiobiological effect in seas and oceans at expected or existing levels of the radiocontamination of sea-water.

The way in which one studies the effect of radionuclides on the vital processes of marine organisms will be affected by the parameter under consideration. Thus, one can study the effect of various concentrations of radionuclides on the mitotic rate of marine microphytes (POLIKARPOV and LANSKAYA [1961]), or on the embryonic development of sea fishes (POLIKARPOV and IVANOV [1961, 1962a, b], IVANOV [1965b]) or on other parameters. There have been few attempts to study the reactions of marine biocoenoses under natural conditions to the effect of radioactive substances in nuclear weapon test areas (ODUM [1956], BLINKS [1952]). A powerful radiobiological effect under natural conditions on any species possessing high radiosensitivity may adversely affect resources that are economically important to man, for example the reproduction of fish stocks when the early stages of development of sea fishes are affected either immediately or gradually by the elimination or significant modification of one or more links in the intricate structure of the food and competitive relations of marine communities, although the links affected may not in themselves be of practical interest to man.

The three aspects of radioecological research here enumerated are, therefore, intimately related and complementary. Progress in any one will promote the development of the others. In other words, it is possible to make a full statement of the problems and achievements of marine radioecology: the science of the interrelation between a radioactive environment and marine organisms (POLIKARPOV [1963, 1964]).

SUMMARY OF PART I

Nuclear weapon explosions and the development of the nuclear power industry have led to the appearance of a new radioecological factor in nature. The concentration of long-lived fission products in the marine environment is continually rising and is, for example, 10^{-12} or even 10^{-11} Cur/l in terms of strontium-90 in the Irish Sea, into which radioactive wastes from the British nuclear plants are discharged, and in the Pacific Ocean, which has become a nuclear proving ground. Several countries are producing increasing quantities of radioactive wastes, and more and more attention is being paid to the seas and oceans as sites for disposal of these dangerous waste products. Some countries (the United States, the United Kingdom and France) bury radioactive wastes on the sea floor, in surface waters and in rivers flowing into the sea. The same use of the sea is planned by Sweden, Italy, India, Japan and other countries.

The systematic discharge of radioactive wastes into sea-water from nuclear-powered ships is now a permanent problem (SMITH [1959]). Damage to nuclear reactors at sea and on the sea coast (e.g., the Windscale accident in 1957 and the fate of the American nuclear-powered submarine Thresher in 1963) may have extremely significant consequences.

Artificially radioactive substances have therefore become a permanent radioecological factor in seas and oceans. This has led to development of a new branch of radiobiology: radioecology, whose task it is to study the features of the interaction between a radioactive environment and organisms, and thus to foresee the biological threat to life on earth, and to work out measures for the control of radiocontamination of the biosphere and the elimination of harmful consequences.

Marine radioecology, which is a major division of radioecology, should play an important part in studying the effect of radiocontamination of sea-water on the biological and commercial output of the sea, in predicting the risk to human health of radioactively contaminated marine foodstuffs, and in working out measures for preventing radiocontamination of the seas and oceans.

There are three basic research trends in marine radioecology: 1) systematic study of the content of artificially radioactive substances and their carriers in the most important marine organisms and in the environment in the various seas and oceans (analysis of the radioecological situation in nature), 2) experimental study of the accumulation and release of radioactive substances by plentiful marine organisms (prediction of the levels of accumulation of various radionuclides under natural conditions when they are present in certain concentrations in sea-water) and 3) study of the biological effect of small doses of the radiation of various radionuclides on the vital processes, proliferation and development of marine organisms, on the structure and biomass of their populations and biocoenoses (prediction of the threat to biological productivity and marine life of certain levels of radioactive contamination of sea-water). The first two aspects combined are one side of the theoretical problem of the interaction between marine organisms and a radioactive environment, namely the effect of *aquatic organisms* on a radioactive environment. Study of this matter will make it possible to establish the role of marine populations and biocoenoses in the migration and distribution of radioactive substances in the seas and oceans. The third aspect reflects the other side of this interaction, namely the effect of a *radioactive environment* on marine organisms. The alteration that this entails to the structure and biomas of populations and biocoenoses may, in its turn, lead to modification of the biological migration of radioactive substances.

Marine radioecology is faced with urgent practical problems of the nuclear age and with major theoretical problems. Radioecology, of which marine radioecology is a part, is therefore one of the most important theoretical and practical branches of present-day radiobiology.

PART II

Accumulation of radioactive substances by marine organisms

CHAPTER 2

CONCENTRATION FACTORS
AND THEIR RELATIONSHIP TO THE ENVIRONMENT

2.1. Concentration and discrimination factors

The concentration of a radionuclide is expressed in radioactivity units (curies, picocuries, disintegrations per minute etc.) per unit of weight of the intact organism or of separate organs and tissues (kilograms, grams etc.), or as the ratio of the activity of the nuclide to a unit of weight of a stable carrier, for example in strontium units (picocuries of strontium-90 per gram of calcium) and cesium units (picocuries of cesium-137 per gram of potassium).

The capacity of an organism to accumulate radioactive substances is expressed by the ratio of its radioactivity to that of the aqueous medium or the preceding food link in which the radionuclide was concentrated.

In the first case use is made of the concentration factor* (*facteur de concentration, Anreicherungsfaktor*), which is the ratio of the concentrations of a radionuclide (or corresponding chemical element) in the organism and in the aqueous solution:

$$K = \frac{C}{C'} \tag{1}$$

where C and C' are respectively the concentration of the radionuclide in the aquatic organism and in the aqueous medium.

If the concentration of any nuclide (element) a in the organism and water is expressed not as an absolute concentration but in relation to another nuclide (element) b whose concentration factor K_b will not normally be the same as the concentration factor K_a, the quotient obtained by division of the relative concentrations of the nuclide (element) a in the organism and water will yield a new and important quantity, the *discrimination factor*, D, i.e.,

$$D_{a/b} = \frac{C_a}{C_b} : \frac{C_a'}{C_b'}. \tag{2}$$

For example

$$K_{Sr} = \frac{C_{Sr}}{C_{Sr}'}, \quad K_{Ca} = \frac{C_{Ca}}{C_{Ca}'},$$

----

* In Russian: Koeffitsient nakopleniya.

27

$$D_{Sr/Ca} = \frac{K_{Sr}}{K_{Ca}} = \frac{C_{Sr}}{C'_{Sr}} : \frac{C_{Ca}}{C'_{Ca}} = \frac{C_{Sr} C'_{Ca}}{C'_{Sr} C_{Ca}},$$

or

$$D_{Sr/Ca} = \frac{C_{Sr}}{C_{Ca}} : \frac{C_{Sr}}{C'_{Ca}} = \frac{S}{S'},$$

where S and S' are the content of strontium-90 in the organism and water expressed in strontium units, respectively.

It follows from equation (2) that the quantities K_a, K_b and D are connected by the relation

$$D_{a/b} = \frac{C_a}{C'_a} : \frac{C_b}{C'_b} = \frac{K_a}{K_b}$$

or

$$K_a = K_b D_{a/b} . \tag{2'}$$

The concentration factor is sometimes unjustifiably referred to as the discrimination factor. Thus, LEWIN and CHOW [1961] define the strontium concentration factor as the ratio of strontium to calcium in algal coccoliths to the ratio of strontium to calcium in the aqueous solution. In fact, as we have seen above, this ratio merely describes the degree of discrimination of strontium in the organism and in the aqueous medium in relation to calcium. Since in the study under consideration $D=0.02$, $K_{Sr}=0.02K_{Ca}$. For coccoliths consisting of calcium carbonate $K_{Ca}=1000$. The strontium concentration factor in the algal coccoliths is therefore approximately 20, and not 0.02 as used by Lewin and Chow.

When information concerning the relative concentration of elements in aquatic organisms and in the aqueous medium is available (e.g., in relation to sodium from Fleming's data, cf. STRAKHOV et al. [1954]) it is sufficient to find the concentration factor of the element (sodium) to be in a position to calculate the concentration factors of other elements. The author has made a calculation on the basis of Fleming's figures (table 6). According to A.P. Vinogradov the sodium concentration factor is 0.6 for Calanus finmarchicus, 0.08 for fishes, and 0.7 for Archidoris britannica (VINOGRADOV [1953]).

Taking the percentage distribution of the element between the aquatic organism and the aqueous medium as a basis, expression (1) can be rewritten in the form:

$$K = \frac{C}{C'} = \frac{P'A}{PA'}, \tag{3}$$

where P and P' are respectively the weight of the organisms and of the water; A and A' are activity in the organisms and in the medium expressed as percentages when $A + A' = 100\%$.

Hence,

$$A = \frac{K}{K + P'/P} \times 100\% \qquad (3')$$

Maximum values of the concentration factors can be reliably established for stable isotope carriers and natural radionuclides in any marine organisms, and also for artificial radionuclides in aquatic organisms only when there is an exchangeable fund or in 'fresh' sectors of the structures of a non-exchangeable fund.

Table 5 contains a system of relations linking concentration factors to the other parameters descriptive of the concentration and distribution of chemical elements (radionuclides) in aquatic organisms, in organs and tissues, and in populations and biocoenoses. The derivation of these formulae is given in ZESENKO and POLIKARPOV [1965] and in POLIKARPOV and ZESENKO [1965]. The concentration of an element (radionuclide) in the intact organism and in its parts, or in biocoenoses and populations, may be calculated from the concentration or discrimination factors and on the basis of measurement of only one of the parameters incorporated in these formulae. Similar parameters may be calculated for a biocoenosis and its component parts.

The concentration factor of a mixture of chemical elements (radionuclides) should always have a real and quite definite physical sense. For chemical elements it is the ratio of the concentration of the ash residue in the organism to the concentration of salts in the water, while for radionuclides it is dependent on the quantitative and qualitative composition of the mixture.

When concentration factors are known, use of the formulae enables the number of measurements of radioactivity in aquatic organisms to be reduced to the minimum and restricted for convenience to indicator organisms (or organs). The remaining information may be calculated by the formulae (table 5). It is evident that the practical and theoretical importance of radioecological analysis necessitate every effort to establish reliable concentration factors for chemical elements (radionuclides) in abundant species of aquatic organisms (ZESENKO and POLIKARPOV [1965], POLIKARPOV and ZESENKO [1965]).

FEDOROV [1964] has recently suggested that the following means should be used to express the concentrating capacity of aquatic organisms, retaining the term 'concentration factor', but replacing the usual concept of the con-

Table 5

Formulae for calculation of the concentration factors of elements (radionuclides) in aquatic organisms (organs and tissues) or in biocoenoses (their populations)

Subject	Concentration factor	
Individual (or biocoenosis)	$K_{ju} = \dfrac{C_{ju}}{C_j'}$	$K_{ju} = K_{iju} \dfrac{C_{ju}}{C_{iju}}$
	$K_{ju} = \dfrac{\sum_{i=1}^{n} K_{iju} P_{iu}}{\sum_{i=1}^{n} P_{iu}}$	$K_{ju} = \dfrac{K_{iju}}{q_{iju}} \varrho_{iu}$
	$K_{ju} = K_{j^*u} D_{(j/j^*)u}$	$K_{ju} = K_{iju} \dfrac{K_{mu} S_{ju}}{K_{imu} S_{iju}}$
Organs and tissues of an individual (or elements of a biocoenosis)	$K_{iju} = \dfrac{C_{iju}}{C_j'}$	$K_{iju} = K_{ju} \dfrac{C_{iju}}{C_{ju}}$
	$K_{iju} = K_{ju} \dfrac{q_{iju}}{\varrho_{iu}}$	$K_{iju} = K_{ij^*u} D_{i(j/j^*)u}$
	$K_{iju} = K_{imu} \dfrac{S_{iju}}{S_j'}$	$K_{iju} = K_{ju} \dfrac{K_{imu} S_{iju}}{K_{mu} S_{ju}}$
	$K_{iju} = K_{i^*ju} \dfrac{C_{iju}}{C_{i^*ju}}$	

To avoid use of an excessive number of notations in the formulae we shall apply the same notations to two separate instances: 1) to individuals and 2) to biocoenoses. In the latter instance i denotes a biocoenotic element and n the number of biocoenotic elements.

Additional notations employed: u = organism (or biocoenosis), i = organ (or element of a biocoenosis), ϱ = weight fraction, q = the element (radionuclide) as a proportion of its total amount in the whole organism, s = the element (radionuclide) as a proportion of the total amount of elements (radionuclides), j = chemical element (radionuclide), m = a mixture of elements (nuclides), n = the number of organs (or biocoenotic elements), i and i^* = different organs (or different biocoenotic elements), j and j^* = different chemical elements (radionuclides). *Note:* It is possible to use the concentration factors of a mixture of elements (radionuclides) only knowing quantitative and qualitative composition of the mixture in organisms and water.

centration factor (C/C') by: 1) the ratio of the total radioactivity of an organism to its natural radioactivity and 2) the ratio of the radioactivity of organisms to the initial amount of radioactive substances present in a volume of sea-water adequate for their normal existence (expressed as percentages).

The first relation permits a simple comparison of the concentration of artificial and natural radioactivity in an aquatic organism, but does not incorporate any new or significant information. The second expresses the proportion of a radionuclide concentrated from the environment by aquatic organisms, i.e., A in formula (3). It is evident that extension of the accepted concept of the concentration factor to these different relations is quite unjustified and will merely cause confusion. There is no need to give a detailed analysis of Fedorov's paper, since it has already been the subject of an exhaustive review by HELA* [1963], who notes, with justification, that 'should it be necessary or desirable to use the other related factors, they should be clearly distinguished from the concentration factor in the original sense of the term'.

A is sometimes incorrectly referred to in the literature as the concentration factor (cf., e.g., FEDOROV [1961]). In fact the concentration factor is a measure of the concentrating capacity of the aquatic organisms in relation to radionuclides (elements), whereas the amount of a nuclide extracted by organisms from the surrounding environment A is a function of the concentration factor and of the biomass** of the organisms.

It should be noted that use of expression (3) may in practice lead to great errors if the concentration of the radionuclide in the aqueous solution is not checked, since a considerable amount of the nuclide may pass out of solution on to the walls of the aquarium.

In relation to the objectives of a given investigation the concentration factor may be expressed in relation to the wet (live) weight of the aquatic organism, to dry weight, or to ash weight, each of which has a quite definite meaning. Calculation of the concentration factor in relation to ash weight defines the extent to which the concentration of the element in the mineral residue exceeds its concentration in, on the one hand, the water, and, on the other hand, the mineral residue of the aqueous solution. The figures and the meaning are, of course, different. The concentration factor calculated in relation to dry weight (SCOTT [1954], TIMOFEYEV-RESOVSKIY, TIMOFEYEVA-RESOVSKAYA et al. [1960]) describes the capacity of the organic component to accumulate chemical elements from the aqueous medium.

When there is little or no moisture to remove (e.g., the shells of molluscs) and the concentration factors are very large, the difference between concen-

---

* Fedorov worked in Professor Hela's laboratory.
** The biomass is the amount of living matter in a unit of water volume (1 cubic meter) expressed in weight units (grams).

Table 6

Concentration of chemical elements in relation to sodium concentration taken as unity, discrimination factors, D, and concentration factors, K, in marine animals (calculated from Fleming's data, see STRAKHOV et al. [1954]). For copepods $K_{Na} = 0.60$, for molluscs $K_{Na} = 0.7$ and for fishes $K_{Na} = 0.08$

Element	Relative concentration in terms of sodium				D			$K = K_{Na} D$		
	in copepods a	in molluscs b	in fishes (mean) c	in sea water d	copepods a/d	molluscs b/d	fishes (mean) c/d	copepods	molluscs	fishes (mean)
Cl	1.94	1.80	–	1.80	1.1	1.0	–	0.66	0.7	–
Na	1.00	1.00	1.00	1.00	1.0	1.0	1.0	0.60	0.7	0.08
Mg	0.056	1.56	0.36	0.121	0.46	12.9	3.0	0.28	9.03	0.24
S	0.259	0.071	2.59	0.084	3.1	0.85	3.09	1.86	0.6	0.25
Ca	0.074	2.62	0.52	0.038	1.9	69	13.7	1.14	48	1.10
K	0.537	0.20	3.83	0.036	15	5.5	106	9.0	3.9	8.47
Br	0.017	–	–	0.006	2.8	–	–	1.68	–	–
C	11.13	4.80	41.00	0.0026	4300	1850	15800	2580	1295	1260
Sr	–	0.11	–	0.0012	–	92	–	–	64	–
Si	0.013	–	–	0.00001	1300	–	–	780	–	–
F	–	0.69	–	0.0001	–	6900	–	–	4830	–
N	2.80	1.07	12.76	0.00001	280000	107000	1276000	168000	75000	102000
P	0.241	0.06	2.56	0.000001	241000	60000	2560000	144500	42000	205000
I	0.0004	–	–	0.000005	80	–	–	48	–	–
Fe	0.013	0.0023	0.013	0.000002	6500	1000	6500	3900	700	520
Mn	–	–	0.000008	0.000001	–	–	8	–	–	0.64
Cu	–	0.0043	0.00008	0.000001	–	4300	80	–	3010	6.4

tration factors calculated in relation to dry and wet weight is slight, but otherwise it is fairly large, and may be a more than ten-fold difference.

VERNADSKIY [1929] expressed the ratio of radium concentration in plants and in water in relation to the live weight of the plants. He emphasized that radium concentration in his experiments related to a live plant containing approximately 90% or more of water, and that the dry matter or ash had a correspondingly higher radium content. Nevertheless, Vernadskiy was of the opinion that his method of expressing concentration in relation to live weight was superior when explaining the balance 'radium⇌living organisms in an aqueous medium'.

Concentration factors calculated in relation to wet weight reflect the actual role of living aquatic organisms in concentrating chemical elements from aqueous solutions. Moreover, calculation of internal radiation doses from incorporated radionuclides is possible only from their concentration factors in relation to live weight. It should be noted that concentration factors are almost invariably expressed in relation to wet weight in publications on the accumulation of radionuclides (elements) by marine organisms. SCOTT [1954], who related his figures to dry weight, also gave figures for the water loss of the algae on drying, and it is therefore a simple matter to convert to wet weight. Many students of freshwaters also calculate the concentration factors of radionuclides in relation to the live weight of aquatic organisms (A. N. Marey, M. M. Telitchenko, R. Glaser, R. F. Foster, J. J. Davis and others). I. N. Verkhovskaya, who is a leading specialist on the use of labeled atoms in biology, is of the opinion that concentration factors should always be expressed in relation to the wet rather than the dry weight of the organism (VERKHOVSKAYA et al. [1955], VERKHOVSKAYA [1962]).

Unless otherwise expressly stated, all concentration factors in the present book are calculated in relation to wet (live) weight.

2.2. Dependence of concentration factors on the environment

2.2.1. Concentration of isotopic carriers

According to the laws of radiochemistry (STARIK [1960]) the adsorption of trace amounts of an element on to the surface of a solid is directly proportional to its concentration in a wide range of concentrations up to 10^{-4} to 10^{-3} moles per liter. The adsorption factor remains constant. Further increase of concentration reduces this factor owing to saturation of the capacity of the adsorbent. Within the stated range of concentrations variations of radioisotopic concentration do not affect the adsorption factor of the radioisotope. A similar phenomenon is to be seen in the concentration of radionuclides from aqueous solutions by aquatic organisms.

The concentration factors should clearly be constant for various concentrations of a nuclide, if its addition to the aqueous solution does not affect the concentration of the isotopic carrier. Thus, when the concentration* of

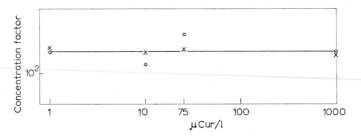

Fig. 5. Stability of the yttrium-91 concentration factor in *Ulva* in the range between 1 and 1000 μCur/l (POLIKARPOV [1961c]).

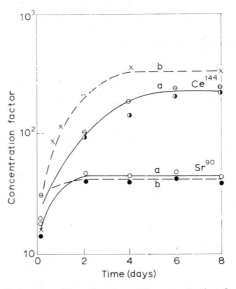

Fig. 6. Dependence of strontium-90 and cerium-144 concentration factors in *Cystoseira*, (a) when the activity level of the nuclides in sea-water decreases, (b) when activity in the solution is maintained constant (POLIKARPOV [1961c]).

---

* The weight of a radionuclide whose activity is one curie: $Q = 7.7 \times 10^{-9} AT$ (T, days) or $Q = 2.8 \times 10^{-6} AT$ (T, yr) where A is atomic weight, T is half-life, Q is the weight in grams without the inactive carrier (BARANOV [1955]).

yttrium-91 is between 10 and 10000 times less than the yttrium concentration in sea-water, its concentration factor remains constant in the green alga *Ulva rigida* (fig. 5).

The same feature is noted if a comparison is made of the concentration factors of strontium-90 and cerium-144 in the brown alga *Cystoseira barbata* when these nuclides were added to the aquarium once (activity decreased during the experiment) and when their concentrations were maintained at a constant level in the sea-water (fig. 6).

The concentration factor of cesium-134 in the red alga *Rhodymenia palmata* is not dependent on the concentration of this nuclide in artificial sea-water between 10^{-8} and 10^{-7} moles per liter (the range investigated by Scott [1954]). Absorption of strontium from artificial sea-water by coccolithophores is directly proportional to concentration between 10^{-4} and 10^{-3} moles per liter, i.e., the discrimination factor; therefore, the concentration factor for strontium remains constant (Lewin and Chow [1961]). For a wide range of zinc-65 concentrations in sea-water (in terms of activity) its accumulation by *Chlamydomonas* cells is directly proportional to concentration (Townsley et al. [1960, 1961]).

One of the best criteria of the stability of concentration factors is if experimental results and results obtained under natural conditions coincide. Table 7 summarizes information on strontium concentration factors, and on calculated and actual strontium-90 concentrations in lobsters at Winfrith (England) in 1960 when strontium-90 concentration in the sea-water was 1.5×10^{-13} Cur/l as a result of atmospheric fallout.

Table 7

Strontium content in lobsters (Templeton [1962])

Object	Sr content in relation to wet weight (mg/kg)	Concentration factor	Calculated ⁹⁰Sr content (10^{-12} Cur/kg)	Measured amount of ⁹⁰Sr (10^{-12} Cur/kg)
Muscles	8	1	0.15	< 1
Shell	1450	180	30	30
Other tissues	270	33	5	6

For the brown alga *Fucus serratus* the expected strontium-90 content was $(2 \text{ to } 5) \times 10^{-12}$ Cur/kg. This quantity was obtained from a concentration factor of 16 to 30 and a strontium-90 concentration in the sea-water of 1.5×10^{-13} Cur/l. The measured amount was $(2 \text{ to } 9) \times 10^{-12}$ Cur/kg. The mean cesium-137 level in sea-water in the same area in 1960 was 1.7×10^{-13} Cur/l, the calculated cesium-137 activity in *Fucus serratus* was 6×10^{-12} Cur/kg, and the measured activity was 9×10^{-12} Cur/kg (TEMPLETON [1961]).

The concentration factors obtained for stable nuclides under natural conditions and for radionuclides of the same elements* in experiments by various methods are also very similar. This led SPOONER [1949] to suggest the use of the tracer method for precise quantitative estimates of the concentration factors of chemical elements taken up from the water by marine organisms. The concentration factors of strontium-90 (in experiments at a concentration of 0.87×10^{-5} Cur/l) and of stable strontium (determined by flame spectrometry) in *Fucus serratus* were practically the same at 40 and 35, respectively.

The author's data concerning the concentration factors of strontium-90 under experimental conditions and of stable strontium under natural conditions for Black Sea plants (table 8) show that they are identical (PARCHEVSKIY *et al.* [1965]).

Table 8

Stable strontium and radiostrontium concentration factors in Black Sea plants (PARCHEVSKIY *et al.* [1965])

Aquatic organisms	Concentration factors	
	^{90}Sr in experiments	Sr under natural conditions
Ulva rigida	2	2
Cystoseira barbata	43	46
Zostera marina	3	5

In their detailed study PENDLETON and HANSON [1958] demonstrated that when cesium-137 concentration in the aquarium was 0.6×10^{-5} Cur/l the concentration factor of this nuclide in *Oedogonium* sp. was 1 200, which was similar to the value of 1 500 to 4 000 obtained under natural conditions in a pond where the concentration of the nuclide was $(1 \text{ to } 0.6) \times 10^{-10}$ Cur/l (for three species of green algae *Oedogonium* sp., *Rhizoclonium crassipelitum* and *R.* sp.). The figures for bulrush seeds were 400 in the experiment and 70 in the pond.

---

* Provided, of course, that they are in the same physical and chemical state. This rule needs no qualification for elements such as strontium and cesium and their radionuclides which occur in solution mainly in an ionic state and with a single valency.

In the Baltic Sea, where salinity is reduced, the concentration factors of stable strontium and strontium-90 were found to be 2.0 and 3.5, respectively, in the roach *Rutilus rutilus*, and 30 and 22 in the pike *Esox lucius* (AGNEDAL *et al.* [1958]).

It was found in experiments with ruthenium-106 that there was no relationship between the concentration factors (calculated in relation to dry weight) in *Elodea canadensis* and *Myriophyllum spicatum* (table 9 and fig. 7) and its activity when the concentration of the nuclide in the water was investigated between 1×10^{-7} and 4×10^{-5} Cur/l.

POVELYAGINA and TELITCHENKO [1959] established similar relationships for strontium-(89+90) (table 10).

The concentration factors of cesium-137 for *Euglena* and *Chlorella* monocultures were not dependent on concentration in the water at 1.5×10^{-10} and 1×10^{-9} moles per liter. Significant reduction of the concentration factors commenced at a concentration of 4×10^{-5} to 1.5×10^{-4} moles per liter (WILLIAMS and SWANSON [1958]).

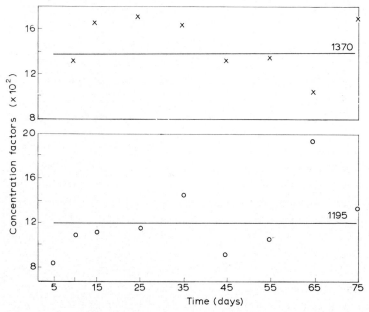

Fig. 7. Fluctuation of the concentration factors of ruthenium-106 in *Elodea* (calculated in relation to dry weight) (upper line, mean concentration factor 1 370) and in *Myriophyllum* (lower line, 1 195), when ruthenium-106 concentration in the water is maintained constant (TIMOFEYEVA-RESOVSKAYA *et al.* [1959]).

Table 9

Mean concentration factors (in relation to dry weight) of ruthenium-106 in plants for various concentrations in water (TIMOFEYEVA-RESOVSKAYA et al. [1959])

Plant	Concentration factor for a concentration of Cur/l		
	1×10^{-7}	2×10^{-6}	4×10^{-5}
Elodea	1 080	1 600	1 140
Myriophyllum	910	1 494	1 240

Table 10

Concentration factors of strontium-(89+90) in pearly mussels (*Unionidae*) for various concentrations of these nuclides in the water (POVELYAGINA and TELITCHENKO [1959])

Object	0.74×10^{-6} Cur/l	0.7×10^{-8} Cur/l
Foot	5.9	7.1
Mantle	257.7	205.7
Internal organs	81.5	45.7
Shell	2.8	4.2
Mantle edge	14.2	7.1
Gills	204.8	388.5

In GLASER's [1961a] experiments lake water contained 10^{-8} moles per liter of iodine (3.50 ± 0.234 μg/l). When iodine was added to this water, which contained the molluscs *Dreissensia polymorpha*, at a rate of between 10^{-7} and 10^{-6} moles per liter the concentration factor of iodine-131 fell in this range by 4.4 times. The concentration factors of iodine-131 in these molluscs differed by 1.6 times in the control water when 10^{-7} moles per liter of potassium iodide were added to it (fig. 8). Glaser emphasized that the addition of radioiodine to the solution did not affect the chemical nature of the water: the amount of iodine-131 added was quite insignificant by comparison with the stable iodine present in the water. In Glaser's opinion the mass numbers of iodine-127 and radioactive iodine-131 are too high for there to be any possibility of the animals reacting differently to them.

TITLYANOVA and IVANOV [1960] noted direct proportionality in the absorption of cesium by *Elodea, Lemna* and *Ceratophyllum* from solutions in which the concentration of the nuclide varied between 10^{-9} and 10^{-3} g equiv/l; this also indicates that the concentration factor of the element is constant in the range of concentrations indicated (fig. 9).

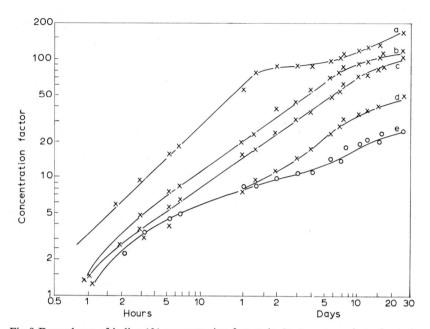

Fig. 8. Dependence of iodine-131 concentration factors in *Dreissensia* on isotopic carrier concentration (GLASER [1961a]). (a) not added, (b) 10, (c) 20, (d) 50, (e) 100 μg of stable iodine per liter of water.

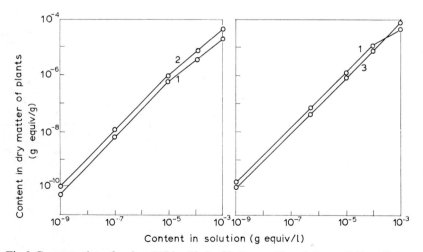

Fig. 9. Concentration of cesium-137 by *Elodea* (1), *Lemna* (2) and *Ceratophyllum* (3) from stable cesium solutions of various concentration (TITLYANOVA and IVANOV [1960]).

Table 11

Concerning the 'reciprocal' dependence of concentration factors on radioisotopic activity

Carp fry (LEBEDEVA [1961])		Yearling carp (SAUROV [1957])		Fishes (AGRANAT [1958])		Copepoda (FEDOROV [1964])	
^{90}Sr (Cur/l)	Concentration factor	^{90}Sr (Cur/l)	Concentration factor	^{210}Po (Cur/l)	Concentration factor	^{89}Sr (Cur/l)	Concentration factor
1.4×10^{-10}	5871 (100%)	5×10^{-10}	28 (100%)	5×10^{-14}	500 (100%)	3.7×10^{-10}	61 600 (100%)
				1×10^{-13}	1900 (380%)		
1.4×10^{-8}	564 (9.6%)	5×10^{-9}	40 (143%)	0.8×10^{-11}	737 (147%)	3.7×10^{-9}	6000 (9.7%)
				3×10^{-11}	600 (120%)		
1.4×10^{-6}	31 (0.5%)	5×10^{-7}	68 (243%)	1.5×10^{-10}	13 300 (2660%)	3.7×10^{-8}	5000 (8.1%)
				3×10^{-10}	2130 (425%)		
				0.6×10^{-9}	4000 (800%)		

Mention is made in some papers on radiation hygiene (MAREY *et al.* [1958] and other papers by Marey and his colleagues) of a 'reciprocal' relationship between the concentration of radioactive substances (in terms of activity) and the concentration factors in aquatic organisms.

If it is borne in mind that the amounts of radionuclides employed in experiments are usually 'weightless', in the sense that they do not affect the natural concentrations of their isotopic carriers, it is impossible to reconcile a reciprocal relationship with the laws of radiochemistry or with any of the material here cited concerning the agreement between the concentration factors of radionuclides (both in experiments and under natural conditions) and the corresponding chemical elements in aquatic organisms. It is note-worthy that the existence of a reciprocal relationship is not borne out by analysis of the information given in the papers in which the claim is made. The concentration factors in these papers reveal the most varied dependence on radionuclide concentration (in terms of activity) in the water (directly proportional, inversely proportional, or without any correlation), and clearly indicate the effect of extraneous factors (table 11).

It can only be suggested that undetermined amounts of stable elements, sufficient to have significantly modified the concentrations of these elements in the natural water, may on occasion have been added to the freshwater on introduction of the radionuclides.

One other probable explanation of error is failure to allow for the content of potassium-40 and artificial radioactive substances (in ash), present as contaminants in experiments involving very low activities.

It has been demonstrated in experiments with marine algae that the yttrium-91 concentration factor still remains constant at very low concentrations of this nuclide of between 10^{-7} and 10^{-11} Cur/l (fig. 10). These experiments were carried out in 1960 when contamination of natural sea-water by short-lived fission products was least. Potassium-40 was the major component of the radioactivity of *Ulva* ash at this time. If no account is taken of potassium-40 content and the radiocontamination of aquatic organisms in experiments involving very low concentrations of radionuclides, the impression may be created that there is an increase (and, moreover, a considerable increase) in the concentration factors as the radioactivity of the added nuclide decreases (fig. 11) (PARCHEVSKIY *et al.* [1965]). The higher the radiocontamination of the experimental material, the higher will be the concentration at which the seeming curve of the inverse relation begins to rise, and the steeper will be its slope.

There are probably a number of explanations for the results claimed by

FEDOROV [1960], who stated that the percentage of extraction of fission products from solution by zooplankton (the 'concentration factor') was increased when their concentration in sea-water was reduced from 10^{-7} to 10^{-12} Cur/l (and, moreover, was again decreased at 10^{-13} Cur/l). The same author obtained the opposite picture for nanoplankton*: the percentage of extraction decreased from a concentration of 10^{-9} Cur/l to one of 10^{-12} Cur/l.

Let us now consider the assertion by FEDOROV [1964] that concentration factors (in the generally accepted sense) are dependent on biomass: that as biomass decreases the aquatic organisms increasingly utilize the radioactivity of the environment, and that this, in Federov's own words, leads to greater concentration of radioactive substances by the organisms and to an increase

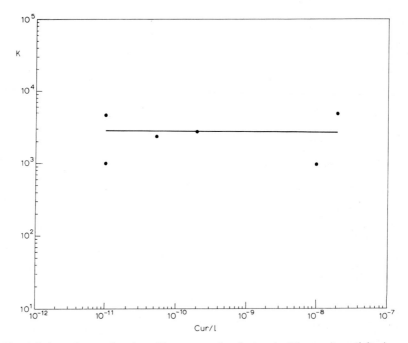

Fig. 10. Independence of yttrium-91 concentration factors in *Ulva* on its activity in sea-water. Experimental points and mean line (PARCHEVSKIY *et al.* [1965]).

---

* The nanoplankton consists of certain plants and many protozoans up to 50 μm in diameter.

in the concentration factors (in the usual sense). This line of reasoning is seen to be impossible if consideration is given to the actual processes involved in the exchange of chemical substances between aquatic organisms and the environment. When a radioisotope such as of strontium is added to sea-water, the effect is to label the stable strontium present in the water. The concentration factors of stable and radioactive strontium are equal both in experiments and under natural conditions (table 8). This is, however, the best criterion of the independence of the concentration factors of chemical elements (radionuclides) from biomass and radioactivity levels. Fedorov, moreover, contradicts himself by assuming that the concentration factors of a radionuclide should decrease as its concentration increases. Concentration factors of 226 000 in *Calanus* are obtained by calculation employing Fedorov's

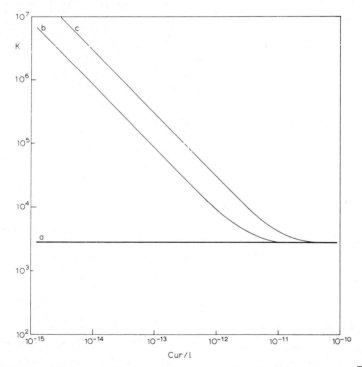

Fig. 11. Seeming increase of yttrium-91 concentration factors in *Ulva* disregarding (from 10^{-10} to 10^{-11} Cur/l and below) the content of radiocontaminants (c) and of potassium-40 (b); (a) yttrium-91 concentration factors in *Ulva* (after deduction of the activity of radio-contaminants and potassium-40). Radiocontamination of *Ulva* in 1959 taken as an example (PARCHEVSKIY *et al.* [1965]).

own data ($A = 12.4\%$ for strontium-90 in copepods). His 'experimental' value of the concentration factor was 61 600. It follows from this that pure strontium content in *Calanus* is $226\,000 \times 0.008$ g/l (or $61\,600 \times 0.008$ g/l), i.e., 1 800 g in 1 kg of these copepods (or 500 g/kg). It would therefore appear that, according to Fedorov, *Calanus* consists of metallic strontium. It is readily apparent from what has been said that such phenomena are impossible.

In fact, if radionuclides especially in concentrations that are insignificant fractions of the natural concentrations of the corresponding carriers present with them in an identical state and under conditions of complete isotopic exchange, were to be accumulated by organisms in relation to their concentration (in terms of activity) in the water, this would indicate that the radioactive and stable isotopes of the same element possessed different chemical properties or, at least, that living organisms are capable of 'discriminating' between the radioisotope and the stable isotope of an element. To accept this would involve rejection of the existing concept of a 'chemical element' as an 'artificial' category, and would entail rejection of the tracer method, because it would be impossible to treat radioactive indicators as indicators of the other isotopes of an element by virtue of their independent chemical behaviour. Science has clearly advanced beyond the stage at which such conclusions might have been suggested.

Therefore, the concentration factors of radionuclides, whose addition to an aqueous medium will not, other conditions being equal, affect the concentrations of the corresponding chemical elements and will not, of course, have a biological effect as a result of radiation, will not be dependent on the concentration of the isotopes (in terms of activity) in balanced accumulation in aquatic organisms.

When an isotopic carrier is present in low concentration or not at all this feature may hold for radioisotopic concentrations of up to 10^{-6} to 10^{-3} moles per liter.

2.2.2. *Concentration of non-isotopic carriers*

Scott [1954] demonstrated that when potassium was replaced by an equivalent amount of sodium *Fucus vesiculosus* made good the deficit of potassium by absorption of cesium. At the same time variation of phosphate and nitrate concentrations, in imitation of their seasonal fluctuations, had practically no effect on the absorption of cesium-134 by *Fucus*. Similar results were obtained on other brown algae (*F. serratus, Ascophyllum nodosum, Laminaria digitata* and *L. saccharina*), and on the green alga *Ulva lactuca* and the red alga *Polysiphonia fastigiata*.

At high concentrations potassium reduces the absorption of cesium-137 by freshwater protozoa. When the concentration of stable cesium is up to 1.5×10^{-4} moles per liter potassium has relatively little effect on the absorption of cesium-137. Thus, the concentration factor of this nuclide in *Chlorella pyrenoidosa* is 154 when potassium concentration in the water is 2×10^{-7} moles per liter and 100 even at 10^{-3} moles per liter. Sodium does not affect the absorption of cesium-137 by *Chlorella*. When present in large amounts it affects concentration in *Euglena intermedia*, apparently because its permeability is affected. It will be clear from the last paragraph why the concentration factors of cesium-137 are dependent on the cesium-potassium ratio. They are determined by stable cesium (WILLIAMS and SWANSON [1958], WILLIAMS [1960]).

RICE [1956] established that calcium in concentrations of between 0.75×10^{-2} and 3.75×10^{-3} moles per liter (0.300 to 0.150 g/l) in artificial sea-water did not affect absorption of radiostrontium by *Carteria* sp. On the other hand, absorption of calcium by the cells of this alga was not dependent on strontium concentration in the range 1.3×10^{-4} to 0.0 moles per liter (0.011 to 0.000 g/l).

In sea-water of the Swedish coast where salt content is reduced to a tenth (Studsvik) the concentration factors of strontium-90 were found to be considerably higher than in ordinary sea water (AGNEDAL *et al.* [1958]).

As calcium concentration in freshwater increased, absorption of radiostrontium by goldfish decreased (PROSSER *et al.* [1945]). The concentration factors of strontium-90 were reduced to a fifth or a sixth in the scales and bones of freshwater fishes, and halved in the muscles and internal organs when the calcium content in the water was increased from 40 to 60 to 280 to 300 mg/l. There was a simultaneous reduction in the rate of absorption of radiostrontium by fishes (MAREY *et al.* [1958]) and by water plants (OWENS *et al.* [1961]).

According to PICKERING and LUCAS [1962] the radiostrontium concentration factor has no real meaning if calcium concentration in the aqueous solution is unknown (table 12).

Therefore, small variations (twice or even more) of the considerable concentrations of potassium and calcium in sea-water do not significantly affect the concentration factors of cesium-137 and strontium-90 in those species of marine organism that have been studied.

Table 12

Dependence of calcium-45 and strontium-90 concentration factors in the filamentous alga *Rhizoclonium hieroglyphicum* on stable calcium concentration in the water (PICKERING and LUCAS [1962])

Ca++ concentration		Logarithm of calcium concentration	Logarithm of strontium concentration factor	Strontium concentration factor	Logarithm of calcium concentration factor	Calcium concentration factor	$\dfrac{CF^*Sr}{CF\,Ca}$
10^{-3} g/l	10^{-6} moles/l						
60.0	1.5	0.176	1.985	97	1.995	98.9	0.98
40.0	1.0	0	2.17	148	2.16	145	1.03
4.0	0.1	−1	3.22	1660	3.095	1250	1.33
0.4	0.01	−2	4.27	18600	4.03	10700	1.74

* CF = concentration factor

2.2.3. *State of isotopes in solution and the* pH *of the medium*

Tables 13, 13a and 14 summarize information on the physical state of various chemical elements whose radioisotopes are found in fission products.

Very little is as yet known concerning the physical and chemical states of the various elements absorbed by aquatic organisms. In experiments with

Table 13

Probable physical state of some elements in sea-water (GREENDALE and BALLOU [1954])

Element	Content (%) in the form of		
	ions	colloids	particles
Cs	70	7	23
I	90	8	2
Sr	87	3	10
Sb	73	15	12
Te	45	43	12
Mo	30	10	60
Ru	0	5	95
Ce	2	4	94
Zr	1	3	96
Y	0	4	96
Nb	0	0	100

Table 13a

Observed physical state of fission products in sea-water (FREILING and BALLOU [1962])†

Elements	Fraction (%)					
	soluble			colloidal	particulate	
^{89}Sr	99*	–	–	11*	0*	4**
^{95}Zr	3*	0–8**	42***	16*	38*	34–66**
^{95}Nb	0*	0–8**	–	39*	45*	34–66**
^{99}Mo	60*	26–68**	65***	6*	28*	–
^{103}Ru	–	–	–	25*	60*	82±5**
^{132}Te	–	–	0***	18*	60*	64±9**
^{140}Ba	99*	–	35***	1*	0*	5±1**
Rare earths	3*	3**	34***	14*	83*	67±3**
^{237}U	4*	–	71***	11*	2*	46±5**
^{239}Np	2*	–	50***	46*	47*	50–93**

† Table prepared by S. A. Patin, Trudy Inst. Oceanol., vol. 82, in press [1966].
* Explosion on the bottom of shallow lagoon.
** Explosion in not great depth.
*** Deep-water explosion.

Table 14

Valence state of certain metallic ions in sea-water (KRAUSKOPFF [1956])

Elements	Ions	Elements	Ions
Zn	Zn^{++}, $ZnCl^+$	Au	$AuCl_4^-$
Cu	Cu^{++}, $CuCl^+$	Cr	CrO_4^-
Bi	BiO^+	V	$H_2VO_4^-$, $H_3V_2O_7^-$
Cd	$CdCl^+$, $CdCl_2$	Mg	Mg^{++}
Ni	Ni^{++}, $NiCl^+$	Ca	Ca^{++}
Co	Co^{++}	Sr	Sr^{++}
Hg	$HgCl_4^-$	Ba	Ba^{++}
Ag	$AgCl_2^-$		

radioiron Goldberg has shown that marine diatoms assimilate only colloidal particles of iron hydroxide and do not absorb complexes of iron with citric, ascorbic and humic acid. Zirconium, titanium and other anions are extracted from sea-water by iron hydroxide particles and may enter plants in company with iron (REVELLE et al. [1956]). Iron hydroxide also forms an insoluble complex with ruthenium-106 from sea-water, and it is possible that differences in the uptake of nitrosyl ruthenium-106 and of ruthenium-106 from a solution obtained in the separation of plutonium fission products, and from a solution formed in the separation of uranium decay products may be explained by different physical and chemical forms of the ruthenium which they contain (JONES [1960]).

Direct deduction of the concentration factors of a chemical element from those of its radioisotope and vice versa is impossible for those elements that readily change their valence and the form of their chemical bonds. Preliminary investigation of their chemical and physical state is necessary in such instances. Although this is a major problem, it is one that must be solved*. What has been said is particularly applicable to ruthenium and a number of other elements that have a great many different states and compounds.

The exchange of an element, mitosis and other processes may be affected by the pH of the medium. It should be borne in mind that whereas the pH of sea-water is highly stable, that of freshwater varies considerably: when the biomass is large the pH is 7.6 to 8.1 and may reach 9 to 10 when there is an abundant 'bloom' (ZERNOV [1949]).

It has been noted that the absorption of nitrosyl ruthenium-106 from sea-

---

* Recently excellent works were carried out by S.Kečkeš et al. in Yugoslavia (Ann. Rept. on Res., Contr. 210/RI/RB [1964–1965]).

water by the red alga *Porphyra* decreases as the *p*H of the medium falls (fig. 12). This apparently indicates that cation exchange reactions occur on the surface of the algae (JONES [1960]). On the other hand, it has been noted that absorption of iron-59 by the red alga *Ceramium* and the green alga *Enteromorpha* is not dependent on the *p*H of sea-water (TAYLOR and ODUM [1960]).

Lack of carbon dioxide may explain why the red alga *Rhodymenia palmata* ceases to concentrate cesium-134 when the *p*H of the medium is raised over a period of a few days from the normal (7.9 to 8.0) to the maximum (9.1) (SCOTT [1954]).

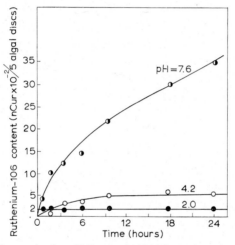

Fig. 12. Effect of *p*H on ruthenium-106 concentration from sea-water by *Porphyra* (JONES [1960]).

According to RICE [1956] the effect of *p*H on accumulation of radiostrontium from sea-water in *Carteria* sp. is also an involved relationship. The concentration factor was 975 at *p*H 8.2 (control), 223 at *p*H 6.0, and 85 at *p*H 9.0. In the last instance the cells were larger and more mobile than the controls, but the number of mitoses (and the number of cells) was correspondingly lower. Rice is of the opinion that the reduction of concentration factors when the *p*H deviates from the normal is due not only to the effect of hydrogen ion concentration on the availability of strontium, but also to reduction in the mitotic rate, since it has been shown by special experiments that the strontium concentration factor is directly related to the number of cell divisions in *Carteria*.

The great stability of hydrogen ion concentration in sea-water is one of the prerequisites for the stability of the concentration factors of various radionuclides for marine organisms.

2.2.4. Water temperature and illumination

Very little has been published concerning the effect of temperature on concentration of radionuclides. GLASER [1961a] carried out experiments to reveal the effect of temperatures between 3 and 20°C on the concentration

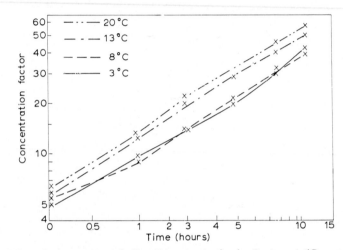

Fig. 13. Effect of temperature on iodine-131 concentration by *Dreissensia* (GLASER [1961a])

Table 15

Iodine-131 concentration factors in molluscs at various temperatures (GLASER [1961a])

Period since commencement of experiment	Concentration factors at 3°C	Upper limit 3 σ*	Concentration factors at 13°C
7 hours	4.96±0.23	5.65	5.88
1 day	9.45±0.57	11.16	12.46
2 days	13.86±0.88	16.50	19.32
4 days	19.72±1.34	23.74	27.8
7 days	29.0±1.94	34.82	39.8
11 days	41.5±3.96	53.68	49.4

* σ = standard deviation.

factors of iodine-131 for *Dreissensia polymorpha* (fig. 13). The difference between 3 and 20°C was only 1.4. The information in table 15 shows that there are no statistically significant differences between the concentration factors at 3 and 13°C. Normal temperatures therefore have a relatively minor effect on the concentration factors of iodine-131 in *Dreissensia*.

It is evident that temperature affects the uptake of fission products from an aqueous medium both directly and indirectly – through its effect on the rate of biological processes. Thus, 24 hr after strontium-89 had been added to sea-water the amount concentrated by plaice fry at 10°C was only a third of the amount concentrated at 20°C (BOROUGHS, CHIPMAN *et al.* [1957]).

Temperature has an even more involved effect on the concentration of radioactive substances under natural conditions. During the cold months, fish in the Columbia River consume little plant food and consequently contain less radioactivity. The same applies to aquatic insects, crustaceans and molluscs. The dormant stages of insects contain less radioactivity than the actively feeding stages (DAVIS and FOSTER [1958]).

Light has a great influence on the concentration of certain chemical elements. Concentration of the elements of the first group of the periodic system, or at least of potassium and cesium, by plants is related to light conditions, which regulate photosynthesis and related processes.

It has long been known that light and carbon dioxide promote accumulation of potassium. The connection between uptake of radiocesium from sea-water and illumination was first demonstrated for *Rhodymenia palmata* by SCOTT [1954], who used radioautography, and later for *Ulva rigida* by POLIKARPOV [1963], who used radiometry.

Scott did not discover any significant translocation of cesium-137 through the plant from the photic zone. Light filters were used to establish the part played by various portions of the visible light spectrum in inducing the concentration of cesium by *Rhodymenia*. It was established by comparison of the total quantity of light transferred to the zone of cesium-134 concentration that no part of the spectrum was particularly effective in 'stimulating' accumulation of this nuclide by the algae. According to the present author's data (table 16) the concentration factor of cesium-137 in *Ulva* after two days of exposure in active sea-water (10^{-5} Cur/l) was six units (or 300%) under conditions of normal illumination, and two units (100%) in darkened conditions.

There is no record of light affecting concentration of the elements of the main subgroup of the second group in Mendeleyev's periodic system by aquatic macrophytes. There are no significant differences between the concentration of calcium-45 by *Fucus vesiculosus* in the dark and in the light (SWIFT and TAYLOR [1960]). A similar phenomenon has been described for the accumulation of strontium-90 in *Fucus serratus* (SPOONER [1949]) and in *Ulva rigida* (table 16) (POLIKARPOV [1963]), and also in freshwater plants (MAREY [1962], OWENS et al. [1961]).

A more intricate relation has been noted in experiments with unicellular marine algae (RICE [1956]). A culture of *Carteria* sp. that had been kept in darkness for seven days, and had ceased to divide, was divided into two cultures which were placed in solutions containing radiostrontium, one kept in the light and the other in the dark. The mean amount of radiostrontium per cell was 64 times greater in the light than in the dark, i.e., dividing cells concentrate this radionuclide far more readily than non-dividing cells. Rice related this to the formation of new protoplasm.

In another experiment one culture in a radiostrontium solution was reared from a dark-adapted culture, and another from cells kept in a medium without phosphates but under ordinary light conditions. The cells of the first culture divided more rapidly and concentrated more of the nuclide than those of the second culture. When conditions of illumination were the same, the concentration of radiostrontium was proportional to the mitotic rate. The cells of the older culture always contained more radiostrontium whatever the mitotic rate.

Concentration of zinc-65 – a member of the auxiliary subgroup of the second group of the periodic system – is so intimately related to photosynthesis that the radiocarbon method of determining the primary productivity of water plants is now being augmented by development of a new method employing labeled zinc (BACHMANN and ODUM [1960]).

In the third group of the periodic system cerium-144 (and also, apparently, all the lanthanides and yttrium) are accumulated by *Ulva* with the same concentration factors whatever the lighting conditions, 287 in the light and 279 in the dark (table 16).

The fourth and fifth groups of Mendeleyev's table contain the main biogenous elements: carbon and phosphorus. Carbon is assimilated in the process of photosynthesis, in which it participates from the beginning, and phosphorus is incorporated in more distant links. Different data concerning the connection between the rate of accumulation of phosphorus-32 by plants and their illumination have therefore been obtained in different studies. The present author failed to detect any connection in experiments with *Ulva* lasting 32 days (POLIKARPOV [1960d]).

To the best of the author's knowledge there is no available information concerning concentration of elements of the sixth and seventh groups of the periodic system by aquatic organisms under various light conditions. From the eighth group there is some information concerning iron and cobalt. Thus, no connection was established between the concentration of iron-59 in chloride form by *Ceramium* and the presence or absence of light. The concentration followed the same pattern in both cases (TAYLOR and ODUM [1960]). On the other hand cobalt-60 was concentrated far more readily in light (by 250%) than in dark by *Ulva* (table 16).

Table 16

Effect of light on concentration factors of certain elements in *Ulva* (POLIKARPOV [1963])

Element	Concentration factors (wet weight)		$\dfrac{a}{b} \times 100\%$
	In light (a)	In darkness (b)	
^{60}Co	10	4	250
^{90}Sr	1	1	100
^{137}Cs	6	2	300
^{144}Ce	287	279	103

2.2.5. Complexing agents

Complexing agents, which are also known as complexons and addenda, are substances which form complex compounds with cations. There are very many of them in the organism, ranging from simple acids to protein polymers. Each complex compound has its own stability constant, which may be very high.

There are two aspects to the study of the effect of complexing agents on concentration of radionuclides by aquatic organisms. First, to establish the role of natural complexing agents and biological complexing agents in the accumulation of various chemical elements (and their radioisotopes) by aquatic organisms and, second, to solve the important practical problem of the concentration of major radionuclides by the aquatic biocoenoses of artificial ponds to purify contaminated water.

Most studies of the effect of complexing agents on the concentration factors of various radioactive substances have been based on ethylenediaminetetraacetic acid (EDTA), which is the most effective. In concentrations equimolar to the salt composition of lake water (400 mg/l) EDTA affected the concentration factors of various elements differently. There was a marked reduction (between 10 and 100 times) in the concentration factors of the radioisotopes of iron, cobalt, zinc, yttrium and cerium (calculated in relation to dry weight) in 12 species of freshwater plant and animal studied. On the average the concentration factors of calcium, zirconium, niobium, ruthenium and iodine were slightly reduced while the concentration factors of rubidium, strontium and cesium were scarcely affected. Complexes of the first five elements with EDTA have been found to have very high stability constants, and to be readily soluble and mobile. The stability constant of calcium is higher than that of strontium with EDTA. Calcium therefore becomes more mobile in the presence of EDTA (its concentration factors are reduced), and strontium is retained in the organism in large amounts as a calcium analog.

Timofeyev-Resovskiy has suggested that reduction of the concentration factors of zirconium, niobium and ruthenium, and increase of those of rubidium and cesium can probably be explained not by the direct effect of EDTA on these elements, but by disruption of calcium metabolism and therefore of the general mechanism of mineral metabolism (TIMOFEYEV-RESOVSKIY, TIMOFEYEVA-RESOVSKAYA et al. [1960], GETSOVA et al. [1960], TIMOFEYEVA-RESOVSKAYA and TIMOFEYEV-RESOVSKIY [1960], TIMOFEYEVA-RESOVSKAYA [1958a], TIMOFEYEV-RESOVSKIY and TIMOFEYEVA-RESOVSKAYA [1959]).

The concentration factors of yttrium-91 in Chlorella fell sharply when EDTA and citrate were added to freshwater (GLASER and SPODE [1960]).

It was found in experiments in sea-water that iron-59 was not adsorbed on the surface of a vessel in the presence of EDTA, and that the concentration factors and rate of exchange between Ceramium and Enteromorpha and the environment were noticeably reduced. It is noteworthy that when the aquaria contained Fucus the radioactivity of iron-59 in sea-water fell less rapidly than when these algae were not present. It has been suggested that Fucus liberates substances into the water which prevent adsorption of this element (TAYLOR and ODUM [1960]).

JONES [1960] is of the opinion that the difference between ruthenium-106 concentration in Laminaria, Ulva and Porphyra is related to differences in the chemical composition of the cell surface of these algae due to differences in the content of extracting alginic acid which, like its sodium salt, forms a soluble complex with ruthenium-106.

In experiments on the absorption of ruthenium-106 by sand from sea-water without calcium the presence of sodium alginate (0.09 to 0.08%) reduced the amount of ruthenium-106 adsorbed on to the sand to 70 to 80%.

Calcium-free water has been used to prevent deposition of calcium alginate. Iron hydroxide forms an insoluble complex with ruthenium-106.

This information will suffice to show that studies of the use of complexing agents in marine radioecology are still in their infancy.

2.3. Kinetic features of radionuclide exchange between aquatic organisms and the water

The method usually employed in experimental determination of radionuclide concentration factors is to trace the curves of variation of concentration until a stable level, known as the plateau or balanced (steady) state of accumulation, is reached.

Some authors (Davis and Foster [1958], Glaser [1962a,b]) are of the opinion that the exponential law may be employed in examination of the kinetics of radionuclide concentration by aquatic organisms (calcium-45, strontium-90, iodine-131). It should, however, be pointed out that the exponential law is suited only to the description of some elementary physicochemical laws, and is by no means universal in its application. In particular it is not amenable to quantitative description of the decay of a mixture of radioactive products (Way and Wigner [1948]), to adsorption on to nonuniform surfaces and to the kinetics of involved chemical reactions (Roginskiy [1948,1956], Trapnell [1955]), or to the concentration and exchange of radionuclides by algae (Barinov [1965a,b]). An account will be given below of the results obtained by Barinov in kinetic studies of algal concentration and exchange of a number of radionuclides including carbon-14, phosphorus-32, calcium-45 and strontium-90.

When there is steady state distribution of an element between aquatic organisms and the environment and the rate of exchange of the element is constant, the concentration and elimination of a radionuclide is determined by the phenomena of isotopic exchange, in which the total content of the element in the organism is not affected, but there is replacement of the stable isotopes by its radioisotope or vice versa.

By analogy with the concentration factor (K), use may be made of the concept of the elimination factor $(\hat{K})$ when studying elimination of a radionuclide from the organism. This coefficient is the ratio of the concentrations of a radionuclide in an aquatic organism and in the surrounding water. The elimination factor is the ratio of the concentration of a radionuclide eliminated from the organism to its concentration in the medium in which accumulation took place. If the functions of accumulation and elimination of a radionuclide are known it is possible to calculate K and $\hat{K}$, and also the time taken to reach a balanced or, to be more precise, a steady state (t_s). It is evident that the values of K and $\hat{K}$ will coincide in isotopic exchange. It is now thought (ROGINSKIY and SHNOL' [1963]) that there may be considerable non-exchangeable funds of an element in the organism. If this is so, the ratio $K_{\text{isotope}}/K_{\text{element}} = \alpha$ will describe the degree of exchangeability of the element *in vivo*. When concentration of an isotope has occurred from early ontogeny, K will be the same for the isotope and the element, but K may not equal $\hat{K}$, and if not $\alpha = \hat{K}/K$. In principle these coefficients should, therefore, be determinable by chemical analysis, and it should be possible to calculate the rates of the processes from the concentration or elimination function of the radionuclide. Simultaneous study of concentration and elimination is, in general, essential for determination of K, $\hat{K}$ and t_s. The time taken for concentration of a radionuclide is assumed to equal the elimination time. The form of the concentration and elimination functions of the radionuclide must be established for calculation purposes.

It is now thought that the kinetics of isotopic exchange in involved systems (ROGINSKIY [1956]) are not subject to the exponential law, although it remains in force for description of the separate elementary processes that go to make up the involved process. The kinetics of involved isotopic exchange are, in the main, determined by the form of the statistical distribution function of the velocity constants of the separate elementary processes. Taking into consideration the intricate multi-phase microheterogeneous structure of organisms, the universality of statistical laws, and the subordination of elementary biochemical processes to the exponential law (BRAY and WHITE [1957], TROSHIN [1956]), it may be assumed that Roginskiy's theory of involved isotopic exchange will be applicable to biological systems, and that the function of isotopic exchange between the organism and the water will be defined by the statistics of the distribution of the exchange constants for the separate homogeneous microfunds constituting the general exchange budget of the element in the organism. According to Roginskiy the main statistical distribution functions are uniform, exponential and power functions. These distributions correspond to logarithmic, power and power-logarithmic isotherms. In the last analysis the kinetics of isotopic exchange are experimentally determined. The constant terms of the equation are calculated by processing the data by the method of least squares. The extent to which the calculated values of K and $\hat{K}$ agree with the experimentally derived values may serve as a criterion of selection of the correct formula. As in all experiments concerned with the exchange of radionuclides between an organism and the environment, care must be taken to ensure that the total concentration of the element in the medium remains practically unaltered after addition of the radionuclide to the solution.

Research already carried out on the concentration and elimination of carbon-14, strontium-89, yttrium-91, cesium-137 and cerium-144 by Black Sea macrophytic algae has shown that Roginskiy's statistical kinetic theory is applicable to a hydrobiological system and that calculations based on this

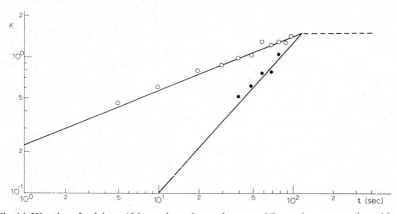

Fig. 14. Kinetics of calcium-45 isotopic exchange between *Ulva* and sea-water (logarithmic coordinates). ○ uptake of isotope, ● elimination of isotope. The dashed line denotes a steady level of isotopic exchange. (BARINOV [1965b].)

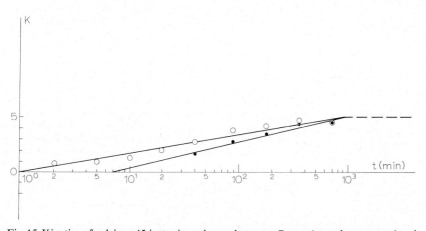

Fig. 15. Kinetics of calcium-45 isotopic exchange between *Cystoseira* and sea-water (semi-logarithmic coordinates). ○ uptake of isotope, ● elimination of isotope. The dashed line denotes a steady level of isotopic exchange. (BARINOV [1965b].)

theory are possible. Thus, for example, it was found that exchange of calcium-45 by the green alga *Ulva rigida* was described by a power function (fig. 14), whereas exchange of the same radionuclide by the brown alga *Cystoseira barbata* was described by a logarithmic function (fig. 15).

The form of an isotopic exchange function is affected by the position of the element in the periodic table as well as by the biological features of the organism. Whereas the kinetics of calcium-45 exchange in *Cystoseira* are described by a logarithmic function (fig. 15), those of strontium-89 in the same alga are described by a power-logarithmic isotherm (fig. 16) corresponding to the power distribution of strontium microfunds in this alga.

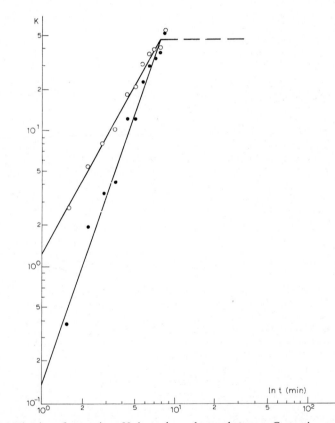

Fig. 16. Kinetics of strontium-89 isotopic exchange between *Cystoseira* and sea-water (logarithmic coordinates). ○ uptake of isotope, ● elimination of isotope. The dashed line denotes a steady level of isotopic exchange. (Barinov [1965b].)

Table 17

Calculation formulae for the main exchange parameters of isotopes and elements in exchange between an aquatic organism and the environment in relation to the form of the function $K(t)$ (BARINOV [1965b])

Form of the function $K(t)$	K_s	t_s	$K'(t)$	V_e
1	2	3	4	5
Logarithmic $K = K_I + n \ln t$	$\exp\dfrac{\hat{n}K_I - n\hat{K}_I}{\hat{n} - n}$	$\exp\dfrac{\hat{K}_I - K_I}{n - \hat{n}}$	$\dfrac{n}{t_s}$	$\dfrac{C_i n}{t_s}$
Power $K = K_I t^n$	$\exp\dfrac{\hat{n}\ln K_I - n\ln\hat{K}_I}{\hat{n} - n}$	$\exp\dfrac{\ln\hat{K}_I - \ln K_I}{n - \hat{n}}$	$\dfrac{nK_s}{t_s}$	$\dfrac{C_i nK_s}{t_s}$
Power-logarithmic $K = K_I(\ln t)^n$	$\exp\dfrac{\hat{n}\ln K_I - n\ln\hat{K}_I}{\hat{n} - n}$	$\exp\exp\dfrac{\ln\hat{K}_I - \ln K_I}{n - \hat{n}}$	$\dfrac{nK_s}{t_s \ln t_s}$	$\dfrac{C_i nK_s}{t_s \ln t_s}$

Notations: K_s = steady state concentration factor, t_s = time taken to reach a steady state, K_I and $\hat{K}_I$ = isotope concentration and elimination factors in the time I, n and $\hat{n}$ = coefficients characterizing the rate of concentration and elimination respectively, C_i = concentration of element in the environment, $K'(t)$ = derivative of the function and V_e = rate of exchange of the element.

The total exchange fund of calcium, strontium and some other elements in marine algae is, therefore, not uniform but is made up of a large number of microfunds. The statistical distribution of the exchange constants of these microfunds largely determines the form of the isotopic exchange function.

The calculated values of K for calcium-45 were 1.5 in *Ulva* and 5 in *Cystoseira* (50 in *Cystoseira* for strontium-89). The calculated values are also very similar to the experimental data of other authors (e.g., POLIKARPOV [1964]). BARINOV (1965a) has given data on the kinetics of cesium-137 and cerium-144 exchange. The rate of exchange of an element between an organism and the environment and the size of the exchange fund of the element in the organism may also be calculated from established isotopic exchange functions.

Table 17 sets out BARINOV's formulae [1965b] for calculation of the various parameters of radioisotopic exchange in relation to the form of the function $K(t)$.

Established features of isotopic exchange may therefore be used to forecast levels of concentration and the extent of deactivation of aquatic organisms, as well as to determine the rate of these processes in relation to specific radioecological and experimental conditions.

The experiments carried out by Barinov to establish the nature of exchange processes for calcium and strontium have demonstrated that when algae (*Ulva* and *Cystoseira*) are killed in boiling alcohol they continue to exchange these elements with the environment at the same rate as live algae. Calcium and strontium apparently occur in the exchange funds of the algae in a diffused and sorbed state. Non-exchangeable calcium and strontium are probably of different origin, and may have formed as a result of enzymatic reactions.

The kinetics of carbon-14 and phosphorus-32 concentration and exchange for the alga *Cystoseira barbata*, which are described by a power function, are related to enzymatic catalysis, since the alga practically ceases to concentrate these elements when killed in boiling alcohol (BARINOV [1965c]).

CHAPTER 3

CONCENTRATION OF RADIONUCLIDES OF THE FIRST GROUP OF ELEMENTS IN THE PERIODIC SYSTEM

3.1. General characteristics of the most important radionuclides and their carriers in water

Cesium-137, which is produced in considerable quantities in the fission of heavy nuclei (180000 Cur per megaton of a thermonuclear bomb), is the most important radionuclide of the first group. Cesium-137 yield in fission of uranium, thorium and plutonium nuclei by thermal and fast neutrons is between 6.15 and 6.63% (KATCOFF [1960]). Radiobiologically it is a hazard both as an emitter that accumulates in the organism and as an external source of gamma-radiation.

Very little is known concerning cesium as an element in living matter, because of its extreme dissemination in the earth. Studies of the behavior of its radioisotope cesium-137 (or of its other radioisotopes) will, therefore, also contribute directly to our knowledge of the biogeochemistry of this element.

The following figures have been given for cesium content in sea-water: in all seas and oceans (various data) $5 \times 10^{-7}, 4 \times 10^{-7}$ to $1.3 \times 10^{-6}, 2 \times 10^{-6}$ g/l (VINOGRADOV [1953], REVELLE et al. [1956], KRUMHOLZ, GOLDBERG et al. [1957]), in the Barents Sea 1.3×10^{-6} g/l and in the Black Sea 0.8×10^{-6} g/l (BUROVINA et al. [1964]). It is thought that of the cesium in sea-water 70% is in an ionic state, 23% adsorbed on particles and 7% in a colloidal state (table 13).

Potassium, which is the non-isotopic carrier of radiocesium, is one of the main components of sea-water, 3.8×10^{-1} g/l. The natural radionuclide potassium-40 produces an activity of 3.3×10^{-10} Cur/l in sea-water, or 90% of the total beta-activity of sea-water due to natural radionuclides and radioactive elements. The rubidium content of sea-water is 2×10^{-4} g/l and the activity of its natural radioisotope, rubidium-87, is 0.6×10^{-11} Cur/l, or approximately 1.6% of total natural radioactivity and approximately 1.8% of the radioactivity of potassium-40 (REVELLE et al. [1956]). According to other data, rubidium concentration in the seas and oceans is 1.2×10^{-4} g/l (KRUMHOLZ, GOLDBERG et al. [1957]). Valuable information has recently been obtained concerning concentrations of the elements of the main sub-

Table 18

Concentration factors for the stable isotopes of lithium, sodium, potassium, rubidium and cesium (from BUROVINA *et al.* [1964])

Aquatic organisms	Sea	Concentration factors				
		Li	Na	K	Rb	Cs
Coelenterates:						
Cyanea arctica						
muscle zone	Barents Sea	1.1	0.7	1.8	0.8	2.0
Metridium senile						
muscle wall	Barents Sea	0.7	0.4	6.2	8.9	3.7
Worms:						
Arenicola marina						
musculocutaneous sac	Barents Sea	0.4	0.2	7.6	10.9	10.2
Crustaceans:						
Eriphia spinifrons						
muscles	Black Sea	1.4	0.4	14.6	17.3	–
Carcinus maenas						
muscles	Black Sea	1.4	0.4	15.0	12.7	28.6
Hyas araneus						
muscles	Barents Sea	0.8	0.2	9.5	8.9	17.0
Molluscs:						
Mytilus edulis						
adductor muscle	Barents Sea	0.3	0.3	5.8	11.0	8.2
M. galloprovincialis						
adductor muscle	Black Sea	0.3	0.2	5.5	14.0	3.2
Pecten islandicus						
adductor muscle	Barents Sea	0.1	0.2	8.5	26.5	5.6
Rapana besoar						
muscles	Black Sea	0.4	0.1	12.3	11.1	15.0
Echinoderms:						
Cucumaria frondosa						
muscles	Barents Sea	0.5	0.3	8.4	13.7	7.0
Fishes:						
Trygon pastinaca						
muscles	Black Sea	0.4	0.2	16.1	23.8	13.3
Raja clavata						
muscles	Black Sea	0.8	0.2	17.7	16.7	22.5
Odontogadus euxinus						
muscles	Black Sea	0.2	0.2	20.0	14.6	13.8

Table 18 (continued)

Aquatic organisms	Sea	Concentration factors				
		Li	Na	K	Rb	Cs
Melanogrammus aeglefinus						
muscles	Barents Sea	0.1	0.02	11.0	7.6	10.9
Myoxocephalus scorpius						
muscles	Barents Sea	0.3	0.06	8.5	–	13.0
Gobiidae						
muscles	Black Sea	0.5	0.1	16.4	23.4	–
Trachurus trachurus						
muscles	Black Sea	0.5	0.1	16.4	15.4	15.0
Sea-water (g/l)	Barents Sea	150×10^{-6}	10.0	0.390	10×10^{-5}	1.3×10^{-6}
Sea-water (g/l)	Black Sea	90×10^{-6}	5.5	0.220	6×10^{-5}	0.8×10^{-6}*
Sea-water (g/l)	All seas and oceans	200×10^{-6}	10.5	0.340	12×10^{-5}	$(0.4–1.3) \times 10^{-6}$

* 0.18×10^{-6} (calculations by A. A. Bachurin).

group (lithium, sodium, potassium, rubidium and cesium) in the waters of the Barents and Black seas (table 18). The potassium content of freshwaters varies greatly, but is usually very low, 10^{-3} g/l.

As already noted (WILLIAMS and SWANSON [1958], WILLIAMS [1960]), there is a definite cesium-potassium metabolic ratio in living cells, which is disturbed only on death. Living cells accumulate potassium and exclude sodium, but cesium may replace potassium. In this connection a cesium unit has been suggested, i.e., one pCur of cesium-137 per gram of potassium.

Of the elements of the subsidiary subgroup, copper is represented by the neutron-induced radioisotope copper-64, which is of radioecological importance because it finds its way into waters with waste water from reactors. The concentration of stable copper in sea-water is 10^{-5} to 10^{-6} g/l (REVELLE et al. [1956]) or 3×10^{-6} g/l (KRUMHOLZ, GOLDBERG et al. [1957]). The content of silver is considerably lower at 3×10^{-7} g/l, and of gold even less at 6×10^{-9} to 4×10^{-9} g/l (REVELLE et al. [1956], KRUMHOLZ, GOLDBERG et al. [1957]).

3.2. Concentration and elimination of nuclides

It has been established that cesium-137 is rapidly concentrated by marine plants and animals, and that a balance is reached within a period of between a few days and two weeks (fig. 17) (CHIPMAN [1960], POLIKARPOV [1961b],

BRYAN [1961], BARINOV [1965a]). It has also been established that a state of equilibrium is reached in a similar period for sodium-22 in *Chlamydomonas* sp., *Artemia salina* and *Lingula reevi* (TOWNSLEY *et al.* [1960, 1961]).

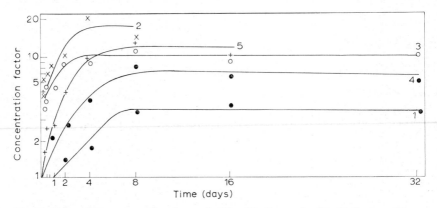

Fig. 17. Cesium-137 concentration factors in *Ulva* (1), *Cystoseira* (2), *Phyllophora* (3), sea anemones (4) and the body of mussels (5) over a period (POLIKARPOV [1961b]).

In order to study the state of accumulated cesium-137, *Chlorella* and *Euglena* cells were washed in inactive water after an experiment and fragmented. It was found that between 60 and 70% of the cesium-137 was firmly bound to the cell structures, and that the remainder consisted of free ions, or was bound to soluble substances in the cell fluids. Therefore, exchange of cesium-137 takes place not only between the environment and the cell, but also between ions in the solution and cesium atoms in a combined state. Young and dividing cells contain little combined cesium, but high activity due to combined cesium-137 is a feature of old *Euglena* cells and their detritus. A considerable proportion of the uncombined intracellular cesium is incorporated in cell structure as the cell ages (WILLIAMS [1960]).

Over a period of several weeks there was no noticeable loss in pure seawater of cesium-134, concentrated by *Rhodymenia palmata*. Elimination was intensified when stable cesium was added to the water up to a concentration of 10^{-4} g/l. When the algae were treated with boiling organic solvents or placed in distilled water there was rapid loss of cesium-134. Since this alga needs light and carbon dioxide to concentrate cesium-134, SCOTT [1954] assumes that the retention of cesium by live algae can be satisfactorily explained by the formation of complex cesium compounds with the intermediate products of photosynthesis.

The rate of elimination of cesium-137 and potassium-42 from crabs (*Carcinus maenas*) is slower than the rate of concentration (BRYAN [1961]).

Cesium-137 passes fairly rapidly from freshwater snails, and caddisfly and mosquito larvae into pure freshwater. Of the cesium concentrated, approximately 8% remained in the molluscs (body and shell) 12 days later, approximately 6% in the mosquito larvae, approximately 3% in the cases of caddisfly larvae after 32 days, and approximately 44% in the body (total for caddisfly larvae as a whole up to 33%). The addition of EDTA to the water intensified migration of cesium-137 into the environment (GETSOVA [1960]).

For the first four days elimination of cesium-137 from the fry of freshwater fishes is described by an exponential relation; 8% of the original amount of nuclide remains in the fry at the end of 20 days (WILLIAMS and PICKERING [1961]).

3.3. Concentration factors

3.3.1. Cesium

Published information on the concentration factors of stable cesium in marine organisms under natural conditions, and of its radioisotopes in experiments (of the order of 10^{-5} Cur/l), and in nature (10^{-13} Cur/l) is given in table 19. It is noteworthy that cesium-137 concentration factors for marine oozes are much higher than for marine biota.

If the figures in the three columns are compared it is seen that the concentration factors obtained by various methods are in good agreement. The lowest concentration factors for cesium and its radioisotopes (between zero and a few units) are a feature of green algae, diatoms, *Zostera*, lower crustaceans and the shells of molluscs. The concentration factors in brown and red algae, the muscles of higher crustaceans, the bodies of molluscs and the muscles of fishes range from one to several tens of units. It is of importance that, with few exceptions, the data are similar, although the investigations were carried out by various authors in the Pacific and Atlantic oceans, and in the Irish, Black and Barents seas. It is noteworthy that cesium-137 concentration factors for bottom material in marine shallows are measured in hundreds of units (BARANOVA and POLIKARPOV [1965]).

A large team (BUROVINA et al. [1964]) have recently obtained valuable information concerning the concentration factors of lithium, sodium, potassium, rubidium and cesium in the muscles of marine animals ranging from coelenterates to fishes in the Barents and Black seas (table 18). It is particularly noteworthy that the concentration factors for each element coincided in the related animals for the two seas. Moreover, the concentration factors of lithium and sodium were very low (less than or around unity), whereas those of potassium, rubidium and cesium reached 10 or more. It should be emphasized that studies of this type covering entire groups of chemical elements are of the greatest importance.

Table 19

Cesium concentration factors in marine organisms

Aquatic organisms	Concentration factors		stable cesium	Source
	radiocesium		stable cesium	
	in experiments	in nature	in nature	
Green algae:				
Chlamydomonas sp.	1	–	–	a
Carteria sp.	1	–	–	a
Chlorella sp.	2	–	–	a
Pyramimonas sp.	3	–	–	a
Nannochloris atomus	3	–	–	a
Ulva rigida	4	–	–	b
U. sp.	6	–	–	c
Monostroma sp.	1.2	–	–	d
Diatoms:				
Nitzschia closterium	1	–	–	a
Nitzschlia sp.	2	–	–	a
Amphora sp.	2	–	–	a
Brown algae:				
Fucus vesiculosus	30*	–	–	e
	–	–	44–50	f
F. serratus	–	53	–	g
	–	–	36	f
Cystoseira barbata	27	–	–	b
Sargassum natans	10	–	–	i
S. fluitans	12	–	–	i
Ascophyllum nodosum	–	–	38–42	f
Laminaria digitata	–	–	20–26	f
L. saccharina	–	–	16–28	f
Scytosiphon lomentarius	2.0	–	–	d
Red algae:				
Porphyridium curentum	1	–	–	a
Gracilaria confervoides	1.6	–	–	d
Porphyra umbilicalis	50	–	–	j
	–	–	16.5	f
Phyllophora nervosa	10	–	–	b
Rhodymenia palmata	–	–	22–28	f

Table 19 (continued)

Aquatic organisms	Concentration factors			Source
	radiocesium		stable cesium	
	in experiments	in nature	in nature	
Algae	–	70	–	g
Protozoa				
Cephidium crispum	10*	–	–	t
Flowering plants:				
Zostera marina	2	–	–	b
Coelenterates:				
Cyanea arctica				
muscle zone	–	–	2	l
Metridium senile	10.1–11.6	–	–	t
muscle wall	–	–	3.7	l
Actinia equina	7	–	–	a
	–	–	2.7	h
Worms:				
Arenicola marina				
musculocutaneous sac	–	–	10.2	l
Nereis diversicolor	6.3	–	–	t
Crustaceans:				
Tigropus californicus	1*	–	–	m, n
Artemia salina	2*	–	–	m, n
Centropages	–	–	0.1–1	o
Calanus helgolandicus	13*	–	–	t
Palaemonetes pugeo	25	–	–	m, n
Leander sp.				
exoskeleton	15	–	–	d
soft tissues	15	–	–	d
Carcinus maenas				
whole	8	–	–	p
muscles	18	–	–	p
	–	–	19.1–26.8	h,l
Homarus vulgaris	8	–	–	q
Hyas araneus				
muscles	–	–	17.0	l
Portunus ruber				
whole	7	–	–	p
muscles	13	–	–	p
P. depurator				
whole	7	–	–	p
muscles	12	–	–	p
Polybius henslowi	5	–	–	p

Table 19 (continued)

Aquatic organisms	Concentration factors		stable cesium	Source
	radiocesium			
	in experiments	in nature	in nature	
Cancer pagurus				
whole	6	–	–	p
muscles	11	–	–	p
Corystes cassivelsunus				
whole	4	–	–	p
muscles	11	–	–	p
Molluscs:				
Mytilus galloprovincialis				
shell	0	–	–	b
body	12	–	–	b
adductor muscle	8.4–9.8	–	3.2	l,t
M. edulis				
adductor muscle	7.8–8.8	–	8.2	l,t
Venus mercenaria				
body	6*	–	–	a
Venerupis philippinarum				
gills	8.1	–	–	d
adductor muscle	8.6	–	–	d
viscera	8.5	–	–	d
Meretrix meretrix				
gills	9.3	–	–	d
adductor muscle	8.6	–	–	d
viscera	6.8	–	–	d
Pecten irradians				
body	8*	–	–	a
body without adductor				
muscle	8	–	–	m
adductor muscle	10*	–	–	m
P. islandicus				
adductor muscle	–	–	5.6	l
Rapana besoar				
muscles	–	–	15.0	l
Echinoderms:				
Cucumaria frondosa				
muscles	–	–	7.0	l
Strongylocentrotus pulcherrimus				
digestive tract	9.3	–	–	d
gonad	2.0	–	–	d
test	0.41	–	–	d

Table 19 (continued)

Aquatic organisms	Concentration factors		stable cesium	Source
	radiocesium			
	in experiments	in nature	in nature	
Invertebrates:				
soft tissues	–	–	10	k
Fishes:				
Rhombus maeoticus				
eggs (before hatching)	8.8	–	–	r
larvae (just hatched)	10.3	–	–	r
larvae (48 hr old)	14.1	–	–	r
Engraulis encrasicholus ponticus				
eggs (before hatching)	9.0	–	–	r
larvae (just hatched)	9.4	–	–	r
larvae (42 hr old)	26.1	–	–	r
Fundulus heteroclitus				
body muscles	35*	–	–	m
Micropogon undulatus				
heart, spleen and liver	13*	–	–	a
brain	10*	–	–	a
muscles	6*	–	–	a
Trygon pastinaca				
muscles	–	–	13.3	l
Raja clavata				
muscles	–	–	22.5	l
Melanogrammus aeglephinus	–	–	10.9	l
Pleuronectes platessa				
whole	6	–	–	j
muscles	7	–	5	j
Chasmichthys gulosus				
muscles	5–10	–	–	c
Paralichthys dentatus	10	–	–	m
Leiostomus				
muscles	23	–	–	m
Trachurus trachurus				
muscles	–	–	15.0	l
Acanthogobius flavimanus				
digestive tract	12	–	–	d
heart and spleen	7.3	–	–	d
liver and gallbladder	5.9	–	–	d
gill	3.0	–	–	d
skin with scales	2.8	–	–	d

Table 19 (continued)

Aquatic organisms	Concentration factors			Source
	radiocesium		stable cesium	
	in experiments	in nature	in nature	
vertebra	3.5	–	–	d
muscles	25	–	–	d
Odontogadus euxinus				
muscles	–	–	13.8	l
Myoxocephalus scorpius				
muscles			13.0	l
Vertebrates:				
soft tissues	–	–	10–42	k,s

* Balance concentration not reached.

a   BOROUGHS, CHIPMAN *et al.* [1957].
b   POLIKARPOV [1961a, b].
c   ICHIKAWA [1961], calculated from the data of SHIMIZU and HIYAMA [1959].
d   HIYAMA and SHIMIZU [1964].
e   SCOTT [1954].
f   SMALES and SALMON [1955].
g   TEMPLETON [1962].
'h   NESTEROV and SKUL'SKIY [1965].
i   POLIKARPOV, ZAYTSEV *et al.*, in the press.
j   TEMPLETON [1959].

k   KRUMHOLZ, GOLDBERG *et al.* [1957].
l   BUROVINA *et al.* [1964].
m   CHIPMAN [1958a, b].
n   CHIPMAN [1960].
o   KETCHUM and BOWEN [1958].
p   BRYAN [1961].
q   BRYAN and WARD [1962].
r   IVANOV [1965a, b].
s   FUKAI and YAMAGATA [1962].
t   BRYAN [1963].

A completely different pattern is revealed by work on freshwater organisms (tables 20–22). First, the concentration factors of radiocesium are considerably (two or three orders) higher in freswater organisms than in marine organisms. Second, the figures differ greatly from locality to locality. Thus, the cesium-137 concentration factors in aquatic organisms from the Columbia River and from the Ohio River, which is a tributary of the Mississipi River, are in general far higher than those for aquatic organisms from Lake Bol'shoye Miassovo in the Urals. When comparing the data in tables 20–22, it should be remembered that the concentration factors in tables 21 and 22 have been calculated in relation to dry weight, which clearly makes exact comparison difficult. A rough conversion factor of 10 may be employed for conversion from dry to wet weight (for a 90% water content), although values range between three in organisms with a high mineral content and 30 in those with a low mineral content. Wet and dry weight are, obviously, the same or similar for the shells of molluscs.

The chemical composition of sea-water is fairly constant, whereas that of freshwater is highly variable and is affected by the geological nature of the surrounding rocks and the geochemical composition of the waters supplying a lake or reservoir. Because chemical content is rarely similar in any two bodies of freshwater, TEMPLETON [1962] is justifiably of the opinion that each situation has to be treated as a separate case. In the United States, for example, sodium concentration in freshwaters varies between 10^{-3} and 2×10^{-1} g/l, that of calcium between 2×10^{-3} and 2×10^{-1} g/l, of silicon between 3×10^{-3} and 2×10^{-2} g/l, of phosphorus between 10^{-6} and 1.5×10^{-3} g/l and of iron between 10^{-5} and 6×10^{-3} g/l (KRUMHOLZ and FOSTER [1957]). In Lake Bol'shoye Miassovo in the Urals the concentrations of the main cations are 10^{-2} g/l for sodium and potassium, 0.9×10^{-2} g/l for magnesium, and 2.5×10^{-2} g/l for calcium (TITLYANOVA and IVANOV [1960]).

3.3.2. Potassium and sodium

The explanation for the difference between the concentration factor of potassium in the marine alga *Valonia* (50) and in the freshwater algae *Nitella* (1 000) and *Hydrodyction* (4 000) (BLINKS and NIELSEN [1940]) will be apparent from what has already been said. Potassium concentration in freshwater is approximately 10^2 times lower than in sea-water. This great difference in potassium content undoubtedly explains why cesium concentration factors are high in freshwater organisms and low in marine organisms. When information on cesium-137 concentration in marine organisms (POLIKARPOV [1961a]) was compared with information on its accumulation in freshwater organisms (TIMOFEYEV-RESOVSKIY [1957], TIMOFEYEVA-RESOVSKAYA, POPOVA et al. [1958], POLIKARPOV [1958], TIMOFEYEVA-RESOVSKAYA and TIMOFEYEV-RESOVSKIY [1958], TIMOFEYEVA-RESOVSKAYA et al. [1959], TIMOFEYEV-RESOVSKIY and TIMOFEYEVA-RESOVSKAYA [1959], TIMOFEYEV-RESOVSKIY, TIMOFEYEVA-RESOVSKAYA et al. [1960]) it was found that the concentration factors of this nuclide were, on average, one order of magnitude lower in the marine organisms.

Sodium-24 concentration factors are also fairly high in freshwater organisms: 500 in phytoplankton and filamentous algae, and 100 in insect larvae and fishes in the Columbia River (KRUMHOLZ and FOSTER [1957]) and 30 in goldfish under experimental conditions (PROSSER et al. [1945]). FRETTER [1955] studies some questions connected with uptake of sodium-23 by marine worms.

Table 20

Concentration factors for radioactive cesium in freshwater organisms

Aquatic organisms	Radiocesium concentration factors		Source
	in experiments	in nature	
Mixed plankton	1 000–25 000	–	a
Flagellata:			
Euglena intermedia	706	–	b
Green algae:			
Chlamydomonas sp.	52	–	b
Gonium pectorale	138	–	b
Chlorella pyrenoidosa	154	–	b
Oedogonium vulgare	790	–	b
O. sp.			
Rhizoclonium crassipelitum	1 500–4 000	1 200	a
R. sp.			
R. hierogliphicum	1 530	–	b
Oocystic elliptica	670	–	b
Spirogyra sp.	400	–	a
S. ellipsospora	341	–	b
S. cummunis	220	–	b
Flowering plants:			
Elodea canadensis	1 000	–	a
Ceratophyllum demersum	400	–	a
Potomogeton pactinatus	700	–	a
Azolla filiculoides	250	–	a
Lemna minor	500	–	a
Scirpur americanus			
stem	50	–	a
seeds	300	–	a
S. acutis			
stem	90	–	a
seeds	400	70	a
Typha latifolia			
leaves	600	–	a
seeds	100	–	a
Polygonum lapathifolium			
seeds	–	240	a
Crustaceans:			
Hyalella azteca	11 000	–	a
Aselus aquaticus	60–80	–	d
Insects (nymphs):			
Ischnura sp.	800	–	a
Erythemis callocata	800	–	a

Table 20 (continued)

Aquatic organisms	Radiocesium concentration factors		Source
	in experiments	in nature	
Molluscs:			
Radix japonica	600	–	a
Fishes:			
Cyprinus carpio			
muscles	3000	–	a
Lepomis gibbosus			
mature fish, whole	7500	–	a
muscles	9500	–	a
L. macrochitus			
muscles	900	–	c
bones	600	–	c
Macropterus salmoides			
muscles	1200	–	c
bones	500	–	c
Ictalurus natalis			
muscles	1200	–	c
bones	800	–	c
Amphibians:			
Rana catesbyana			
mature	8000	–	a
tadpoles			
whole	2600	–	a
viscera	4500	–	a
body	1000	–	a
Scaphiopus hammondi intermontanus			
tadpoles	6000	–	a
Waterbirds:			
Fulica americana			
muscles	–	1800	a
liver	–	2200	a
bones	–	800	a
Anas platyrhynchos			
muscles	–	2000	a
liver	–	2500	a
bones	–	700	a
Oxyura jamaicensis rubida			
muscles	–	2200	a
liver	–	2800	a
bones	–	900	a

a   PENDLETON and HANSON [1958].   c   HARVEY [1964].
b   WILLIAMS and SWANSON [1958].   d   BRYAN [1963].

Table 21

Concentration factors for 20 radionuclides in various species of freshwater plants calculated in relation to dry weight (TIMOFEYEVA-RESOVSKAYA [1963])

Aquatic organisms	^{32}P	^{35}S	^{45}Ca	^{51}Cr	^{59}Fe	^{60}Co	^{65}Zn	7
Aquatic bacteria	–	–	–	–	–	–	–	–
Bacteria from crustose lichens	–	5	–	–	555	370	1 555	–
Scenedesmus quadricauda Bret.	–	500	70	–	9 000	390	775	–
Scenedesmus acuminatus Chodat.	–	400	45	–	35 000	250	–	–
Cladophora fracta Kütz.	76 750	2 600	330	–	26 800	8 750	6 110	–
Cladophora glomerata Kütz.	135 000	165	335	–	31 500	1 985	3 900	–
Mougeotia sp.	2 100	575	140	–	34 000	238 000	36 600	–
Spigogyra crassa Kütz.	1 750	650	190	–	1 660	17 000	1 950	–
Spirogyra sp.	–	795	175	–	3 120	5 640	31 500	–
Chara sp.	–	85	480	–	–	–	4 600	–
Chara aspera Wild.	–	45	345	–	–	–	2 270	–
Chara fragilis Desw.	–	60	330	–	15 500	7 425	2 860	–
Ricciocarpus natans L.	–	200	285	–	–	–	19 000	–
Fontinalis sp.	–	–	–	–	–	–	–	–
Ranunculus conferoides Fries.	–	460	285	–	–	–	2 460	1
Myriophyllum spicatum L.	–	40	220	695	530	3 500	3 980	2
Ceratophyllum demersum L.	5 750	165	215	470	4 510	4 665	4 740	■
Lythrum sp.	–	–	–	–	–	–	–	
Cicuta virosa L.	–	–	–	–	–	–	–	
Utricularia vulgaris L.	4 400	30	185	–	–	11 650	3 900	–
Typha angustifolia L.	–	–	–	–	–	–	–	
Calta palustris L.	–	–	–	–	–	–	–	
Lemna minor L.	–	200	235	–	9 050	3 900	8 950	
Lemna trisulca L.	11 000	75	250	480	–	14 000	28 300	
Lemna polyrrhyza L.	–	95	290	–	–	–	6 100	
Potamogeton natans L.	7 200	–	–	–	–	–	–	
Potamogeton compressus L.	–	–	–	–	–	–	–	
Potamogeton filiformis Pers.	–	50	425	–	–	–	8 060	
Potamogeton perfoliatus L.	–	65	485	–	–	–	5 800	
Elodea canadensis Rich.	4 400	135	500	480	4 735	3 490	3 110	
Vallisneria spiralis L.	–	–	–	–	–	–	–	
Stratiotes aloides L.	–	–	–	–	–	4 900	–	
Hydrocharis morsus ranae L.	11 000	205	410	260	–	5 430	6 900	
Carex sp.	1 730	95	330	–	6 865	4 595	7 245	
Mean	23 735	335	290	475	12 775	18 665	8 690	

Rb	^{90}Sr	^{91}Y	^{95}Zr	^{95}Nb	^{106}Ru	^{115}Cd	^{131}I	^{137}Cs	^{144}Ce	^{147}Pm	^{203}Hg	Mean
	440	–	–	–	855	–	–	100	11140	–	–	3160
	205	–	–	–	100	–	–	200	740	–	–	465
0	130	22050	5650	–	3000	–	–	280	20700	–	–	5245
	105	1530	13240	–	760	–	–	390	38500	–	–	9020
280	1910	119625	32300	5800	2550	16200	–	1230	35600	–	5880	17865
275	900	40000	19520	6950	1280	17400	–	1565	31000	–	5410	10945
	190	15600	4370	4130	9300	2810	–	–	5400	–	–	29260
	235	6860	6815	20900	2330	3270	–	285	41200	–	–	7950
25	550	13990	71250	–	3200	–	–	1920	12800	–	–	12270
	350	3790	3840	8020	815	–	–	200	4120	–	–	2630
0	280	5825	6920	12850	1200	–	–	180	6800	–	–	3670
	400	6335	14150	4885	3750	–	–	365	19510	–	–	6295
75	590	20425	8560	20500	2430	–	–	715	9800	–	–	8250
	360	–	–	–	3320	–	–	1020	9650	–	–	3585
05	465	–	6790	11430	1550	–	–	765	7880	2900	–	3195
0	445	4710	1965	6000	885	2720	410	295	4100	4850	6420	2335
0	510	920	5280	2340	1900	985	370	300	11250	5300	7700	3025
	50	–	–	–	–	–	–	185	–	–	–	115
	130	–	–	–	165	–	–	140	–	–	–	145
60	665	6935	2800	16500	1885	700	835	340	6050	–	6700	4240
	25	–	–	–	–	–	–	20	–	–	–	22
	45	–	–	–	210	–	–	70	–	–	–	110
30	400	4425	6930	6670	2810	–	285	2405	15000	–	–	4660
80	315	6170	3660	15890	2920	1530	615	940	10500	6400	6350	6435
00	590	8435	6180	5200	3325	–	–	1050	5520	–	–	3680
	670	–	–	950	1055	–	–	205	1370	–	–	1910
0	1020	–	–	3800	1190	–	–	115	2435	–	–	1710
5	585	11670	2445	725	1990	–	–	185	3255	–	–	2940
0	685	4740	1725	1110	750	–	105	195	1925	–	–	1600
5	805	2120	3600	5040	1125	1510	535	285	5300	2650	4700	2380
	220	–	–	–	270	–	–	360	7100	–	–	1985
	615	–	–	–	375	–	240	375	–	–	–	2235
50	415	5000	1535	2925	885	1070	660	535	2485	3500	4180	2530
0	160	6535	4925	5440	–	–	–	670	6730	–	–	3330
60	470	14440	10195	7660	1875	4820	405	545	11660	4265	5915	–

Table 22

Concentration factors for various radionuclides in certain species of freshwater animals calculated in relation to dry weight (TIMOFEYEVA-RESOVSKAYA [1963])

Aquatic organisms	^{32}P	^{35}S	^{45}Ca	^{51}Cr	^{59}Fe	^{60}Co	^{65}Zn	7
Worms:								
Herpobdella sp.	1 070	4	–	–	330	275	125	–
Molluscs:								
Anadonta cellensis Schröter.	165	–	–	–	–	1 090	–	–
Limnaea stagnalis L.	2 530	–	–	–	125	325	155	7
Radix auricularia L.	1 740	–	–	–	–	925	220	–
Radix ovata Drap.	2 400	–	–	–	–	–	–	–
Galba palustrix Müll.	20 800	–	–	–	–	1 160	–	–
Bithynia tentaculata L.	3 180	–	–	–	–	560	345	–
Aplexa hyphorum L.	35 600	35	–	–	720	1 380	165	–
Anisus vortex L.	13 200	–	–	–	–	870	–	–
Planorbis planorbis L.	–	–	–	–	–	800	–	–
Crustaceans:								
Diaptomus graciloides	–	–	–	–	–	–	–	–
Mecocyclops sp.	–	–	–	–	–	–	–	–
Chydorus sphaericus	–	–	–	–	–	–	–	–
Rivulogammarus lacustris Sars.	85	85	–	–	350	1 100	1 820	6
Larvae and insects:								
Culex pipiens pipiens L.	–	292	155	–	–	4 010	16 400	–
Theobaldia alascensis Ludl.	–	–	–	–	–	–	–	–
Halesus interpunctatus Lett.	–	7	–	240	410	8 500	450	–
Leptocerus sp.	–	–	–	–	–	–	–	–
Phryganea grandis L.	–	–	–	–	–	–	–	–
Aeschna sp.	–	2	–	–	–	635	1 930	–
Lestes sp.	–	–	–	–	–	–	–	–
Eristalis sp.	–	8	–	–	–	390	–	–
Tendipes sp.	–	–	50	–	680	375	–	–
Vertebrates:								
Cyprinus caprio L. × *C. c. haematopterus* Tem. et Schl.	–	14	35	–	–	60	–	6
Rana sp. (tadpoles)	–	–	–	–	–	–	–	6
Mean	8 075	55	80	240	435	1 405	2 400	1

[86]Rb	[90]Sr	[91]Y	[95]Zr	[95]Nb	[106]Ru	[115]Cd	[131]I	[137]Cs	[144]Ce	[147]Pm	[203]Hg	Mean
180	8	145	90	480	30	–	50	10	125	34	–	195
–	85	–	455	275	170	–	–	90	370	–	–	338
260	3100	2020	270	420	130	–	115	215	2180	325	–	670
–	60	–	–	–	100	–	70	290	1050	365		535
–	320	–	–	–	295	12	–	390	7130	–		1750
370	585	–	–	–	100	–	–	–	2300	780	590	3335
–	170	–	–	–	340	2	–	690	3300	360	–	995
–	505	–	530	–	290	–	–	510	7220	1250	–	4380
230	305	–	1660	–	–	–	265	60	–	1920	–	2315
–	210	–	–	–	–	–	350	60	–	1080	–	500
–	90	–	–	–	1670	–	–	1270	6460	–	–	2370
–	170	–	–	–	710	–	–	530	–	–	–	470
–	370	–	–	–	2250	–	–	1020	–	–	–	1215
290	400	–	120	–	440	–	–	235	1580	–	–	550
–	355	–	–	–	1140	–	–	245	9050	–	–	3955
–	–	–	–	–	10	–	–	40	12	–	–	20
–	12	–	300	–	400	40	–	120	1100	270	1100	995
–	20	–	–	–	63	–	–	20	360	–	–	115
–	18	–	–	–	500	–	–	680	2430	–	–	905
–	7	–	–	–	125	–	–	100	2600	300	–	710
–	4	–	–	–	90	–	–	315	325	–	–	185
–	60	–	–	–	130	–	–	35	805	1060		350
–	105	123	–	–	500	–	–	530	1020	–	–	425
135	160	–	60	–	80	–	10	55	820	–	–	60
–	210	–	–	–	1510	–	1275	–	70100	–	–	905
245	305	765	435	390	480	25	305	325	5730	705	845	–

The difference between the concentration factors of potassium and other macroelements for freshwater and marine algae is depicted in fig. 18 (SABININ [1955]).

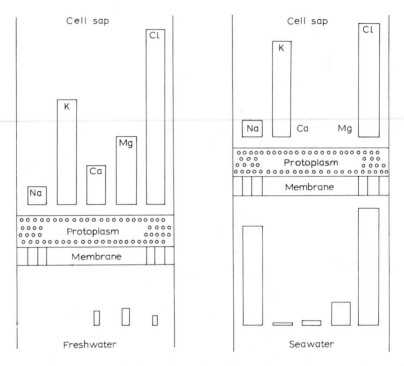

Fig. 18. Ion concentration in the surrounding medium and in the cell sap of *Nitella* (left) and *Valonia* (right). The scale of the diagram showing the concentration of substances in the cell sap of *Nitella* is 10 times that depicting the concentration of substances in the cell sap of *Valonia* (HOAGLAND and BROYER [1936], cited by SABININ [1955]).

3.3.3. Copper, silver and gold

The concentration factors of stable copper are fairly high in marine organisms: 100 in noncalcareous algae, 5000 in soft tissues and skeletons of invertebrates and 1000 in soft tissues and skeletons of vertebrates (KRUM-HOLZ, GOLDBERG *et al.* [1957]).

The Columbia River copper-64 has been found to accumulate with a concentration factor of 2000 in phytoplankton, 500 in filamentous algae and insect larvae, and 50 in fishes (KRUMHOLZ and FOSTER [1957]).

Crab shell accumulates silver-110 with a concentration factor of 150 (Zesenko [1965]).

Contents of gold and other microelements in marine organisms were studied by Fukai and Meinke [1962].

Gold-198 concentration factors are hundreds of units in marine green, brown and red algae, flowering plants, crustaceans and molluscs (Polikar-pov [1963]), while that of stable gold in the soft tissues of fishes is 60 under natural conditions (Pillai and Ganguly [1961]).

3.4. Distribution of radionuclides by organs and tissues

It has been established by radioautography that the distribution of cesium-134 is almost uniform in *Rhodymenia palmata*. The pattern was similar in *Laminaria digitata* with cesium-134 uniformly distributed throughout the thallus, with some increase of activity, as in *Rhodymenia*, in the meristematic zone and, to a lesser extent, at the free end of the thallus (Scott [1954]).

In land animals, muscles and tissues with a high potassium content are the depot for cesium-137 (Zakutinskiy [1959]). This also holds for aquatic organisms (tables 18 and 19), but with a number of specific distribution features associated with the nature of the habitat.

According to data from the author's laboratory, the concentration factors of cesium-137 in crabs (*Carcinus maenas*) are highest in the muscles, but the greater part of the cesium-137 (in terms of total content in the crab) is to be found in the shell (approximately 50%). The activity of the muscles accounts for approximately 20%, of the liver for less than 10%, and of the remaining tissues and organs for less than 1%.

This distribution pattern is apparently related to the high specific activity (radionuclide/radionuclide plus carrier) of cesium-137 in sea-water and, therefore, to its enhanced capacity for adsorption. As an element which imitates potassium, cesium is incorporated in considerable quantity and with the highest concentration factor in muscle tissue, but since it is a readily adsorbed element present in the water in insignificant concentration, it is even more significantly concentrated by the surface of the shell.

In the freshwater mollusc *Anodonta celensis* the largest amount of cesium-137 (approximately 40%) is concentrated in the shell, and there is approximately 20% in the visceral mass, 15% each in gills and muscles and approximately 10% in the mantle. If only the body activity is considered the figures are approximately 37% in the visceral mass, 24% each in gills and muscles and 15% in the mantle (Getsova *et al.* [1964]).

When young tunny were fed on cesium-137 it was rapidly incorporated in the liver, heart and kidneys. These organs lost cesium-137 throughout the following week. The rate of concentration of the radionuclide in the muscles, gonads, brain and skin was more rapid than its rate of elimination from them. A study has also been made of the uptake of cesium-137 by sea fishes from a solution maintained at a constant level of activity by daily changing. On the 28th day, concentration had still not reached a stable level in the heart, spleen, liver, brain and muscles. The cesium-137 concentration factor in the first three organs was more than twice its concentration factor in the muscles (BOROUGHS, CHIPMAN et al. [1957]).

The relative concentration of cesium-137 by various organs is approximately the same for different species of fish. Concentration was of the same order of magnitude when cesium-137 was taken up with food in sea-water containing the nuclide, and when cesium-137 was directly extracted from sea-water without feeding (BOROUGHS, CHIPMAN et al. [1957]).

Silver-110 concentration factors are 10 to 20 in the heart, gonads and muscles of the crab *Carcinus maenas*, 80 in the liver and gills, and 150 in the shell (ZESENKO [1965]).

CHAPTER 4

CONCENTRATION OF RADIONUCLIDES OF
THE SECOND GROUP
OF ELEMENTS IN THE PERIODIC SYSTEM

4.1. General characteristics of the most important radionuclides and their carriers in water

In addition to being the main representative of artificial radionuclides of the second group, strontium-90 is one of the most important and biologically harmful components of fission products. Although the concentration of stable strontium in sea-water is fairly high $(0.81 \times 10^{-2}$ g/l), there has been very little biological study of this element.

The biogeochemistry of strontium will throw some light on its connection with calcium in various aquatic organisms. The calcium content of sea-water is high at 4×10^{-1} g/l. As already reported (table 13) the strontium in sea-water is mainly in an ionic state (87%), 10% is adsorbed on to particles, and only 3% is represented by colloids. The connection between strontium and its non-isotopic carrier (calcium) provided the basis for definition of a unit of strontium-90 concentration in organisms and in the environment known as the strontium unit or sunshine unit, which is one pCur of strontium-90 per gram of calcium. Another unit, known as a moonshine unit, is defined as the ratio of one pCur of strontium-90 to the total amount of strontium in the organism.

There is no biogeochemical information concerning beryllium (MAUCHLINE [1961]). The concentration of magnesium in sea-water is 1.272 g/l, that of barium 0.62×10^{-5} g/l and that of radium-226 only 2×10^{-15} to 3×10^{-14} g/l. The radioactivity of the last-named element in sea-water is 2.7×10^{-13} Cur/l, i.e., approximately 0.07% of the total natural beta-activity of sea-water, or approximately 0.08% of the radioactivity of potassium-40 (REVELLE and SCHAEFFER [1957], KRUMHOLZ, GOLDBERG et al. [1957]).

Calcium concentration varies widely in freshwater lakes, but not in sea-water: the range in the United States is between 2×10^{-3} and 2×10^{-1} g/l (KRUMHOLZ and FOSTER [1957]), while calcium content in the waters of Lake Bol'shoye Miassovo is 2.5×10^{-2} g/l (TITLYANOVA and IVANOV [1960]).

The following figures have been given for radium concentration in freshwaters: 0.7×10^{-13} Cur/l (mean) in USA rivers, $(0.2$ to $1.0) \times 10^{-12}$ in tap water in Sweden, 0.4×10^{-13} (mean) and 7.0×10^{-12} (maximum) in tap water in the USA, 1.0×10^{-12} (mean) in the USSR, 0.6×10^{-12} at Gastein in Austria, $(1.7 \text{ to } 3.1) \times 10^{-13}$ at Frankfurt on Main in Germany; and in particular zones in spring water in the USA at Boulder, Colorado, 3×10^{-7}, in Japan 0.7×10^{-6}, at Jachymov in Czechoslovakia 0.5×10^{-6}, at Gastein in Austria 10^{-7}, and in France $(3 \text{ to } 4) \times 10^{-11}$ Cur/l (TEMPLETON [1962]).

The most important radionuclide from among those of the elements of the secondary subgroup of the second group is zinc-65, which is formed in the environment as a neutron-induced product.

Stable zinc is a typical biogenous trace element incorporated in many enzymes and hormones and involved in a number of metabolic links in animals and plants. It has already been stated that it is directly related to photosynthetic activity. The concentration of zinc has been found to be 1×10^{-5} g/l and that of cadmium 5.5×10^{-8} g/l in sea-water (KRUMHOLZ, GOLDBERG et al. [1957]).

4.2. Concentration and elimination of nuclides

4.2.1. Uptake of strontium

Concentration of radiostrontium from sea-water reaches a balanced accumulation within a few days in marine algae and invertebrates, and within weeks in fishes, except in the case of mineralized tissues of slow growth, in which the concentration of this element increases over a long period (BOROUGHS et al. [1956a], CHIPMAN [1960], POLIKARPOV [1960c, 1961d]).

According to the author's data, concentration of strontium-90 by marine algae and sea-grass reaches a stable level within four to eight days from commencement of an experiment (POLIKARPOV [1960c, 1961a]). This time is 2 to 4 days for Cystoseira, but within a minute for Ulva (BARINOV [1965b,c]). In the unicellular Carteria accumulation of this nuclide is directly dependent on mitotic rate. The strontium-90 content of nondividing cells is $1/64^{th}$ that of a dividing culture 72 hr after the introduction of radioactivity (calculated for a single cell). The higher the mitotic rate of Carteria, the greater is the concentration factor of this radionuclide. RICE [1956] is therefore of the opinion that Carteria cells are incapable of avoiding the penetration of strontium or of eliminating it, and that therefore the amount of strontium in a single cell or in a gram of cells increases in course of time.

Artemia salina larvae concentrate strontium-$(89+90)$ to a balanced accumulation in a few hours. Molluscs, shrimps and crabs accumulate strontium-$(89+90)$ rapidly from sea-water (BOROUGHS, CHIPMAN et al.

[1957]). Strontium-90 concentration continues to build up in mussel shells as late as two months after the commencement of an experiment, but reaches a stable level in the body of the mussel within 12 hr (POLIKARPOV [1960c]).

Mature fishes (*Tilapia mossambica*) concentrate strontium-89 to saturation point in 15 days with a concentration factor of 0.3. In other words, they discriminate against strontium (BOROUGHS *et al.* [1956a]). The strontium-calcium ratio in sea fishes is 1/400 or less (800 moles of calcium to one mole of strontium), while in sea-water this ratio is 1/40, i.e., strontium is either concentrated more slowly or lost more rapidly than calcium (BOROUGHS, TOWNSLEY *et al.* [1957]). It is noteworthy that the retention of strontium is half that of calcium in rats and mice following peroral and intravenous administration of the isotopes of these elements (NORRIS and KISIELESKI [1948]).

In freshwater organisms strontium-90 concentration reaches a stable level within a few hours of commencement of an experiment in *Scenedesmus* sp. and *Daphnia*, within 10 to 130 days in *Elodea*, the body of molluscs, and the internal organs and muscles of fishes, and within 2 to 3 months in the shells of molluscs, and the scales and bones of fishes (MAREY *et al.* [1958]). According to the data of other authors accumulation of various nuclides reaches a stable level in freshwater plants and animals in the main within periods of between a few days and two weeks (TIMOFEYEV-RESOVSKIY, TIMOFEYEVA-RESOVSKAYA *et al.* [1960]).

4.2.2. Elimination of strontium

BARINOV [1965a,b] has demonstrated that if strontium-89 were to accumulate in *Cystoseira* until a state of kinetic equilibrium was reached, this nuclide would be completely returned to the environment when the alga was placed in pure sea-water, and that return would take the same time (would occur at the same rate) as concentration of radiostrontium in this alga with its maximum level of concentration factor. According to Barinov's data the same feature applies to the radioisotopes of calcium, yttrium, cerium and cesium when they are exchanged by mature macrophytes.

The cells of *Carteria* sp. hold on tenaciously to the strontium-90 which they have concentrated. Thus, when placed in inactive artificial sea-water after 12 days in a radiostrontium solution they had lost less than 2% of the nuclide even after 48 hr, but in distilled water had lost 35% after 65 hr.

A complexing agent (EDTA sodium salt) intensifies elimination of strontium-90 in proportion to its concentration (RICE [1956]). The rate of elimination of strontium-90 from *Cystoseira barbata* is dependent on the time for which the thallus has previously been in active sea-water. The strontium-90 output curve is exponential within 24 hr in pure sea-water. When the thallus had previously been in a strontium-90 solution for three hr, the elimination constant of the nuclide (the rate of elimination of strontium per unit of its concentration) was 3.7×10^{-2}, and when it had been in the solution for two days the elimination constant was 2.3×10^{-2} (POLIKARPOV and TEN [1962]). When strontium-89 was fed to mobile pelagic fishes in gelatin capsules it was rapidly eliminated: within 24 hr no more than approximately 2% of the amount of radionuclide administered remained in the fishes. The relatively immobile bottom-dwelling species *Tilapia mossambica* loses up to 5% of strontium-89 in four days. In other words the loss of strontium-89 from the organism is directly related to metabolic rate in fishes.

Soft visceral tissues eliminate strontium-89 more rapidly than bones, gills, the integument and muscles (BOROUGHS, CHIPMAN et al. [1957]). When strontium-89 is administered by intramuscular injection it is retained by fishes for a long period: only 30 to 40% of the activity was lost in 14 days (BOROUGHS et al. [1956a]).

EDTA in the solution retards the elimination of strontium-90 from mosquito and caddisfly larvae as a result of formation of a strontium-EDTA complex, which is less stable than the calcium-EDTA complex. Considerable quantities of calcium are eliminated in the form of the calcium complex compound; this disrupts calcium metabolism and, as a result, intensifies the concentration and retention of its chemical analogue strontium, which is more available under these conditions (GETSOVA et al. [1960], TIMOFEYEV-RESOVSKIY and TIMOFEYEVA-RESOVSKAYA [1959]).

After the molluscs *Unio pictorum* had been kept in a strontium-90 solution, the strontium-90 was redistributed when they were kept in pure freshwater. At the end of 44 days its content was reduced in the gills, foot, mantle and internal organs (42 to 89%), but had increased in the whole shell and the shell edge to 120% and 458%, respectively. It is assumed that strontium is eliminated via the gills (POVELYAGINA and TELITCHENKO [1959]). Two hundred days after radioactive water had been replaced by pure water, the muscles of freshwater fishes still contained 10%, and the bones 50% of the strontium-90 (MAREY et al. [1958]).

4.2.3. Uptake of calcium

There is rapid exchange of calcium-45 between sea-water and *Fucus*, and a stable level of accumulation is reached within an hour (SWIFT and TAYLOR [1960]). This time is six hr for *Cystoseira* and one to two min for *Ulva*

(BARINOV [1965b]). In the mantle of a mantle-shell oyster preparation a stable level of calcium-45 accumulation was reached within four hr of commencement of the experiment. Calcium-45 was rapidly deposited in the shell of a preparation, especially in its rear part, at a nearly linear mean rate. Only a small part of the mantle calcium (no more than 2.4%) was exchangeable. Of the 1.34 mg of calcium which passed through the mantle in the first 24 hr, 1.27 mg was subsequently deposited in the shell (JODREY [1953]).

Calcium uptake by a coral was studied by GOREAU and BOWEN [1955].

Intact sea urchin eggs concentrate far less calcium-45 than eggs without the jelly-like membrane. The presence of this nuclide has been demonstrated by radioautography. Since calcium-45 is also present in the cytoplasm, there is exchange between calcium-45 and calcium-40 in the cytoplasm of sea urchin eggs (HSIAO and BOROUGHS [1958]).

In artificial sea-water concentration of calcium-45 reaches a stable level in the fishes *Tilapia mossambica* by the 11th day, reaching a concentration factor of 0.6. Since, according to VINOGRADOV [1953], the concentration factor of stable calcium in sea fishes is of the order of 20, the exchangeable calcium in their body is no more than two to three % in the present case. The exchange of bone calcium is apparently extremely slow, and the greater part remains in the fishes throughout their life. Townsley and his colleagues are justified in thinking that sea fishes concentrate calcium directly from sea-water and do not need a source of supply of this element in food (BOROUGHS, TOWNSLEY *et al.* [1957]).

When the water was experimentally isolated for the head and tail areas of *Tilapia mossambica*, it was found that salinity did not affect calcium-45 and strontium-85 uptake (TOWNSLEY *et al.* [1960]).

At a water temperature of 17 to 20°C only 10 to 15 hr are needed for total replacement of calcium in the body of the freshwater worm *Rhizodrilus limasus* by calcium in the water (76.8 mg/l). By contrast to sea fishes, the rate of uptake of calcium-45 by the worms is little dependent on variation of calcium concentration in the water. EDTA slows down the concentration of calcium-45 by the worm (TOMIYAMA, KOBAYASHI *et al.* [1956b]).

Uptake of calcium-45 from freshwater by carp has been studied when the water around the head was separated from that around the tail by an impermeable partition. In the first portion of the experiment the nuclide was added to one half of the apparatus (in the presence of 260 mg of calcium per liter of water), and in the second to the other half. When the head end of the carp was exposed to a calcium-45 solution concentration of the nuclide in it was higher than in the tail end. Thus, far more calcium-45 (59 times more) accumulated in the gills in the first portion than in the second, 2.5 times more in the scales, six times more in the skin, 48 times more in the internal organs, 49 times more in the backbone, six times more in the remaining organs, and 6.4 times more in the body as a whole. The gills, therefore, provide the best access for calcium-45 to freshwater fishes (TOMIYAMA, ISHIO *et al.* [1956a]).

4.2.4. Elimination of calcium

As already stated, the time taken for elimination of calcium from algae is the same as the time needed for concentration to a steady state: six hr for *Cystoseira* and one to two min for *Ulva* (BARINOV [1965b]).

In the marine alga *Fucus vesiculosus* 90% of calcium-45 is replaced by stable calcium within the first two hr in pure sea-water (SWIFT and TAYLOR [1960]). This demonstrates the intensive metabolism of the element and is in good agreement with the author's data for the rate of elimination of strontium-90 from another brown alga, *Cystoseira barbata* (POLIKARPOV and TEN [1962]). In non-radioactive sea-water loss of calcium-45 from the edge of the mantle of the oyster *Crassostrea virginica* was rapid (0.5 to 2 hr), with an exponential loss curve and an elimination half-period of 16.5 min (JODREY [1953]).

The rate of elimination of calcium-45 from the freshwater worms *Rhizodrilus limasus* is comparable to the rate of uptake, and increases as calcium concentration in the medium rises. EDTA, on the other hand, reduces its elimination from these worms in an aqueous solution (TOMIYAMA, KOBAYASHI *et al.* [1956c]). In the opinion of these authors, calcium-45 is excreted from goldfish via the kidneys, the gall bladder and, possibly, the gills. In pure water the content of this nuclide decreased in the alimentary canal, the kidneys, the gall bladder, the gills, the internal organs excluding the alimentary canal, the skin and the muscles and increased in the scales, the fins and the vertebral column (TOMIYAMA, KOBAYASHI *et al.* [1956a]). Here the process taking place was similar to that described by POVELYAGINA and TELITCHENKO [1959] for strontium-(89+90) in freshwater molluscs. Following intramuscular administration of strontium-90 to carp, up to 84% was eliminated via the kidneys, and the remainder via the gills (TOMIYAMA, KOBAYASHI *et al.* [1956e]).

4.2.5. Uptake of zinc

It has already been noted that concentration of zinc-65 is affected by illumination and, therefore, by the rate of photosynthesis in waterplants. In experiments with the marine unicellular alga *Carteria* it was found that there was almost total absorption of zinc-65 from sea water containing a normal concentration of stable zinc (10^{-5} g/l) in half a day (BOROUGHS, CHIPMAN et al. [1957]). Zinc-65 was rapidly accumulated (hours or days) by the marine unicellular alga *Chlamydomonas* sp., by the branchiopod *Artemia salina*, and by the fish *Chaetodon militaris* (TOWNSLEY et al. [1960]), but accumulation had still not reached a balanced level in the body and shell of marine molluscs by the 6th day (GONG et al. [1957]). A significant amount of zinc-65 was, however, found in various organs and tissues of the marine mollusc *Pecten irradians* within two hr (BOROUGHS, CHIPMAN et al. [1957]).

CHIPMAN et al. [1958] demonstrated that *Nitzschia closterium* cells (4.3×10^8 cells per liter of water) concentrated up to 80% of zinc-65 in the first hour and practically the whole of the nuclide (to below the sensitivity of the method) within 24 hr from a solution with an initial activity of 27 μCur/l. Non-dividing cells of this alga in darkness extracted 87% of the initial amount of zinc-65 from sea water in 24 hr.

It should be noted that stable zinc begins to have a toxic effect on cell cultures at approximately one order of magnitude (2.5×10^{-4} g/l) higher than the normal concentration of this element in the sea, which is 2.8×10^{-6} to 1.46×10^{-5} g/l, with a mean of 0.96×10^{-5} g/l.

When a complex compound was formed between zinc-65 and EDTA (0.05 g/l) concentration of the nuclide by the cells practically ceased. Nevertheless the cells still divided at the normal rate, despite the extremely low concentration of available zinc. Concentration of zinc-65 in whole oysters and in their isolated gills reaches a stable level rapidly, within 42 to 48 hr.

4.2.6. Elimination of zinc

When *Nitzschia closterium* cells are washed in non-radioactive sea-water they lose only a small amount of previously concentrated zinc-65 (up to 15%). The addition of the isotopic carrier increases the output of zinc-65 into sea-water to 45% at a stable zinc concentration in the medium of 10^{-4} g/l. Washing the cells with distilled water has scarcely any effect. Live *Carteria* cells therefore give up little of this element to the environment.

The pattern is different in animals. *Crassostrea virginica* loses zinc-65 rapidly in pure sea-water. When kept for four days in a solution of this nuclide and then placed in non-radioactive sea-water, the sea fishes *Lagodon rhomboides* lost approximately 80% in one day, and 94% of total activity in two days; the loss curve was exponential in form. The remaining portion of the radioactivity (approximately 6%) was, however, retained until the end of the experiment, the 26th day (CHIPMAN et al. [1958]).

According to the data of other authors, elimination of zinc-65 from the molluscs *Meretrix meretrix lusoria* was 40% in two days in pure sea-water. There was little or no concentration of zinc-65 in the hard tissues of these molluscs (MORI and SAIKI [1956]).

4.3. Concentration factors

4.3.1. Strontium

Table 23 gives concentration factors of the stable and radioactive isotopes of strontium for marine organisms under experimental and natural conditions. In all cases strontium-90 was in equilibrium with yttrium-90. The concentration factors are moderate (tens of units) and high (hundreds) for diatoms and brown algae, and for the shells of crustaceans and mussels. The unicellular alga *Carteria* sp. stands out from all other organisms in its capacity to concentrate strontium-90. Its concentration factors are between 560 and 1600. According to calculations by the author, based on SCHREIBER's data [1960] the concentration factors of stable strontium in whole *Acantharia* were approximately 110, but in their spicules were 60000. Note that strontium-90 concentration factors in marine oozes are approximately unity (BARANOVA and POLIKARPOV [1965]).

It emerges from comparison of concentration factors obtained under natural and experimental conditions for radioactive and stable strontium that all the strontium is exchangeable in all the groups of organisms. It was established by spectral analysis that the strontium content was always the same in the soft tissues of marine molluscs as in the sea-water; no higher concentration was discovered. Data obtained by the use of radionuclides pointed to the same conclusion. For obvious reasons the concentration factors of strontium-90 are far lower than those of stable strontium in the skeleton of invertebrates and fishes, because of the extremely slow and limited exchange in mineralised tissues. Strontium content is lowest in muscles.

The concentration factors were correspondingly higher in the less saline water from the Baltic Sea (salinity lowered to a tenth or less) being 194 in the brown alga *Fucus vesiculosus*, 630 in the mollusc *Macoma baltica*, 3 in the muscles of fishes and 150 in their bones (AGNEDAL et al. [1958]).

It is evident that strontium-90 is concentrated to a greater extent in freshwater organisms than in marine organisms.

The discrepancies between the data of various authors are as high for radiostrontium and stable strontium in freshwater organisms as for cesium (table 24, also tables 21 and 22). This can be related to differences of stable

Table 23

Strontium concentration factors in marine organisms

Aquatic organisms	Concentration factors			Source
	radioactive Sr		stable Sr	
	in experiments	in nature	in nature	
Flagellata:				
Coccolitophoridae				
coccoliths	–	–	20	a
Peridinium trochoideum	6	–	–	b
Prorocentrum micans	6	–	–	b
Green algae:				
Chlamydomonas minima	4	–	–	b
Carteria sp.	560	–	–	c
rapidly dividing culture	1 600	–	–	c
Ulva lactuca	1	–	–	d
U. rigida	2	–	2	e, b
Monostroma sp.	6.7	–	–	g
Enteromorpha comressa	1	–	–	e, b
Diatoms:				
Nitzschia closterium	17	–	–	c
Brown algae:				
Laminaria digitata	14	–	–	d
	–	–	14–90	h
Ascophyllum nodosum	22	–	–	d
	–	–	16	h
Pelvetia canaliculata	–	–	20	h
Dictyota fasciola	18	–	–	e
Padina pavonia	19	–	–	e
Fucus serratus	40	–	35	d
	–	16–60	16–30	j
	–	–	11	h
F. vesiculosus	23	–	–	d
	–	–	18	h
young	35	–	–	d
F. spiralis	–	–	8	h
Cystoseira barbata	40	55	46	e, b, f
Sargassum thunbergii	6.5	–	–	g
S. natans	41	–	–	i
S. fluitans	35	–	–	i
Red algae:				
Chondrus crispus	2	–	–	d
Gigartina stellata	2	–	–	d
Rhodymenia palmata	1	–	–	d

Table 23 (continued)

Aquatic organisms	Concentration factors			Source
	radioactive Sr		stable Sr	
	in experiments	in nature	in nature	
Corallina rubens	4	–	–	e
Ceramium rubrum	1	–	–	e
Polysiphonia elongata	1	–	–	e
Phyllophora nervosa	8	–	–	e
Porphyra umbilicalis	0.05–0.3	–	0.1–0.3	k
P. tenera	–	–	0.2	l
Laurencia obtusa	1	–	–	e
Algae	–	40	–	j
Noncalcareous algae	–	–	20	m
Flowering plants:				
Zostera marina	3	–	5	e, b
Protozoa:				
Acantharia				
whole cells	–	–	110	n
spicules	–	–	60000	n
Coelenterates:				
Actinia equina	1	–	–	e
Crustaceans:				
Tigropus californicus	1	–	–	o, p
Artemia salina	0.2	–	–	o, p
	0.1	–	–	q
nauplii	0.7	–	–	q
Calanus	–	–	0.28	r
Euphausia	–	–	0.3	r
Decapoda				
plankton larvae	–	10	–	j
Leander squilla	8*	–	–	e
L. sp.				
viscera	1.3	–	–	g
muscles	0.2	–	–	g
exoskeleton	15	–	–	g
Metapenaeus monoceros				
muscles	–	–	0.4	l
Penaeus japonicus				
muscles	–	–	0.1	l
Panulirus japonicus				
muscles	–	–	1	l
Paralithodes camtschatics				
muscles	–	–	0.9	l

Table 23 (continued)

Aquatic organisms	Concentration factors		stable Sr	Source
	radioactive Sr			
	in experiments	in nature	in nature	
Neptunus trituberculatus				
muscles	–	–	0.9	l
Pachygrapsus marmoratus	3*	–	–	e
Homarus vulgaris				
exoskeleton	–	180	180	j
muscles	–	–	1	j
remainder	–	40	33	j
Molluscs:				
Mercenaria mercenaria				
(young) shell	10*	–	–	o
body	0.9	–	–	o
Mytilus galloprovincialis				
shell	6*	–	–	e, b
body	0.6	–	–	e, b
Venerupis philippinarum				
gills	0.7*	–	–	g
adductor muscle	0.4*	–	–	g
mantle	0.75*	–	–	g
viscera	1.2*	–	–	g
Meretrix meretrix lusoria				
gills	1.5	–	–	s
	1.2	–	–	g
mantle	1.0	–	–	s
	1.9	–	–	g
viscera (without kidneys)	1.0	–	–	s
liver	0.8	–	–	s
siphon	0.7	–	–	s
foot	0.5	–	–	s
adductor muscles	0.5	–	–	s
	0.7	–	–	g
muscles	–	–	0.2	l
Venus mercenaria				
shell	9	–	–	t
body	1	–	–	u
Crassostrea				
soft tissues	1	–	–	v
Ostrea gigas				
muscles	–	–	0.2	l
Turbo cornutus				
muscles	–	–	0.2	l

Table 23 (continued)

Aquatic organisms	Concentration factors			Source
	radioactive Sr		stable Sr	
	in experiments	in nature	in nature	
Haliotis gigantea				
muscles	–	–	0.1	l
Ommatostrephes	–	–	0.3	r
Octopus fangsiao				
muscles	–	–	0.4	l
Echinoderms:				
Stichopus japonicus				
muscles	–	–	0.7	l
Strongylocentrotus pulcherrimus				
digestive tract	0.6	–	–	g
gonad	0.43	–	–	g
Anthocidaris crassispina				
ovary	–	–	2	l
Chaetognatha:				
Sagitta	–	–	70	r
Invertebrates:				
whole	–	25	–	j
skeleton	–	–	1 000	m
soft tissue	–	–	10	m
Fishes:				
pelagic eggs	0.8	–	–	w
Scorpaena porcus				
eggs (before hatching)	1.8	–	–	x
larvae (just hatched)	1.7	–	–	x
larvae (24 hr old)	3.6	–	–	x
Rhombus maeoticus				
eggs (before hatching)	1.6	–	–	x
larvae (just hatched)	1.8	–	–	x
larvae (96 hr old)	4.3	–	–	x
Engraulis encrasicholus ponticus				
eggs (before hatching)	0.8	–	–	x
Odontogadus merlangus ponticus				
eggs (before hatching)	4.4	–	–	x
Mullus barbatus ponticus				
eggs (before hatching)	0.8	–	–	x
larvae (56 hr old)	2.1	–	–	x
Mugil cephalus				
young	5*	–	–	o, p

Table 23 (continued)

Aquatic organisms	Concentration factors			Source
	radioactive Sr		stable Sr	
	in experiments	in nature	in nature	
Fundulus heteroclitus				
bones	3*	–	–	o, p
muscles	0.1	–	–	o, p
Tilapia mossambica	0.3	–	–	y
Pleuronectes platessa				
whole	1.0	–	4.2–5.5	k
muscles	0.15	–	1.5	k
ray				
soft tissues	–	–	0.15	k
Thunnus thynnus				
muscles	–	–	0.1	z
Prognichtys agoo				
muscles	–	–	0.1	z
Theragra chalcogramma				
muscles	–	–	0.3	z
Rudarius ercodes				
gills	0.4	–	–	g
scales	1.2*	–	–	g
vertebra	0.5*	–	–	g
muscles	0.03	–	–	g
head and fins	0.4*	–	–	g
viscera	0.45	–	–	g
Pterogobius elapoides				
gills	2	–	–	g
scales	8.4	–	–	g
vertebra	1.4*	–	–	g
muscles	0.08	–	–	g
head and fins	1.5*	–	–	g
viscera	1.7	–	–	g
Acanthogobius flavimanux				
gills	0.8	–	–	g
scales	9.8	–	–	g
vertebra	0.7*	–	–	g
muscles	0.04	–	–	g
head and fins	0.8*	–	–	g
viscera	1.1*	–	–	g
gallbladder	0.16	–	–	g
liver	0.046	–	–	g
digestive tract	1.6	–	–	g

Table 23 (continued)

Aquatic organisms	Concentration factors			Source
	radioactive Sr		stable Sr	
	in experiments	in nature	in nature	
Trachurus japonicus				
gills	2.4	–	–	g
scales	13	–	–	g
vertebra	1.5*	–	–	g
head and fins	1.7*	–	–	g
viscera	0.37	–	–	g
muscles	0.06	–	–	g
	–	–	0.2	z
T. mediterraneus ponticus				
eggs (before hatching)	0.9	–	–	x
larvae (just hatched)	0.8	–	–	x
Argyrosomus argentatus				
muscles	–	–	0.1	z
Vertebrates:				
skeleton	–	–	200	m
soft tissues	–	–	1	m

* Stable concentration not achieved.

a   LEWIN and CHOW [1961].
b   POLIKARPOV [1963].
c   RICE [1956].
d   SPOONER [1949].
e   POLIKARPOV [1960b, 1961a].
f   KULEBAKINA, in the press.
g   HIYAMA and SHIMIZU [1964].
h   BLACK and MITCHELL [1952].
i   POLIKARPOV, ZAYTSEV et al., in the press.
j   TEMPLETON [1962].
k   TEMPLETON [1959].
l   ICHIKAWA [1961], calculated from the data of KAGAWA [1959], and THOMPSON and CHOW [1955].
m   KRUMHOLZ, GOLDBERG et al. [1957].

n   SCHREIBER [1960].
o   CHIPMAN [1958a].
p   CHIPMAN [1958b].
q   BOROUGHS, TOWNSLEY et al. [1958].
r   KETCHUM and BOWEN [1958].
s   SUYEHIRO et al. [1956].
t   CHIPMAN [1960].
u   GONG et al. [1957].
v   BOROUGHS, CHIPMAN et al. [1957].
w   POLIKARPOV and IVANOV [1962b].
x   IVANOV [1965a, b].
y   BOROUGHS et al. [1956a].
z   ICHIKAWA [1961], calculated from the data of ASARI and OSHIMA [1949].

Table 24

Strontium concentration factors in freshwater organisms

Aquatic organisms	Concentration factors		stable Sr in nature	Source
	^{90}Sr			
	in experiments	in nature		
Phytoplankton	–	75 000	–	a
Flagellata:				
Euglena sp.	16	–	–	b
Green algae:				
filamentous algae	–	500 000	–	a
	–	808	1 050	c
Oedogonium sp. and				
Cladophora sp.	–	808	–	c
Flowering plants:				
Myriophyllum sp.				
whole	–	270	350	c
tops of stems	–	230	–	c
middle of stems	–	287	–	c
roots	–	106	–	c
Aquatic plants:				
leaves and stems	–	280	–	d
Crustaceans:				
Leander paucidens	20	–	–	e
Insects:				
Chironomidae				
larvae	–	9	–	c
insect larvae	–	100 000	–	a
Molluscs:				
shells	–	–	3 820	f
clams				
soft tissues	–	730	–	d
Unio pictorum				
foot	13	–	–	g
mantle	232	–	–	g
internal organs	64	–	–	g
shell	4	–	–	g
shell edge	11	–	–	g
gills	297	–	–	g
Dreissena polymorpha				
shell	11	–	–	g
body	24	–	–	g
Anodonta cygnea				
foot	7	–	–	g
mantle	167	–	–	g

Table 24 (continued)

Aquatic organisms	Concentration factors			Source
	^{90}Sr		stable Sr	
	in experiments	in nature	in nature	
internal organs	14	–	–	g
shell	4	–	–	g
shell edge	26	–	–	g
gills	177	–	–	g
Planorbis corneus				
shell	34	–	–	g
body	26	–	–	g
Fishes:				
Cyprinus carpio				
yearlings	45	–	–	h
Carassius auratus	10	–	–	e
	150	–	–	i
Phoxinus sp.	100	–	–	i
Leuciscus erythrophtalmus				
muscles	–	2.6	7.5	c
skin	–	6.3	–	c
scales	–	208.6	–	c
bones	–	71.0	750	c
intestines	–	10.5	–	c
contents of intestines	–	74.9	–	c
parasites (cestodes)	–	12.6	–	c
kidneys	–	6.0	–	c
gills	–	96.9	–	c
liver	–	5.7	–	c
eyes	–	5.5	–	c
heart	–	0.5	–	c
remainder	–	48.6	–	c
skin and scales	–	134.9	–	c
*Esox lucius**	–	22	30	j
*Clupea harengus**	–	3	1.4	j
*Rutilus rutilus**	–	3.5	2	j
Cyprinus carpio				
eggs in female	0.3	–	–	k
Salmo salar				
eggs	30	–	–	l
fry	417**	–	–	m
S. trutta				
fry	417**	–	–	m
S. t. f. fario				
yearlings and two year old				

Table 24 (continued)

Aquatic organisms	Concentration factors		stable Sr	Source
	^{90}Sr			
	in experiments	in nature	in nature	
whole	600	–	–	n
muscles	150	–	–	n
bones	1 000	–	–	n
yellow perch				
bones	–	1 840–4 200	–	d
soft tissues	–	5	–	d
minnows				
whole	–	950	–	d
Fishes (species not stated)	–	20 000–30 000	–	a
Birds:				
tame ducks				
muscles	50	–	–	o
bones	420	–	–	o
internal organs	85	–	–	o
Mammalia:				
mink				
bones	–	820	–	d
beaver				
bones	–	1 300	–	d
muskrat				
bones	–	3 500	–	d

* In the Baltic Sea.
** Balanced concentration not reached.

a	KRUMHOLZ and FOSTER [1957].	i	PROSSER et al. [1945].
b	WILLIAMS and PICKERING [1961].	j	AGNEDAL et al. [1958].
c	TEMPLETON [1959].	k	TELITCHENKO [1961].
d	OPHEL [1963].	l	BROWN and TEMPLETON [1964].
e	SUYEHIRO et al. [1956].	m	BROWN [1962].
f	NELSON [1963].	n	TEMPLETON [1962].
g	POVELYAGINA and TELITCHENKO [1959].	o	MAREY et al. [1958].
h	SAUROV [1957].		

strontium content in different waters and, possibly, to the determination procedure. Concentrations of stable strontium in water were obtained by dividing the concentration of stable strontium in green algae, insect larvae and fishes by the corresponding strontium-90 concentration factors in them in the Columbia River and in White Oak Lake (KRUMHOLZ and FOSTER [1957]). The values obtained were very low being of the order of no more than

10^{-8} to 10^{-9} g/l. In a detailed study by TEMPLETON [1959], whose results can be accepted with confidence, the strontium-90 concentration factor in filamentous algae differed from that obtained by the American authors for the same plants by almost 1000 times, the factor for insect larvae by 10000 times, the factor for fishes by approximately 1000 times. Templeton supported his investigations by information concerning the concentration factors of stable strontium in a number of objects. It should be emphasized that Templeton found good agreement between the concentration factors of strontium-90 and of stable strontium, with the quite understandable exception of fish bones, for which the former was approximately 10 % of the latter.

Templeton also gives interesting information concerning strontium-90 concentration factors calculated in relation to dry weight. The values obtained were 182 to 359 in bottom sediments (sands and silts), 245 in plankton detritus and 1507 in suspended matter. If the accumulation factors of filamentous algae and flowering plants are also calculated in relation to dry weight, thus increasing them approximately ten-fold, it is found that strontium-90 accumulation factors are lower in bottom sediments and detritus, and even in suspended particles, than in live plants.

4.3.2. Calcium

It is a simple matter to calculate the calcium concentration factor in the coccoliths of Coccolithophoridae (Flagellata), which consist of calcium carbonate (LEWIN and CHOW [1961]); it is equal to 1000. The figure of 0.6 obtained experimentally for the concentration factor of calcium-45 in whole *Tilapia mossambica* by BOROUGHS, TOWNSLEY *et al.* [1957] is approximately 3 % of that of stable calcium in whole fishes (20) calculated from VINOGRADOV's data [1953].

If the concentration factors of stable strontium and calcium are compared, it is found that they are 20 and 10, respectively, in noncalcareous marine algae, 10 in the soft tissues and 1000 in the skeleton of marine invertebrates, and one in the soft tissues and 200 in the skeleton of marine vertebrates (KRUMHOLZ, GOLDBERG *et al.* [1957]). In other words the ratio of these elements that characterizes sea-water (1/50) is not apparently disturbed. There is however a wide range in particular cases (WISEMAN, [1955]), e.g., 1/10 in *Laminaria* and 1/200 in the shells of molluscs. Wiseman is of the opinion that calcium is removed from sea-water by biogeochemical processes, and partly by chemical precipitation on to the sea floor, and that strontium is concentrated in sea-water. There is other interesting information that has a bearing on this matter (AGNEDAL *et al.* [1958]), e.g., the concentration factors of stable strontium and calcium in the aquatic organisms of the Baltic Sea, where salt content is reduced, are 49 and 5, respectively, in bottom sediments, 290 and 4300 in the littoral fauna, 630 and 4730 in the mollusc *Macoma baltica*, 10 and 8 in *Phragmites communis*, 194 and 102 in the brown alga *Fucus vesiculosus*, 3 and 22 in fish muscles and 150 and 1700 in fish bones.

Some authors are of the opinion that fairly similar strontium and calcium concentration factors are a feature of freshwater organisms. For example, the factors are 1050 and 980 in the filamentous algae *Oedogonium* sp. and *Cladophora* sp., 350 and 327 in spiked milfoil (*Myriophyllum spicatum*), 7.5 and 6.7 in the muscles of rudd and 750 and 600 in their bones (TEMPLETON [1959]).

Other authors found great differences in the concentration factors of these elements being 2 and 1500 in *Spirogyra*, 0.2 and 300 in insect larvae, and 0.3 and 3000 in minnows (KRUMHOLZ and FOSTER [1957]). The calcium concentration factor for the guppy is 1000 (PROSSER *et al.* [1945]).

These conflicting data suggest that the differences arise from the fact that the analyses were made in freshwaters with greatly differing contents and ratios of these and other elements. Rivers and lakes should, therefore, always be considered separately; this is not the case for seas and oceans*. PICKERING and LUCAS [1962] have correctly stressed that strontium-90 concentration factors for aquatic organisms have no meaning unless the calcium content of the water is known.

4.3.3. Barium and radium

The concentration factor of radiobarium is fairly high being reported as 150 in goldfish (PROSSER *et al.* [1945]).

There is as yet little information concerning the capacity of aquatic organisms to accumulate radium-226. The data of TURNER *et al.* [1958] yield concentration factors of 210 for cockles, 69 for mussels, 23 for sardines (tinned) and 6 for plaice. Marine organisms apparently concentrate radium-226 to a greater extent than strontium. The radium concentration factor is, for example, 100 in marine macrophytes, and 100 to 2750 in plankton

---

* It is highly probable that the considerable differences between the data obtained by various investigators in freshwaters may sometimes be due not only to the effect of environmental factors, but also to the extent to which exchange equilibrium has been established between the radionuclides and to the level of isotopic exchange, as well as to the experimental method and determination procedure, neither of which have as yet been standardized, and are not precise in some studies, especially biological studies.

(table 39). Concentration factors are very similar for plaice: up to 5.5 for strontium (table 23) and 6 for radium-226 (*op. cit.*).

VERNADSKIY [1929] found considerable differences between the concentration factors of radium-226 in various species of duckweed from different waters in the USSR: 14 to 47 for *Lemna minor* from the ponds at Peterhof and 200 to 477 for *L. gibba* and *L. trisulca* from ponds at Kiev. The explanation for these differences is at present unknown.

4.3.4. Zinc

Zinc is a trace element essential to life that is incorporated in many enzymes and hormones in cells.

The capacity of marine organisms to concentrate zinc is considerable (table 25). Most of the data in table 25 relate to the concentration factors of stable zinc under natural conditions. Many of the figures were calculated (MAUCHLINE [1961], ICHIKAWA [1961]) from published biogeochemical material (VINOGRADOV [1953] and other papers). Almost all the concentration factors of stable zinc are of the order of thousands of units, reaching tens of thousands for some molluscs (mussels and oysters), but for sponges, worms, and a number of echinoderms they are only tens or even units. The zinc-65 concentration factor is extremely high in *Nitzschia* cells, for which a value of 50000 was established (a balanced concentration was not reached). The concentration factors of zinc-65 were considerably lower than those of stable zinc for all the marine organisms investigated. Possible explanation will be considered below.

Only zinc-65 concentration factors are known for freshwater organisms (table 26; also tables 21 and 22), and it is only for *Chlorella* that there is information concerning the concentration factor of stable zinc. If it is remembered that many of the figures given are based on dry weight, it would appear that the concentration factors of zinc in freshwater organisms are tens or hundreds of units (in relation to wet weight). There are various concentration factors for accumulation of this isotope in trout in relation to calcium concentration in the water: 39 at 2×10^{-3} g/l, 6.3 at 2×10^{-2} g/l and 3.6 at 6.4×10^{-2} g/l under conditions of prolonged exposure (50 days).

4.4. Distribution of radionuclides by organs and tissues

4.4.1. Strontium

Most of the strontium-90 (80 %) in male crabs (*Carcinus maenas*) was in the shell, and only 13 % in the lower part of the cephalothorax (chitin and mus-

Table 25

Zinc concentration factors in marine organisms

Aquatic organisms	Concentration factors		Source
	^{65}Zn in experiments	Zn in nature	
Green algae:			
Ulva rigida	127*	–	a
U. pertusa	290	–	b
Diatoms:			
Nitzschia closterium	50 000*	–	c
Brown algae:			
Cystoseira barbata	186*	–	a
Fucus vesiculosus	–	420	d
	–	1 100	e
F. serratus	–	600	e
Ascophyllum nodosum	–	1 400	e
Pelvetia canaliculata	–	1 000	e
Laminaria saccharina	–	420	d
L. digitata	–	400–1 000	e
Red algae:			
Phyllophora nervosa	839*	–	a
Gelidium gracilaria	–	80	f
Flowering plants:			
Zostera marina	336	–	a
Sponges:			
Helichondra sp.	–	30	d
Coelenterates:			
Cyanea capillata	–	1 600	d
Metridium dianthus	–	300	d
Actinia equina	69*	–	a
Worms:			
Nereis japonicus	–	6	d
Crustaceans:			
Copepoda	200*	–	g
Idothea sp.	–	12 000	d
Gammarus sp.	–	15 000	d
Eupagurus sp.	–	9 400	d
Palaemon vulgaris muscles	–	1 900	d, f
Leander squilla	37*	–	a
L. sp. muscles	40*	–	b

Table 25 (continued)

Aquatic organisms	Concentration factors		Source
	^{65}Zn in experiments	Zn in nature	
exoskeleton	150*	–	b
viscera	500*	–	b
Pachygrapsus marmoratus	26*	–	a
Callinectes hastatus			
muscles	–	4400	f
Molluscs:			
Limacina	–	30000	h
Littorina littorea	–	4900	d
L. obtusata	–	5600	d
Patella vulgata	–	4300	d
Buccinum undatum	–	5000	d
Nucella lapillus	–	4600	d
Mya arenaria	–	7700	d
Tellina crassa	–	8000	d
T. tenius	–	16000	d
Mactra corallina	–	5200	d
Dosinia exoleta	–	7400	d
Lutraria lutraria	–	9500	d
Cardium edule	–	1400–4600	d
Venus verrucosa	–	4700	d
V. mercenaria			
body	24*	–	i
shell	35*	–	i
Pecten irradians			
body	20*	3500	j
kidneys	138*	–	j
liver	24*	–	j
gills	22*	–	j
testis or ovaries	14*	–	j
foot	13*	–	j
heart	11*	–	j
adductor muscle	10*	–	j
mantle	9*	–	j
P. japobaeus			
muscles	–	17000	f
P. maximus	–	5900	d
Venerupis philippinarum			
viscera	68	–	b
shell	10*	–	b

Table 25 (continued)

Aquatic organisms	Concentration factors		Source
	^{65}Zn in experiments	Zn in nature	
Meretrix			
soft tissue	0.25–1.75	–	k
Mytilus edulis	–	2100–7500	d
M. galloprovincialis			
shell	49*	–	a
body	629*	–	a
Crassostrea virginica			
body	210*	17000	c
Ostrea edulis	–	20000–31000	d
muscles	–	40000	f
Haliotis tuberculata	–	5200	d
muscles	–	10000	f
Loligo vulgaris			
muscles	–	5700	f
Sepia officinalis			
muscles	–	2600	f
Ommatostrephes	–	20000	h
Octopus vulgaris			
muscles	–	11000	f
Echinoderms:			
Stichopus tremulus			
muscles	–	1400	f
Astropecten sp.	–	25	d
Asterias rubens	–	56	d
Strongylocentrotus pulcherrimus			
digestive tract	200*	–	b
gonad	25	–	b
test	15*	–	b
Tunicates:			
Ciona intestinalis	–	6600	d
Fishes:			
Squalus acanthias	–	15500	d
Torpedo torpedo	–	330	d
Mugil cephalus			
muscles	–	540	f
Anguilla anguilla			
muscles	--	4200	f
Clupea harengus	–	5000	d
muscles	–	4400	f
C. pilchardus	–	2900	d

Table 25 (continued)

Aquatic organisms	Concentration factors		Source
	^{65}Zn in experiments	Zn in nature	
Engraulis encrasicholus	–	1 500	d
Oncorhynchus nerka	–	280	d
O. tschawytscha	–	800	d
Chaetodon miliaris	3–4	–	1
Chasmichthys gulosus			
viscera	22	–	b
digestive tract	20	–	b
gills	15*	–	b
skin	7.6	–	b
vertebra	3*	–	b
muscles	1.4*	–	b
head and fins	3*	–	b
Gadus morrhua	–	1 800	d
Merluccius vulgaris	–	5 500	d
Pleuronectes flesus	–	1 400	d
P. sp.			
muscles	–	2 900	f

* Stable concentration not achieved.

a POLIKARPOV [1963].
b HIYAMA and SHIMIZU [1964].
c CHIPMAN *et al.* [1958].
d MAUCHLINE [1961], calculated from VINOGRADOV's data [1953].
e BLACK and MITCHELL [1952].
f ICHIKAWA [1961], calculated from VINOGRADOV's data [1953].
g CHIPMAN [1958a].
h KETCHUM and BOWEN [1958].
i GONG *et al.* [1957].
j BOROUGHS, CHIPMAN *et al.* [1957].
k MORI and SAIKI [1956].
l TOWNSLEY *et al.* [1961].

Notes from table 26 page 105:

* 2 mg of stable calcium per liter of water.
** 20 mg of stable calcium per liter of water.
*** 64 mg of stable calcium per liter of water.

a KRUMHOLZ and FOSTER [1957].   d TEMPLETON [1962].
b KNAUSS and PORTER [1954].   e COFFIN *et al.* [1949].
c WHITTAKER [1953].   f HANSON and KORNBERG [1956].

Table 26

Zinc, phosphorus and iron concentration factors in freshwater organisms

Aquatic organisms	Concentration factors				Source
	^{65}Zn	^{32}P		^{59}Fe	
	in experiments	in experiments	in nature	in nature	
Phytoplankton	–	–	200000	200000	a
	–	–	150000	–	a
Green algae:					
Chlorella sp.	140	–	–	–	b
filamentous algae	–	–	800000	100000	a
	–	–	100000	–	a
Algae	–	36000	–	–	c
Insect larvae	–	–	100000	100000	a
Molluscs:					
snails					
shell	–	850	–	–	c
body	–	900	–	–	c
Fishes:	–	–	100000	10000	a
	–	–	30000–70000	–	a
	–	3200	–	–	c
Salmo trutta fario	39*	–	–	–	d
	6.3**	–	–	–	d
	3.6***	–	–	–	d
Fundulus sp.	–	13000	–	–	e
Birds:					
swallow, *Petrochiledon* sp.					
adult	–	–	75000	–	f
young	–	–	500000	–	f
seagulls, *Larus* sp.					
adult	–	–	5000	–	f
duck, *Aythya* sp.					
adult	–	–	50000	–	f
merganser, *Mergus* sp.					
adult	–	–	2500	–	f
young	–	–	15000	–	f
duck, *Anas* sp. and goose, *Branta* sp.					
adult	–	–	7500	–	f
young	–	–	40000	–	f
egg yolk	–	–	1500000	–	f

See notes on page 104.

cles), 6% in muscles and 1% in remaining tissues. Strontium-90 distribution was found to be similar in the females of these crabs. It should be noted that it is for the shell that the highest strontium-89 concentration factors have been recorded.

Strontium-89 distribution in the organs and tissues of oysters (*Crassostrea virginica*) was basically the same with 85.3% of total activity concentrated in the shell, 4.1% in the mantle, 3.1% in gills, 2.4% in adductor muscles and 5.1% in other tissues. The shells accounted for 90.1% of total weight, mantle for 2.5%, gills for 1.7%, adductor muscles for 1.9% and other tissues for 3.8%. If attention is confined to the distribution of the radionuclide in the soft tissues and organs, the pattern is as follows: mantle 27.7% of body activity (25% by weight in relation to the weight of the body), gills 21.2% (17.5%), adductor muscles 16.2% (19.2%) and other tissues 34.9% (38.3%) (BOROUGHS, CHIPMAN *et al.* [1957]). Fretter has demonstrated that strontium-90 is not concentrated in the liver but in the pericardial glands of mussels. The mollusc *Acanthodoris pilosa* concentrates the nuclide in the mantle around the numerous calcareous concretions. Strontium-90 that penetrates from the water into the tissues of the snail *Calyptraea sirensis* is concentrated by amoebocytes. In the worms *Platynereis dimerili* this radionuclide is concentrated in the lymphocytic cytoplasm which transports it to the body surface (FRETTER [1953]).

In *Tilapia mossambica*, which absorbs strontium-89 from the aqueous medium, the largest proportion (up to 35 to 40%) is concentrated in the skeleton, followed by 25 to 30% in skin, 18 to 35% in visceral organs, five to eight % in gills and three to five % in muscles. Strontium-89 distribution is similar in fishes that have received the nuclide in gelatin capsules or by intramuscular injection (BOROUGHS *et al.* [1956a]).

Excess of yttrium-90 has been discovered in the liver, gall bladder, heart, kidneys, spleen, gonads, urine, blood clots, scales and fat, and excess of strontium-90 in the gills, stomach, brain, muscles, intestines, eyes, urinary bladder and skin. There is very little strontium-90 on the surface of cellular constituents of blood of fishes, but yttrium-90 is present in the stroma. The whole of the strontium-90 is to be found in the blood plasma (BOROUGHS and REID [1958]). It has been established that the skin of *Tilapia* is only slightly less permeable for strontium-90 (and calcium-45) than the surface of the gills (BOROUGHS *et al.* [1956c]).

The shell accounts for 40% of total strontium-90 activity in the freshwater mollusc *Anodonta cellensis*, i.e., for approximately as much as in the case of cesium-137 (39%). The distribution of these nuclides differs greatly in the marine mollusc *Mytilus galloprovincialis*, in which cesium-137 is not to be found in the shell, and strontium-90 is very slightly concentrated by the soft tissues. There is very little stable strontium in the shell of *Anodonta cygnea* (0.08 %), and far less than in the shell of *Mytilus edulis* (0.3 %) (VINOGRADOV [1953]). This apparently suffices to explain the difference. The shell of *Anodonta cellensis* contains approximately half the strontium-90 (44.1 %), the mantle contains 7.5 %, visceral mass 6.2 % and muscles 2.1 %. If consideration is confined to the body, the gills account for 74 % of the isotope (GETSOVA *et al.* [1964]).

Strontium-90 is laid down mainly in the gills of the carp and only to a small extent in its skin. This nuclide is not assimilated in significant amounts by the digestive tract of the fish. Radioactivity may be detected within 30 min in most tissues apart from muscles and the gall bladder (TOMIYANA, KOBAYASHI *et al.* [1956d]).

TELITCHENKO [1961] demonstrated that, in mirror carp, radiostrontium is concentrated mainly in tissues with a high calcium content. This does not conflict with existing views on the behavior of strontium in the animal organism (ZAKUTINSKIY [1959], TARUSOV [1954]).

As in sea fishes, the distribution of strontium-90 in the carp is similar when the radionuclide is concentrated from the aqueous medium, and when it is administered by intramuscular injection. There is more strontium-90 than yttrium-90 in the tail fin, the scales, the vertebral column and the gills, and also in the blood, the swim bladder, the gall bladder and the alimentary canal. There is more yttrium-90 in almost all the internal organs, the skin and the muscles. The kidneys and spleen have a very high yttrium-90 content, and almost no strontium-90 (TOMIYAMA, KOBAYASHI *et al.* [1956e]).

4.4.2. Calcium

In the extreme case of a skeletal formation consisting almost entirely of calcium carbonate, the concentration factor for concentration of calcium from sea-water (0.4 g/l) by aquatic organisms is 1 000. Usually, however, the figure is lower because of the presence of other elements in the skeleton.

It has been shown by radioautography that calcium-45 is similarly localized in the mantle of freshwater molluscs (*Anodonta*) and marine molluscs (*Venus*). As was to be expected, this nuclide is incorporated in calcium carbonate crystals in developing parts of the shell (BEVELANDER [1952]).

In sea fishes (*Tilapia*) the distribution of calcium-45 is the same as that of strontium-90. At 21 days the skeleton contained 43 % of the calcium-45, skin 24 %, internal organs 21 %, gills eight % and muscles four %. This pattern does not, however, correspond to the distribution of stable calcium in sea fishes, apparently because the concentration of calcium-45 was still in progress. Thus, the bones of tunny contain 70 % of the stable calcium, gills 15 %, skin six %, internal organs, excluding blood, three % and muscles one to two % (BOROUGHS, TOWNSLEY et al. [1957]). The ratio of strontium to calcium, and that of barium to calcium, in various groups of aquatic organisms is a matter of great interest. As already noted, the discrimination factor is the ratio of the concentration factor of one element to that of another. The most clearly defined selective preference for strontium in relation to calcium is found in brown algae where the strontium-calcium discrimination factor is 4.2. In red algae this factor is only slightly greater than unity (1.3), less than unity in green algae (0.5) and in Chiton shells, the skeleton of cephalopods and the shells of other molluscs (0.81, 0.57 and 0.27, respectively). The discrimination factor of 1.4 for corals indicates predominant relative concentration of strontium. There is 50 times more calcium than strontium in sea-water, and 120 to 1000 times more in rocks. Like sodium and magnesium, strontium has therefore been concentrated in the sea in the course of the earth's history, whereas calcium is extracted by marine organisms and in part by chemical precipitation from sea-water and is deposited on the sea floor (BOWEN [1956]). WISEMAN [1955] reaches the same conclusion, and states that the ratio of strontium to calcium is 1/50 in sea-water, 1/10 in laminarians, 1/200 in the shells of molluscs and between 1/100 and 1/1000 in rocks. Discrimination factors for barium are greater than unity in almost all cases. They are 78 in brown algae, 53 in red algae, 16 in green algae, 5.5 in corals, 3.5 in Chiton shells, 5.3 in the skeleton of cephalopods and only 0.7 in the shells of other molluscs (BOWEN [1956]).

Several publications contain data on concentrations of strontium and calcium in aquatic organisms and water in nature as well as in experiments (WEBB [1937], WILSON and FIELDES [1941], ROSENTHAL [1957, 1960], TEMPLETON and BROWN [1963]).

4.4.3. Zinc

In marine molluscs stable zinc is concentrated mainly in the gills and gonads, and least of all in the muscles. Zinc content has been found to be high in the blood of Mytilus edulis (VINOGRADOV [1953]). Zinc-65 may be adsorbed onto the surface of shells (GONG et al. [1957]). The mollusc Meretrix meretrix lusoria concentrates zinc-65 mainly in the gills, mantle, visceral organs and in certain other soft tissues. There is little quantity of the isotope in the shell (MORI and SAIKI [1956]).

In sea fishes stable zinc is concentrated in the liver and spleen, and has an extremely high concentration factor of up to 10000. During the breeding period the quantity present in the sexual products of male fishes increases greatly (VINOGRADOV [1953]).

After 24 hr of exposure in a solution of zinc-65 king salmon fry (*Oncorhynchus tschawytscha*) were kept for two months in running lake water to study changes in the distribution of this nuclide in the organism. Immediately after removal from the active solution, 4.1% of the total zinc-65 activity was concentrated in the vertebral column of the fry, and at the end of the experiment in pure water 8.2%. Radioactivity in the head also increased in the course of the investigation from 25.9% on the 8th day to 31.2% on the 63rd day. There was also an increase in the visceral organs (heart, kidneys, liver, spleen, pancreas and alimentary canal) from 28.3 to 40.4%. In the remaining tissues (mainly the muscles, skin, scales and fins) the proportion of zinc-65 decreased during the same period from 41.4 to 20.2%.

The growing fry retained almost all the previously accumulated zinc-65 for the two months spent in pure water. The only changes were redistribution throughout the organs and tissues (JOYNER and EISLER [1961]). Comparative accumulation of zinc-65 in young rainbow, cutthroat and brook trout was studied by SLATER [1961].

When carp (*Cyprinus carpio*) were kept for 22 days in zinc-65 solution, concentration of this nuclide was highest in the gills and kidneys. When the nuclide was administered by intramuscular injection, concentration was also highest in the kidneys and far lower (in decreasing order) in the pancreas, heart, intestines, gills, scales, tailfin, swim bladder, skin, vertebral column and finally, muscles (MORI and SAIKI [1956]).

When sea fishes (*Micropogon undulatus*) were fed zinc-65 in gelatin capsules, the distribution by organs and tissues 12 hours after administration was as follows: muscles 44.7%, bones 21.1%, gills 7.6%, liver 11.4%, gonads 2.4%, kidneys 4.3%, heart 1.0% and spleen 0.9%. 6.7% of the total radioactivity was concentrated in the skin, scales, alimentary canal, blood, brain, eyes and other parts. Although the concentration of the nuclide was highest in the internal organs, they accounted for only a small part (only approximately 2%) of overall weight. The muscles (the main component of the fish by weight, 91%) therefore concentrated 44.7% of all the zinc-65, and together with the bones accounted for 66% of the total activity of the fish (CHIPMAN et al. [1958]).

CHAPTER 5

CONCENTRATION OF RADIONUCLIDES OF THE THIRD GROUP OF ELEMENTS IN THE PERIODIC SYSTEM

5.1. General characteristics of the most important radionuclides and their carriers in water

The radionuclides of the rare earths and of yttrium are a considerable proportion of the fission products of heavy nuclei. This is shown by their percentage content in a mixture of fission products and variations in content with time.

Nuclide	10 days	30 days	1 yr	10 yr
^{140}La	12.0	12.5	–	–
^{141}Ce	6.3	11.2	–	–
^{144}Ce	–	2.0	26.5	–
^{144}Pr	–	2.0	26.5	–
^{147}Pm	–	–	5.7	15.8
^{90}Y	–	–	1.8	21.8
^{91}Y	3.4	7.6	3.9	–
^{147}Nd	4.8	4.1	–	–
^{143}Pr	10.0	11.2	–	–
^{151}Sm	–	–	–	2.5

(after PALUMBO [1963]).

The radioisotopes of yttrium and cerium (and also, apparently, the other rare earths) occur in sea-water mainly as suspended particles (94% and 96%, respectively), and also as colloids (4% for each element) (table 13). Bottom sediments have a great ability to concentrate yttrium-91 from the sea-water (POLIKARPOV and AKAMSIN [1960]).

Sea-water contains insignificant concentrations of stable cerium (4×10^{-7} g/l), lanthanum (3×10^{-7} g/l) and yttrium (3×10^{-7} g/l) (REVELLE and SCHAEFER [1957]). There has been little study of their concentration in aquatic organisms or of their biological role. The concentration of scandium, an element of the same subgroup, in sea-water is even lower (5×10^{-8} g/l).

Arising from the absence of significant amounts of carriers, specific activity (ratio of a radionuclide to the sum of the radionuclide and the stable carrier, expressed in weight units) is, therefore, always high even at relatively low levels of radioactivity.

The elements of the other subgroup of the third group in the periodic system are not at present of particular radioecological interest.

5.2. Concentration and elimination of nuclides

Concentration of the radioisotopes of yttrium and the rare earths is rapid. Thus, *Carteria* cells adsorb yttrium-90 at such a rate that a balanced concentration is practically reached within 15 min (RICE [1956]). Concentration reaches a stable level in *Artemia* larvae within a few hours. Rapid uptake of this nuclide has also been noted in shrimps, crabs, oysters and scallops (BOROUGHS, CHIPMAN *et al.* [1957]). In the author's experiments, yttrium-91, praseodymium-143 and cerium-144 in equilibrium with praseodymium-144 reached balanced accumulations in green, brown and red marine algae, marine flowering plants, sea anemones, crustaceans and mussels in periods of between a few hours and several days (POLIKARPOV [1960a–c, 1961a–c,e]). According to data obtained in the author's laboratory, a balanced concentration of cerium-144 was reached in three hr in a crab as a whole, and in the principal organs and tissues.

When *Nitzschia* cells were cultured in a radioactive medium and then transferred to pure sea-water, approximately 25% of the cerium-144 in equilibrium with praseodymium-144 passed into the surrounding medium in approximately five hr, but thereafter and until the 42nd day the only further change was biological dilution of this nuclide as a result of cell division. In another experiment 7 to 8% of the cerium-144 was given up by cells when they were centrifuged four times, and the water was changed after each centrifugation. The addition of a complexing agent greatly increased the elimination of this radionuclide from *Nitzschia* cells (RICE and WILLIS [1959]).

Yttrium-91 was rapidly eliminated following its administration in gelatin capsules to the fish *Tilapia mossambica*, up to 60% in the first day, and up to 3% during the second day, thereafter the amount remained constant until the end of the experiment (14 days) (BOROUGHS *et al.* [1956b]).

5.3. Concentration factors

The concentration factors of the radioisotopes of yttrium and cerium are high (hundreds or thousands of units) for marine algae, grasses and most invertebrates (table 27). The radioyttrium concentration factor may reach 10000 in membrane of pelagic eggs. Yttrium and cerium concentration factors

Table 27

Yttrium and cerium concentration factors in marine organisms

Aquatic organisms	Concentration factors				Source
	yttrium		cerium		
	radio-isotopes in experiments	stable isotopes in nature	radio-isotopes in experiments	stable isotopes in nature	
Dinoflagellates:					
Peridinium trochoideum	–	–	340	–	a
Amphidinium klebsi	–	–	4500	–	b
Green algae:					
Platymonas sp.	–	–	2100	–	b
Carteria sp.	–	–	2400	–	b
Ulva rigida	900	–	350	–	c, d
Enteromorpha compressa	630	–	340	–	c, d
Bryopsis plumosa	690	–	640	–	c, d
Monostroma sp.	–	–	100*	–	e
Diatoms:					
Nitzschia closterium	–	–	2000	–	b
Tallassiosira sp.	–	–	3300	–	b
Brown algae:					
Cystoseira barbata	220	–	350	–	c, d
Sargassum thunbergii	–	–	200*	–	e
Red algae:					
Porphyridium cruentum	–	–	3300	–	b
Porphyra umbilicalis	–	–	500	–	f
Corallina rubens	–	–	330	–	c, d, g, h
Ceramium rubrum	160	–	430	–	c, d, g, h
Phyllophora nervosa	–	–	1100	–	c, d, g, h
Laurencia obtusa	–	–	210	–	c, d, g, h
Gracilaria confervoides	–	–	100*	–	e
Algae	–	–	660	–	i
Flowering plants:					
Zostera marina	210	–	130	–	c, d, g, h
Coelenterates:					
Actinia equina	40	–	140	–	c, d, g, h
Crustaceans:					
Idotea baltica	110	–	–	–	c, d, g, h
Gammarus locusta	100	–	–	–	c, d, g, h

Table 27 (continued)

Aquatic organisms	Concentration factors				Source
	yttrium		cerium		
	radio-isotopes in experiments	stable isotopes in nature	radio-isotopes in experiments	stable isotopes in nature	
Leander sp.					
viscera	–	–	200	–	e
muscles	–	–	25	–	e
exoskeleton	–	–	2	–	e
Pachygrapsus marmoratus	–	–	220	–	c, d, g, h
Carcinus maenas	–	–	180	–	c, d, g, h
exoskeleton	–	–	400	–	j
gills	–	–	700	–	j
digestive gland ('liver')	–	–	6	–	j
digestive tract	–	–	4	–	j
testis	–	–	3	–	j
muscles	–	–	5	–	j
Molluscs:					
Mytilus galloprovincialis					
shell	250	–	40	–	c, d, g, h
	–	–	50	–	j
body	12	–	350	–	c, d, g, h
byssus	–	–	1 000	–	j
gills	–	–	20	–	j
mantle	–	–	4	–	j
lock muscle	–	–	4	–	j
viscera	–	–	40	–	j
Venus mercenaria					
body	–	–	280**	–	k
Pecten irradians	–	–	100	–	1
Invertebrates	–	–	1 800	–	i
Fishes:					
pelagic eggs (entire)	100	–	–	–	m
egg membrane	10000	–	–	–	m
Trachurus mediterraneus ponticus					
eggs (before hatching)	196	–	495	–	n
larvae (just hatched)	0.5	–	1.4	–	n
larvae (8 hr old)	92	–	–	–	n
larvae (24 hr old)	–	–	538	–	n
Engraulis encrasicholus ponticus					
eggs (before hatching)	233	–	–	–	n
larvae (just hatched)	5.5	–	–	–	n

Table 27 (continued)

Aquatic organisms	Concentration factors				Source
	yttrium		cerium		
	radio-isotopes in experiments	stable isotopes in nature	radio isotopes in experiments	stable isotopes in nature	
larvae (24 hr old)	250	–	–	–	n
Rhombus maeoticus					
eggs (before hatching)	–	–	308	–	n
larvae (just hatched)	–	–	4.4	–	n
larvae (96 hr old)	–	–	611	–	n
Odontogadus merlangus ponticus					
eggs (before hatching)	–	–	152	–	n
larvae (24 hr old)	–	–	20.4	–	n
Belone belone euxini					
whole egg (after 96 hr)	–	–	22	–	n
egg membrane	–	–	156	–	n
egg content	–	–	12	–	n
Pleuronectes platessa	–	–	12	–	p
whole	–	–	5	–	f
muscles	–	2–10	–	2–25	f
whole egg	10.5	–	–	–	o
Chasmichthys gulosus					
muscles	–	–	0.27	–	e
viscera	–	–	4*	–	e
digestive tract	–	–	20	–	e
gills	–	–	70	–	e
vertebra	–	–	4	–	e
head and fins	–	–	15	–	e

* Concentration did not reach a balanced level.

** The rare earth radionuclides were obtained from a two-month-old mixture of fission products containing two or three times more cerium-144, cerium-141 and elements of the cerium subgroup than yttrium-91.

a POLIKARPOV [1963].
b RICE and WILLIS [1959].
c POLIKARPOV [1960c].
d POLIKARPOV [1961a].
e HIYAMA and SHIMIZU [1964].
f TEMPLETON [1959].
g POLIKARPOV [1961b].
h POLIKARPOV [1961d].
i TEMPLETON [1962].
j ZESENKO [1965].
k GONG et al. [1957].
l CHIPMAN [1958a].
m POLIKARPOV and IVANOV [1962b].
n IVANOV [1965a].
o BROWN and TEMPLETON [1964].
p DANCKWERTE [1956].

are, however, considerably lower for fishes, apparently owing to the low ratio of body surface to volume. The concentration factors of stable yttrium and cerium under natural conditions are amazingly similar to those of their radioisotopes in experiments.

Freshwater organisms also concentrate the radioisotopes of the rare earths in considerable amounts – usually hundreds or thousands of units (tables 28; also tables 21 and 22). It should be noted that variations from water to water

Table 28

Concentration factors for yttrium, rare earths and iodine in freshwater organisms

Aquatic organisms	Concentration factors				Source
	^{90}Y in experiments	rare earths radio-isotopes in nature	^{131}I in experiments	stable I in nature	
Phytoplankton	–	1 000	–	–	a
Green algae:					
Filamentous algae	–	500	–	–	a
Insect larvae	–	200	–	–	
Molluscs:					
Dreissensia polymorpha	–	–	–	1 000	b
	–	–	140	–	c
gills	–	–	70	–	c
mantle	–	–	40	–	c
viscera	–	–	20	–	c
byssus	–	–	400	–	c
shell	–	–	100	–	c
cavity fluid	–	–	6	–	c
Fishes**:	–	100	–	–	a
Salmo trutta and *S. salar*					
whole eggs	30*	–	–	–	d
egg membrane	210	–	–	–	e

* Concentration from a mixture of strontium-90 and yttrium-90, excess yttrium-90 noted in eggs.
** Concentration factor of cerium-144 in fishes from the Clinch and Tennessee rivers (in nature) are 0.13 to 3.2 (FRIEND *et al.* [1965], U.S. Public Health Service Ann. Rept. to U.S. Atomic Energy Commission for Contract AT (49–5)–1288).

a   KRUMHOLZ and FOSTER [1957].   d   TEMPLETON [1962].
b   WILKE-DÖRFURT [1928].   e   BROWN and TEMPLETON [1964].
c   GLASER [1961a].

are not so great for these radionuclides as was found when analysing the concentration factors of cesium and strontium. The decisive factor is apparently that the concentration of the rare earths and yttrium is so low that saturation of surfaces is impossible.

There is a noteworthy difference between the concentration factors of a mixture of strontium-90 and yttrium-90 for eggs in a free state and eggs within the female fish, 30 in the first instance (TEMPLETON [1962], BROWN [1962]) and 0.3 in the second (TELITCHENKO [1961]). The explanation must be sought in the fact that yttrium, unlike strontium, does not penetrate the integument. This is also indicated by the fact that the concentration factor is two for newly-formed salmon and trout larvae (TEMPLETON [1962]). The concentration factor of yttrium-90 in eggs in a free state should undoubtedly be greater than that of a mixture of strontium-90 and yttrium-90, because excess yttrium-90 is concentrated (BROWN [1962]), whereas excess strontium-90, whose concentration reduces the yttrium-90 concentration factor, remains in the solution.

The concentration factor of 0.3 obtained for carp eggs in the female (TELITCHENKO [1961]) relates to strontium-90, because measurements of the same preparations over a period showed that activity increased as a result of the appearance of yttrium-90 with the radioactive decay of strontium-90.

5.4. Distribution of radionuclides by organs and tissues

It has been demonstrated in the author's laboratory that concentration factors for cerium-144 in the gills of male crabs, *Carcinus maenas*, are approximately 1.5 times greater at 550 units than in the shell. In the other organs and tissues the concentration factor of this nuclide is measured in units or tens of units.

A different pattern was revealed when the distribution of cerium-144 was studied in relation to the weight of each organ as a proportion of total weight. Thus, approximately 70% of total activity is accounted for by shell, approximately 10% by underpart of the cephalothorax (shell elements and those muscles that can not readily be separated), approximately 5% by muscles and only 0.8% by gills, although they have the highest concentration factor. Even smaller quantities are accounted for by the remaining organs and tissues. There are some differences in cerium-144 concentration between males and females, the main difference being the greater proportion of activity concentrated in the gills than in the muscles.

Owing to the very high concentration factors, according to the author's data, more than half the cerium-144 activity in *Spicara smaris* (Serranidae) is accounted for by the gills. A considerable amount of the nuclide is concentrated in the fins, head and scales. This pattern reflects the concentration of cerium-144 by adsorption.

For obvious reasons, following peroral administration of yttrium-91 in gelatin capsules to *Tilapia mossambica* a quite different distribution pattern was noted. The activity remaining in the fishes at the end of 14 days (1.3% of the amount administered) was taken as 100%, and it was found that 43% of the nuclide was retained in internal organs, 29% in muscles, 16% in skeleton, 8% in intestines and 4% in gills. Adsorption of yttrium-91 on to the body surface and gills of the fishes was excluded in these experiments (BOROUGHS *et al.* [1956b]).

In *Anodonta cellensis* cerium-144 activity was found mainly in the shell (68.3% of the total activity of the mollusc), whereas the visceral mass accounted for only 26.9%, mantle for 2.4%, gills for 1.4% and muscles for 0.9%. Calculating solely in relation to the body, 85.0% was concentrated in visceral mass, 7.7% in mantle, 4.4% in gills and 2.9% in muscles (GETSOVA *et al.* [1964]). The shell of freshwater snails contains far less cerium-144 than the remainder of the body. Thus, 30% of total activity was concentrated in the shell of *Bithynia leachi*, approximately 20% in *Radix ovata*, 15% in *R. auricularia* and 10% in *Limnaea stagnalis* (TIMOFEYEVA-RESOVSKAYA, POPOVA *et al.* [1958]).

For purposes of comparison let us consider some information on land animals. The mammalian skeleton concentrates up to 60% of yttrium-91 (or yttrium-90) entering the organism, and 25% of cerium-144, which is more prone to concentration in the liver (60%) (SEMENOV and TREGUBENKO [1957], BALABUKHA and FRADKIN [1958], ZAKUTINSKIY [1959]). Also in aquatic organisms rare earths tend to concentrate mainly in mineralized tissues and liver, but the biological surfaces on to which the radionuclides are adsorbed from the aqueous solutions play a most important role.

CHAPTER 6

CONCENTRATION OF RADIONUCLIDES OF THE FOURTH GROUP OF ELEMENTS IN THE PERIODIC SYSTEM

6.1. General characteristics of the most important radionuclides and their carriers in water

Since within a year of fission zirconium-95 accounts for a considerable portion of the fission products of uranium and plutonium, it is important to study its behavior in the biosphere. Unfortunately, however, the biogeochemistry of zirconium has remained almost completely uninvestigated. We do not know its concentration in sea-water, and have no information on its content in aquatic organisms.

It is thought that the radioisotope of zirconium is present in sea-water mainly as particles (96%), with only 3% in a colloidal and 1% in an ionic state (table 13).

All the remaining elements of this secondary subgroup are also present in insignificant amounts in sea-water. Thorium and titanium concentrations are of the order of 10^{-6} g/l while quantitative detection of hafnium in sea-water has not even proven possible.

Because it is one of the most radiogenetically effective factors, the most important radionuclide of the main subgroup is carbon-14 which is induced by thermal neutrons in thermonuclear explosions. A certain amount of this nuclide is continually formed in the atmosphere when atmospheric nitrogen is bombarded by cosmic rays [the $^{14}N(n,p)^{14}C$ reaction]. The dating of remains of biological origin from radioactive decay of carbon-14 (half life 5 600 yr) is based on the secular equilibrium between atmospheric radiocarbon and carbon-14 in living plants and animals. Standards have, however, become a pressing problem in recent years, since trees and bones are now contaminated by carbon-14 from thermonuclear explosions. Germanium-71, which is a short-lived fission fragment, is not a significant component of the mixture of fission products, but use may sometimes need to be made of information on its behavior in the sea, for example, in the planned use of considerable quantities of this nuclide, and especially in relation to nuclear accidents at sea. The stable germanium content of sea-water is extremely low, less than 10^{-7} g/l (VINOGRADOV [1953]), but the concentration of its possible stable carriers is considerably larger with silicon 2×10^{-5} to 4×10^{-3} g/l and tin and lead approximately 10^{-6} g/l. The carbon content of sea-water is 2.8×10^{-2} g/l (REVELLE et al. [1956]).

6.2. Concentration and elimination of nuclides

The author has demonstrated that accumulation of germanium-71 in marine algae and flowering plants reached a stable level between the second and eighth day; however, a stable level was not reached in sea anemones, or in the body or shell of mussels by the end of the experiment, i.e., by the 32nd day (POLIKARPOV [1961b]). Zirconium-95 concentration factors continued to increase throughout the experimental period (16 days) in marine plants and sea anemones investigated by the author (POLIKARPOV [1963]). *Anodonta* also concentrated zirconium-95 throughout the experiment (16 days), and the concentration factors were found to rise continuously both for the mollusc as a whole, as well as its shell, mantle, gills, muscles and visceral mass (GETSOVA et al. [1964]). There is, as yet, scarcely any information concerning concentration of carbon-14 by marine organisms. Carbon concentration factors in marine animals (table 6) and algae (Barinov's data for carbon-14) are of the order of thousands of units.

There is at present little quantitative information on the uptake of carbon by aquatic organisms, and it is therefore not possible to make reliable calculations of the true concentration factors for aquatic organisms, since there are no grounds for assuming that all the carbon of sea-water is available to plants, and for animals we need information on the transformation of organic matter as food and on possible direct utilization of carbon from sea-water in the form of various compounds. In other words, the concentration factors for carbon in marine copepods, molluscs and fishes given in table 6 should not be considered more than a first approximation.

6.3. Concentration factors

Table 29 gives concentration factors for zirconium-95 and the related chemical element titanium, as well as thorium and germanium-71 in marine organisms. The following conclusion can be based on this fairly limited information. The concentration factors for concentration of zirconium-95 and titanium by marine plants range from hundreds to ten thousand units. Allowing for the fact that the data for freshwater organisms (tables 21 and 22) were calculated in relation to dry weight, it is found that the zirconium-95 concentration factors in freshwater plants agree with the figures given above for marine plants and are of the order of hundreds and thousands in relation to wet weight.

Zirconium-95 concentration factors are considerably lower for freshwater animals (tens and hundreds of units when calculated in relation to wet weight). In marine invertebrates the factors are between one and 1000. To the best of the author's knowledge, only one concentration factor has been given for thorium (800 in *Porphyra*). This concentration factor falls within the limits of the concentration factors for zirconium-95 and titanium in algae.

Different values were obtained for elements of the other subgroup (germanium-71), being tens or hundreds of units in water plants and units and tens of units in aquatic animals (table 29).

6.4. Distribution of radionuclides by organs and tissues

Zirconium-95 concentration factors are around unity in the muscles of *Carcinus maenas* and in mussels, hundredths of a unit in fish muscles, tens and hundreds of units in the gills and chitin of crabs, 20 in the gills of mussels and 50 to 60 in their shells (ZESENKO and POLIKARPOV [1965], ZESENKO [1965]).

Almost all the zirconium-95 in *Anodonta* (98.2% of the total activity of the mollusc) is in the shell. When the calculation is based solely on the content of this radionuclide in the body, the greater part (57.5%) is found in the visceral mass, and far less in gills (18.7%), muscles (16.3%) and mantle (7.5%) (GETSOVA *et al.* [1964]).

Table 29

Concentration factors for zirconium, titanium, thorium and germanium in marine organisms

Aquatic organisms	Concentration factors				Source
	^{95}Zr in experiments	stable Ti in nature	Th in experiments	^{71}Ge in experiments	
Green algae:					
Ulva rigida	2050*	–	–	16	a
Brown algae:					
Cystoseira barbata	170*	–	–	66	a
Pelvecia canaliculata	–	2000	–	–	b
Ascophyllium nodosum	–	1000	–	–	b
Fucus spiralis	–	10000	–	–	b
F. vesiculosus	–	2000	–	–	b

Table 29 (continued)

Aquatic organisms	Concentration factors				Source
	^{95}Zr in experiments	stable Ti in nature	Th in experiments	^{71}Ge in experiments	
F. serratus	–	200	–	–	b
Laminaria digitata	–	90–200	–	–	b
Red algae:					
Phyllophora nervosa	2960*	–	–	76	a
Porphyra umbilicalis	900	–	800	–	c
	200–336	–	–	–	d
Noncalcareous algae	–	1000	–	–	e
Algae	1000	–	–	–	f
Flowering plants:					
Zostera marina	1120*	–	–	197	a
Coelenterates:					
Actinia equina	130*	–	–	8*	a
Crustaceans:					
Decapoda					
larvae	1000	–	–	–	f
Carcinus maenas					
exoskeleton	350	–	–	–	g
gills	460	–	–	–	g
digestive gland	3	–	–	–	g
digestive tract	3	–	–	–	g
testis	0.6	–	–	–	g
muscles	2	–	–	–	g
heart	0.7	–	–	–	g
Molluscs:					
Mytilus galloprovincialis					
shell	4	–	–	16*	a
	36	–	–	–	g
byssus	600	–	–	–	g
gills	14	–	–	–	g
mantle	5	–	–	–	g
lock muscle	2	–	–	–	g
viscera	20	–	–	–	g
body	8	–	–	13*	a
Venus mercenaria					
body	13	–	–	–	h
Invertebrates	1400	–	–	–	f
soft tissues	–	1000	–	–	e

Table 29 (continued)

Aquatic organisms	Concentration factors				Source
	^{95}Zr in experiments	stable Ti in nature	Th in experiments	^{71}Ge in experiments	
Fishes:					
Engraulis encrasicholus ponticus					
eggs (before hatching)	14.5	–	–	–	i
larvae (just hatched)	34.1	–	–	–	i
larvae (42 hr old)	152	–	–	–	i
Uranoscopus scaber					
eggs (before hatching)	35	–	–	–	i
larvae (just hatched)	43	–	–	–	i
larvae (96 hr old)	247	–	–	–	i
Belone belone euxini					
whole egg (96 hr old)	24	–	–	–	i
egg membrane	56.7	–	–	–	i
egg content	18.7	–	–	–	i
Diplodus annularis					
liver	0.07	–	–	–	g
gonads	0.12	–	–	–	g
bones	0.08	–	–	–	g
vertebra	0.03	–	–	–	g
digestive tract	11	–	–	–	g
gills	10	–	–	–	g
skin and scales	0.8	–	–	–	g
muscles	0.05	–	–	–	g
head bones	0.4	–	–	–	g
gallbladder	0.1	–	–	–	g
kidneys	0.1	–	–	–	g
brain	0.008	–	–	–	g
heart	0.08	–	–	–	g
eyes	0.2	–	–	–	g
spleen	0.1	–	–	–	g
blood	1	–	–	–	g
Vertebrates:					
soft tissues	–	40	–	–	e

* A stable accumulation was not reached.

a POLIKARPOV [1961b, 1963].
b BLACK and MITCHELL [1952].
c TEMPLETON [1959].
d FOREMAN and TEMPLETON [1955].
e KRUMHOLZ, GOLDBERG *et al.* [1957].

f TEMPLETON [1962].
g ZESENKO [1965].
h GONG *et al.* [1957].
i IVANOV [1965d].

CHAPTER 7

CONCENTRATION OF RADIONUCLIDES OF THE FIFTH GROUP OF ELEMENTS IN THE PERIODIC SYSTEM

7.1. General characteristics of the most important radionuclides and their carriers in water

Niobium-95 makes a significant contribution to the mixture of fission products. The concentration of stable niobium in sea-water, like that of tantalum, remains unknown. The possible stable carriers in sea-water are vanadium $(2 \times 10^{-6}$ g/l) and protactinium $(3 \times 10^{-6}$ g/l). The latter is represented by the neutron-induced isotope protactinium-233.

The radionuclides induced by thermal neutrons in the main subgroup are phosphorus-32, arsenic-76 and antimony-122. The concentration of stable phosphorus in the sea varies considerably between 10^{-6} and 10^{-4} g/l, arsenic is approximately 10^{-6} to 10^{-5} g/l, antimony approximately 10^{-6} g/l and bismuth 10^{-7} g/l. The biological importance of phosphorus as an element needs no stressing. Protactinium-231 is a naturally radioactive element with a half-life of 3.34×10^4 yr.

7.2. Concentration and elimination of nuclides

Concentration of niobium-95 is complete within 2 to 15 days in almost all the marine organisms investigated. Phosphorus-32 accumulates fairly rapidly in marine algae during the first 24 hr and thereafter proceeds at a lesser rate. Molluscs, especially their shells, continue to accumulate phosphorus-32 throughout the experiment (POLIKARPOV [1960a]). The isolated gills of *Mytilus edulis* continuously concentrated phosphorus-32 for 50 hr. When stable phosphate concentration was increased, the rate of accumulation of phosphorus-32 by the gills rose (RONKIN [1950]).

There is considerable spread in the experimental data for concentration of this nuclide by oysters, but a balanced concentration is apparently reached in the soft tissues within 16 hr. Isolated oyster organs and tissues (gills, mantle, muscles and digestive gland) continuously concentrated phosphorus-32 until the end of an experiment after 48 hr (POMEROY and HASKIN [1954]). It has been established that carp concentrate labeled phosphate ions directly from the surrounding water through the gills in the main, rather than via the alimentary canal or through the body surface (TOMIYAMA, KOBAYASHI *et al.* [1956f]).

7.3. Concentration factors

To the best of the author's knowledge, apart from data obtained in his department, the only references in the literature to a concentration factor for niobium-95 are two figures (1 100 and 470) for red laver (*Porphyra*).

All the marine macrophytes investigated by the author exhibited a high capacity to concentrate niobium-95 (up to 2000 units). Although animals do not accumulate it as intensively, the concentration factor of this nuclide may exceed a hundred units. The concentration factors of stable vanadium for marine organisms vary between tens and thousands (table 30).

In general the concentration factors of phosphorus-32 in marine organisms are lower than those of stable phosphorus (tables 6 and 30). At least three factors can be said to account for this: 1) the fact that investigations were carried out by different authors employing different concentrations of phosphorus compounds in sea-water, 2) the fact that the concentration factors of phosphorus-32 continued to rise in a number of marine organisms throughout the experiments and 3) the slow exchange of phosphorus-containing substances, and especially those incorporated in the skeleton. Because determination of antimony in sea-water has been inaccurate and may possibly have been overstated (less than 5×10^{-7} g/l), it is thought that the concentration factors calculated on this basis (table 30) may in fact be considerably higher.

Tables 21 and 22 give values of niobium-95 concentration factors (in relation to dry weight) in freshwater organisms. In general they are slightly lower than zirconium-95 concentration factors for the same aquatic organisms. Comparison of phosphorus-32 concentration factors in experiments and under natural conditions reveals considerable discrepancies between the data of different authors. This may be related to great differences of stable phosphorus content in different bodies of water, and also to methods of determination. A wide range of values for the concentration factors of elements in aquatic organisms from various bodies of freshwater is a regular feature.

Phosphorus-32 concentration factors are as high for waterfowl and swallows as for aquatic organisms (table 26), and always far higher in the young than in adult birds. It is highly indicative that the phosphorus-32 concentration factor in the yolk of wild duck and geese eggs on the Columbia River reaches the colossal value of 1 500 000 units (table 26). It is obvious that this

Table 30

Concentration factors for niobium, vanadium, phosphorus and antimony in marine organisms

Aquatic organisms	Concentration factors					Source
	^{95}Nb in experiments	stable V in nature	^{32}P in experiments	stable P in nature	stable Sb in nature	
Green algae:	–	–	162	–	–	a
Ulva rigida	335	–	328*	–	–	b, c, d
Brown algae:	–	–	30	–	–	a
Pelvecia canaliculata	–	100	–	–	–	e
Ascophyllum nodosum	–	100	–	–	–	e
Fucus spiralis	–	300	–	–	–	e
F. vesiculosus	–	60	–	–	–	e
F. serratus	–	20	–	–	–	e
Laminaria digitata	–	10–30	–	–	–	e
Cystoseira barbata	2038	–	–	–	–	d
Red algae:	–	–	15–257	–	–	a
Porphyra umbilicalis	1100	–	–	–	–	f
	470	–	–	–	–	g
Phyllophora nervosa	1020	–	–	–	–	d
Ceramium elegans	–	–	257	–	–	a
Noncalcareous algae	–	1000	–	10000	–	h
Flowering plants:						
Zostera marina	1094	–	51	–	–	c, d
Sponges:						
Halichondria sp.	–	–	–	–	30	i
Coelenterates:						
Actinia equina	–	–	33	–	–	c
Cyanea capillata	–	–	–	–	30	i
Metridium dianthus	–	–	–	–	90	i
Worms:						
Nereis diversicolor	–	–	1000	–	–	a
Crustaceans:						
Idothea baltica	–	–	100	–	–	a
Pachygrapsus marmoratus	–	–	372	–	–	a
gills	–	–	313	–	–	a
Carcinus maenas	26	–	–	–	–	d
exoskeleton	60	–	–	–	–	d
	100	–	–	–	–	j
gills	89	–	–	–	–	d
	150	–	–	–	–	j

Table 30 (continued)

Aquatic organisms	Concentration factors					Source
	Nb in experiments	stable V in nature	³²P in experiments	stable P in nature	stable Sb in nature	
digestive gland	1.5	–	–	–	–	j
digestive tract	5	–	–	–	–	j
testis	2	–	–	–	–	j
muscles	3	–	–	–	–	j
	1.3	–	–	–	–	d
Molluscs:						
Mytilus galloprovincialis	157*	–	137	–	–	d
shell	165*	–	27*	–	–	c, d
	95	–	–	–	–	j
byssus	3000	–	–	–	–	j
gills	50	–	–	–	–	j
	–	–	751	–	–	a
	–	–	375	–	–	a
mantle	2	–	–	–	–	j
lock muscle	7	–	–	–	–	j
viscera	30	–	–	–	–	j
body	140	–	375	–	–	c, d
Ostrea taurica						
shell	–	–	30*	–	–	d
body	–	–	29*	–	–	d
Echinoderms:						
Asterias rubens	–	–	–	–	70	i
Tunicates:						
Ciona intestinalis	–	–	–	–	30	i
Invertebrates:						
skeleton	–	–	–	10000	–	h
soft tissues	–	100	–	10000	–	h
Fishes:						
Belone belone euxini						
whole egg (after 170 hr)	106	–	5.5	–	–	k, l
egg membrane	256	–	12.5	–	–	k, l
egg content	126	–	4.0	–	–	k, l
Squalus acanthias	–	–	–	–	140	i
Ctenolabrus rupestris	–	–	–	–	100	i
Vertebrates:						
skeleton	–	–	–	2000000	–	h
soft tissues	–	20	–	40000	–	h

* Balanced concentration not reached

a PORA *et al.* [1962].
b POLIKARPOV [1960d].
c POLIKARPOV [1961a].
d POLIKARPOV [1963].
e BLACK and MITCHELL [1952].
f TEMPLETON [1959].
g FOREMAN and TEMPLETON [1955].

h KRUMHOLZ, GOLDBERG *et al.* [1957].
i MAUCHLINE [1961], calculated from
   VINOGRADOV's data [1953].
j ZESENKO [1965].
k ZESENKO and IVANOV [1965].
l IVANOV [1965d].

factor is realistic only for a low phosphorus content in river water.

7.4. Distribution of radionuclides by organs and tissues

Phosphorus moves slowly from the point of uptake in the thallus of *Laminaria agardhii* and *Fucus vesiculosus*, not more than one cm in 12 hr (PARKER [1956]).

Phosphorus-32 has been detected by radioautography on the inner surface of the mantle of *Venus* below the surface of the epithelium (BEVELANDER [1952]). Uptake of labeled phosphate ions by the oysters *Crassostrea virginica* occurs mainly via the gills (POMEROY and HASKIN [1954]). No more than 5% of phosphorus-32 is incorporated in the shell of the freshwater snails *Limnaea stagnalis*, *Radix ovata*, *Galba palustris* and *Bithynia leachi*, and it is only in *Radix auricularia* that the shell accounts for approximately 50% of the total activity (TIMOFEYEVA-RESOVSKAYA, POPOVA *et al.* [1958]). In *Anodonta* 74.2% of the phosphorus-32 is found in the shell, 11.7% in gills, 10.2% in visceral mass and 2.8 and 1.1% in mantle and muscles, respectively. If activity is calculated separately for the body, 45.1% is accounted for by gills, 39.8% by visceral mass, 10.8% by mantle and 4.3% by muscles (GETSOVA *et al.* [1964]). Like calcium-45 and strontium-90 this nuclide is concentrated by the gills, the tail fin and the scales of the carp, but unlike strontium-90 it may accumulate in the blood, heart, liver and kidneys of the fishes. When phosphorus-32 is administered intramuscularly its distribution is the same as in uptake from water. Half the nuclide in the blood is incorporated in the formed elements (TOMIYAMA, KOBAYASHI *et al.* [1956d,g]).

The distribution of niobium-95 in *Anodonta* is as follows: 93.5% of total activity in the shell, 3.11% in gills, 2.03% in visceral mass, 0.82% in mantle and 0.54% in muscles. In the body distribution is more uniform as 48.0% is in gills, 31.2% in visceral mass, 12.5% in mantle and 8.3% in muscles (GETSOVA *et al.* [1964]).

In mammals the critical organs for phosphorus-32 and niobium-95, as for strontium-90, yttrium-91, zirconium-95 and cerium-144, are the mineralized tissue of the bones.

CHAPTER 8

CONCENTRATION OF RADIONUCLIDES OF THE SIXTH GROUP OF ELEMENTS IN THE PERIODIC SYSTEM

8.1. General characteristics of the most important radionuclides and their carriers in water

Of the elements of the secondary subgroup, chromium has a neutron-induced isotope chromium-51 that is of radioecological importance, molybdenum has a fission product molybdenum-99 that is a significant proportion of a fresh mixture of 'fission fragments' and tungsten has the neutron-induced isotope tungsten-185. There are also several isotopes of uranium, neptunium-239 and plutonium-239.

Chromium-51 is one of the principal radioactive corrosion products that form in the cooling circuit of nuclear-powered ships in a concentration of up to 4×10^{-5} Cur/l (Committee on the Effects of Atomic Radiation on Oceanography and Fisheries [1959]). There have been few determinations of stable chromium in sea-water. GOLDBERG [1957] gives a concentration of 5×10^{-8} g/l and BLACK and MITCHELL [1952] give 1.6×10^{-6} g/l. It is thought that chromium is present in sea-water in an ionic state (MAUCHLINE [1961]) in the form CrO_4^{--} (KRAUSKOPFF [1956]).

Statements on molybdenum concentration in seas also vary. KRUMHOLZ, GOLDBERG et al. [1957] give 10^{-5} g/l, while REVELLE et al. [1956] give 5×10^{-7} g/l. It is present in sea-water in all three physical states: 60% particles, 30% ions and 10% colloids (table 13).

Tungsten content in seas has been given as 10^{-7} g/l (KRUMHOLZ, GOLDBERG et al. [1957] and uranium concentration as 3×10^{-6} g/l, or in units of activity 5.4×10^{-12} Cur/l for uranium-238 and 1.6×10^{-13} Cur/l for uranium-235 (TEMPLETON [1962]).

Three elements (radionuclides) of those in the main subgroup are of interest to us: sulphur, which when irradiated by thermal neutrons is converted to sulphur-35, tellurium in the form of the short-lived fission product tellurium-132 and the naturally radioactive element polonium-210, which is the penultimate link in the series of the radioactive family of uranium-238.

The concentration of stable sulphur in sea-water is high at 0.9 g/l, that of tellurium and polonium is unknown, but a figure of 4×10^{-6} g/l has been given for the related element selenium. Almost half (45 %) of the radioisotopes of tellurium are present in sea-water in an ionic state, 43 % in a colloidal state and only 12 % in particles (table 13).

8.2. Concentration factors

All species of marine plants and animals investigated by the author concentrated chromium-51 to a steady state within two to 16 days of commencement of the experiments.

Table 31 gives concentration factors of chromium in brown algae (BLACK and MITCHELL [1952]) based on a chromium concentration in sea-water of 1.6×10^{-6} g/l. All other concentration factors given in the table are based on a concentration in sea-water of 5×10^{-8} g/l. This clearly goes a long way towards explaining the very great difference between the two chromium concentration factors in *Fucus serratus*. In general, marine plants and animals exhibit a considerable capacity to concentrate stable chromium from the

Table 31

Concentration factors for chromium, molybdenum, uranium and plutonium in marine organisms

Aquatic organisms	Concentration factors					Source
	^{51}Cr in experiments	stable Cr isotopes in nature	stable Mo in nature	^{238}U in experiments	^{239}Pu in nature	
Green algae:						
Ulva rigida	70	–	–	16*	–	a, b
Enteromorpha compressa	–	–	–	212*	–	c
Brown algae:						
Cystoseira barbata	–	–	–	129*	–	c
	60	–	–	–	–	b
Pelvetia canaliculata	–	300	8	–	–	d
Ascophyllum nodosum	–	500	14	–	–	d
Fucus spiralis	–	300	15	–	–	d
F. vesiculosus	–	400	4	–	–	d
F. serratus	–	100	3	–	–	d
	–	120000	–	–	–	e
Laminaria digitata	–	200	2–3	–	–	e
Red algae:						
Polysiphonia elongata	–	–	–	49*	–	c
Phyllophora nervosa	80	–	–	–	–	b
Phorphyra umbilicalis	–	–	–	–	1000	f

Table 31 (continued)

Aquatic organisms	Concentration factors					Source
	^{51}Cr in experiments	stable Cr isotopes in nature	stable Mo in nature	^{238}U in experiments	^{239}Pu in nature	
Noncalcareous algae	–	300	10	–	–	g
Flowering plants:						
Zostera marina	40	–	–	–	–	b
Sponges:						
Halichondria sp.	–	800	–	–	–	e
Coelenterates:						
Actinia equina	20	–	–	–	–	b
Cyanea capillata	–	1 600	–	–	–	e
Crustaceans:						
Leander adspersus	37	–	–	–	–	b
Molluscs:						
Mytilus galloprovincialis						
body	18	–	–	–	–	b
shell	2	–	–	–	–	b
Echinoderms:						
Asterias rubens	–	140	–	–	–	e
Echinus esculentus	–	9 000	–	–	–	e
Invertebrates:						
soft tissues	–	–	100	–	–	g
Fishes:						
Squalus acanthias	–	2 000	–	–	–	e
Pleuronectes platessa						
soft tissues	–	–	–	–	0.17	h
intestines	–	–	–	–	1*	h
gills	–	–	–	–	1*	h
liver	–	–	–	–	0.8*	h
kidneys	–	–	–	–	0.6*	h
skin	–	–	–	–	0.4*	h
bones	–	–	–	–	0.1*	h
muscles	–	–	–	–	0.09*	h
Vertebrates:						
soft tissues	–	–	20	–	–	g

* Stable concentration not achieved.

a POLIKARPOV [1960e].
b POLIKARPOV [1963].
c POLIKARPOV and TEN [1961].
d BLACK and MITCHELL [1952].
e MAUCHLINE [1961], calculated from VINOGRADOV's data [1953].

f TEMPLETON [1959].
g KRUMHOLZ, GOLDBERG et al. [1957].
h TEMPLETON [1962].

Table 32

Concentration factors for sulphur-35 and polonium-210 in marine organisms

Aquatic organisms	Concentration factors				Source
	^{35}S in experiments		stable S in nature	^{210}Po in experiments	
	sulphate ions	sulphide ions			
Green algae:					
Ulva rigida	2.6	3.0	–	–	a
Enteromorpha compressa	0.6	0.2	–	–	a
Bryopsis plumosa	0.4	0.4	–	–	a
Brown algae:					
Cystoseira barbata	0.3	1.0	–	–	a
Red algae:					
Ceramium rubrum	0.1	2.0	–	–	a
Gelidium latifolium	0.3	0.5	–	–	a
Porphyra umbilicalis	–	–	–	1 000	b
Noncalcareous algae	–	–	10	–	c
Coelenterates:					
Actinia equina	0.4	–	–	–	a
Molluscs:					
Nassa reticulata					
body	0.3	–	–	–	a
Rapana besoar					
body	0.2	–	–	–	a
Mytilus galloprovincialis					
body	0.4	–	–	–	a
Invertebrates:					
skeleton	–	–	1	–	c
soft tissues	–	–	5	–	c
Fishes:					
Odontogadus merlangus ponticus					
eggs (before hatching)	1	–	–	–	d
larvae (just hatched)	0.6	–	–	–	d
Vertebrates:					
soft tissues	–	–	2	–	c

a   POLIKARPOV [1960f].
b   TEMPLETON [1959].
c   KRUMHOLZ, GOLDBERG et al. [1957].
d   IVANOV [1965a].

surrounding environment. The fact that the concentration factors of chromium-51 are lower than those of stable chromium is possibly due to the fact that chromium-51 was used in the form of chromous chloride (a different chemical form) in the author's experiments, and with a considerable quantity of the isotopic carrier (the chromium content of the sea-water was increased). Moreover the concentration factors given for stable chromium under natural conditions are not completely convincing. This matter has not yet been the subject of special investigation (simultaneous measurements of chromium in aquatic organisms and in the surrounding water under natural conditions).

Molybdenum concentration factors are considerably lower, but still remain fairly large in the soft tissues of invertebrates and as an overall total for marine animals at 100 and 600, respectively.

It emerges from the data in table 31 that the equilibrium value of the plutonium-239 concentration factor in algae is 1000. Nevertheless this nuclide is very poorly concentrated in plaice, although the concentration factor increased slowly but continuously in all tissues except the intestines throughout the 150 days of an investigation.

Uranium concentration factors continued to increase in *Ulva* throughout the experiments. The highest uranium concentration factor in this alga was only one-fifth that of plutonium-239 in *Porphyra*, which also had a very high polonium-210 concentration factor of 1000 (table 32).

The considerable weight concentration of sulphur in sea-water is responsible for the low concentration factors of its radioisotope sulphur-35 in marine plants and animals (table 32). Low concentration factors are also noted for stable sulphur (one to five), except in the case of noncalcareous algae (ten).

Higher concentrations of sulphur-35 form in freshwater organisms than in marine organisms (tables 21 and 22). Since the data in these tables are related to dry weight, the average concentration factor of this nuclide for freshwater plants is approximately 10, and for animals approximately one or slightly above. The lower concentration of sulphur compounds in freshwaters apparently produces conditions that favor slightly higher uptake.

CHAPTER 9

CONCENTRATION OF RADIONUCLIDES OF THE SEVENTH GROUP OF ELEMENTS IN THE PERIODIC SYSTEM

9.1. General characteristics of the most important radionuclides and their carriers in water

Of the radionuclides of elements of the secondary subgroup, only manganese-54 is at present of interest in radioecology. Stable manganese has, in many respects, been the subject of thorough study. This trace element, which is essential to life, growth and development, is incorporated in the enzymes lactase and arginase. Manganese concentration in sea-water is 2×10^{-6} to 4×10^{-6} g/l (VINOGRADOV [1953], GOLDBERG [1957]). Nothing is known of the concentration of the related elements technetium and rhenium in sea-water.

The main subgroup includes iodine and its important fission product iodine-131, which is a significant component of a 20-day-old mixture of 'fission fragments'. Stable iodine is found in the thyroid hormone thyroxine. Iodine concentration in sea-water is 5×10^{-5} g/l, almost all of which (90%) is in an ionic state, with 8% in a colloidal state and only 2% as particles (table 13).

GLASER [1961a] found that iodine content in Lake Stechlin near Berlin in 1959 and 1960 reached $(3.50 \pm 0.234) \times 10^{-6}$ g/l of lake water.

9.2. Concentration factors

Marine plants and animals exhibit a high capacity to accumulate manganese from sea-water. Its concentration factors are measured mainly in thousands of units, but there is a wide range rising from 80 to 550 000 (table 33), i.e., rising to one to two g per kg of live weight for a manganese concentration in sea-water of 2×10^{-6} to 4×10^{-6} g/l (KRUMHOLZ, GOLDBERG et al. [1957], MAUCHLINE [1963]). It may be stated with satisfaction that the concentration factors of radioactive and stable manganese are similar in related aquatic organisms. The behavior of manganese in the lobster was studied by BRYAN and WARD [1965]).

Aquatic organisms concentrate iodine-131 fairly rapidly and reach a balanced concentration in a few days. This has been noted for some forms of

133

marine organisms in the author's investigations, and for freshwater organisms by GLASER [1961a] and by TIMOFEYEV-RESOVSKIY and TIMOFEYEVA-RESOVSKAYA [1959]. In the mollusc *Dreissensia*, iodine-131 accumulation reached a saturation level within approximately 10 days or before (GLASER [1961a]). Several authors studied some questions of accumulation of iodine-

Table 33

Manganese concentration factors in marine organisms

Aquatic organisms	Concentration factors		Source
	^{54}Mn in nature	stable Mn in nature	
Phytoplankton:			
(mainly *Gonyaulax polyhedra*)	2500–6300	–	a
Green algae:			
Ulva lactuca	–	2200	b
	–	1300	c
Enteromorpha sp.	–	4600	b
	–	1500	c
	5500	–	a
Diatoms:			
Rhizosolenia sp.	–	20	b
Coscinodiscus sp.	–	5000	b
Brown algae:			
Pelvetia canaliculata	–	1800	b
	–	6000	d
Ascophyllum nodosum	–	3400	b
	–	4500	d
Fucus serratus	–	20000	b
	–	12000	d
	–	7500	c
F. spiralis	–	11000	d
F. vesiculosus	–	12000	d
Laminaria saccharina	–	350	b
	–	300	c
Macrocystis pyrifera	680	–	a
Red algae:			
Porphyra lanceolata	2000	–	a
Corallina officinalis	–	6000	b
Rhodymenia palmata	–	1000	b
Sponges:			
Halichondria sp.	–	6000	b
Halidona simulans	–	9500	b
Suberites domunculus	–	3000	b

Table 33 (continued)

Aquatic organisms	Concentration factors		Source
	^{54}Mn in nature	stable Mn in nature	
Coelenterates:			
Cyanea capillata	–	120	b
Metridium dianthus	–	4100	b
Crustaceans:			
Amphipoda	930	–	a
Balanus tintinnabulum	2400	–	a
Pollicipes polymerus			
neck	2050	–	a
head	1270	–	a
Crangon vulgaris	–	7500	b
Molluscs:			
Patella vulgata	–	750	b
Littorina littorea	–	14000	b
Buccinum undatum	–	600	b
shell	–	5000	b
Nucella lapillus	–	1300	b
Aplysia punctata	–	750	b
Aporrhais pes-pelicanus	–	29000	b
Turritella communis	–	85000	b
Crassostrea virginica			
shell	–	45000	b
Ostrea sp.			
shell	–	50000	b
O. edulis			
muscles	–	1500	b
Mytilus edulis	830	1200	a, b
shell	–	10000	c
M. californianus	800–830	–	a
Pecten maximus	–	24000	b
P. jacobaens			
muscles	–	10000	b
Haliotis tuberculata			
muscles	–	750	c
Chlamys opercularis	–	3000	b
Tellina crassa	–	3700	b
T. tenuis	–	5800	b
Mactra corallina	–	650	b
muscles	–	620	c
Venus verrucosa	–	1900	b
Dosinia exoleta	–	35000	b

Table 33 (continued)

Aquatic organisms	Concentration factors		Source
	^{54}Mn in nature	stable Mn in nature	
Cardium edule			
shell	–	5 000	b
Mya arenaria	–	400	b
Lutraria elliptica	–	3 000	b
Sepia officialis	–	550 000	b
muscles	–	10 000	c
Octopus vulgaris			
muscles	–	50 000	c
Echinoderms:			
Astropecten sp.	–	3 500	b
Asterias rubens	–	33 000	b
Stichopus regalis			
muscles	–	200	c
Tunicates:			
Cione intestinalis	–	9 500	b
Fishes:			
Pleuronectes sp.			
muscles	–	70	c
Centrophorus sp.	–	1 400	b
Squalus acanthias	–	1 800	b
Gadus sp.			
muscles	–	320	c
Clupea harengus	–	95	b
muscles	–	95	c
C. pilchardus	–	126 000	b
Engraulis encrasicholus	–	350	b
Salmo fario	–	170	b
Osmerus eperlanus	–	130	b
Merluccius merluccius	–	980	b
M. vulgaris	–	1 000	b
Scomber sp.			
muscles	–	80	c
Leuresthes tenuis	850	–	a
Sarda lincolata	240	–	a
Zooplankton			
(minor, from surface layer)	340–1 700	–	a

a FOLSOM *et al.* [1963].
b MAUCHLINE [1961], calculated from VINOGRADOV's data [1953].
c ICHIKAWA [1961], calculated from VINOGRADOV's data [1953].
d BLACK and MITCHELL [1952].

131 in laminaria (ROCHE and YAGI [1952]), invertebrates (ANTHENISSE and LEVER [1956], ROCHE *et al.* [1960]) and *Amphioxus* (THOMAS [1956]).

Iodine concentration factors of hundreds of units (between 19 and 17900) have been obtained for marine algae, while those for flowering plants and animals vary between 1.3 and 5000 (table 34). The figures for stable iodine in algae range between 160 and 10000. Concentration factors for animal material vary considerable in magnitude. A stable iodine concentration factor of 1000 is characteristic of *Dreissensia*, whereas the concentration factor for iodine-131 is only 140. On the whole iodine-131 concentration factors are apparently higher for freshwater plants than for freshwater animals (tables 21, 22 and 28). It is noteworthy that the iodine concentration factor in molluscs is highest in the byssus and considerably lower in the shell.

In pure water the activity of *Dreissensia* that had accumulated iodine-131 fell rapidly by 15% in five hr and then remained at this level until the end of the experiment (30 hr) (GLASER [1961a]).

Table 34

Iodine concentration factors in marine organisms

Aquatic organisms	Concentration factors		Source
	^{131}I in experiments	stable I in nature	
Green algae:			
Ulva rigida	200	–	a
U. pertusa	160–300	–	b
Enteromorpha linza	300*	–	a
E. prolifera	78	–	c
Brown algae			
Cystoseira barbata	300*	–	a
Laminaria sp.	–	3400–6800	d
Undaria pinnatifida	–	400–800	d
Hizikia fusiforme	–	800	d
Sargassum thumbergii	140–270	–	b
Red algae:			
Meristotheca papulosa	205	–	b
Phyllophora rubens	500*	–	a
Ceramium rubrum	500*	–	a
Porphyra sp.	–	160	d
Asparagopsis taxiformes	17900	–	e
Noncalcareous algae	–	10000	f
Flowering plants:			
Zostera marina	30	–	a

Table 34 (continued)

Aquatic organisms	Concentration factors		Source
	^{131}I in experiments	stable I in nature	
Coelenterates:			
Actinia equina	6*	--	a
Crustaceans:			
Centropages	–	6000	g
Calanus	–	100	g
Leander squilla			
with egg	20	–	a
L. pacificus	30*	–	b
Penaeus japonicus			
muscles	–	35	h
Metapenaeus monoceros			
muscles	–	20	h
Panulirus japonicus			
muscles	–	48	h
Homarus vulgaris			
ovaria	5000	–	i
blood	100–300	–	i
Molluscs:			
Limacina	–	50000	g
Mytilus galloprovincialis			
shell	60*	–	a
body	20*	–	a
byssus	2000*	–	a
Ostrea gigas			
muscles	–	71	h
Meretrix meretrix lusoria			
muscles	–	49	h
Venerupis japonica			
muscles	–	41	h
V. philippinarum			
visceral mass	0.98	–	b
gills	1.8	–	b
muscle tissues	0.92	–	b
other viscera	0.39	–	b
shell	2	–	b
Ommatostrephes	–	20000	g
Echinoderms:			
Stichopus regalis			
muscles	–	20–60	j
viscera	–	490–540	j

Table 34 (continued)

Aquatic organisms	Concentration factors		Source
	^{131}I in experiments	stable I in nature	
Strongylocentrotus pulcherrimus			
digestive tract	100*	–	b
gonads	8.5*	–	b
shell	9.2*	–	b
S. droebachiensis			
ovary	–	60	j
Invertebrates:			
soft tissues	–	100	f
skeleton	–	50	f
Fishes:			
Katsuwonus pelamis			
muscles	–	11	h
Sardinops melanosticta			
muscles	–	11	h
Gadus macrocephalus			
muscles	–	12	h
Chrysophrys major			
muscles	–	12	h
Clupea pallasii			
muscles	–	15	h
Chasmichthys gulosus	3.3–8.5	–	b
Acanthogobius flavimanus	1.3	–	b
Girella punctata	7.0	–	b
Pterogobius elapoides	1.3	–	b
Rudarius ercodes	1.6	–	b
Vertebrates:			
soft tissues	–	10	f

* Balanced concentration not achieved.

a   POLIKARPOV [1963].
b   HIYAMA and KHAN [1964].
c   HIYAMA [1962].
d   ICHIKAWA [1961], calculated from the data of ŌTANI *et al.* [1935].
e   PALUMBO [1957].
f   KRUMHOLZ, GOLDBERG *et al.* [1957].
g   KETCHUM and BOWEN [1958].
h   ICHIKAWA [1961], calculated from the data of ŌSHIMA [1949].
i   FONTAINE [1960].
j   ICHIKAWA [1961], calculated from the data of VINOGRADOV [1953].

CHAPTER 10

CONCENTRATION OF RADIONUCLIDES OF THE EIGHTH GROUP OF ELEMENTS IN THE PERIODIC SYSTEM

10.1. General characteristics of the most important radionuclides and their carriers in water

This group contains very important radionuclides, namely iron-55, iron-59, cobalt-60, ruthenium-103 and ruthenium-106. Iron and cobalt are very important trace elements, and their radioisotopes are some of the most important neutron-induced radioactive substances. Even after storage their activity in water from the cooling circuit of the nuclear-powered vessel Savannah was 10^{-5} Cur/l (Committee on the Effects of Atomic Radiation on Oceanography and Fisheries [1959]). Three years after their formation ruthenium-103 and ruthenium-106 are still significant factors in a mixture of fission products.

The radioisotopes of ruthenium occur in sea-water almost exclusively as particles (95%) with only 5% accounted for by colloids (table 13).

Commercially available ruthenium chloride is usually a mixture of the trichloride and the tetrachloride. Eleven nitrosyl ruthenium complexes which may possibly form in radioactive nitrate wastes have been described. AUERBACH and OLSON [1963] have commented with justification that the polyvalent states of ruthenium and the relative ease with which it changes valency and the forms of its various complex compounds place fairly strict limitations on the formulation of experiments relating to its behavior and possible effects and on interpretation of the experimental results.

The iron content of sea-water is between 2×10^{-6} and 2×10^{-5} g/l, that of cobalt and nickel is 5×10^{-7} g/l each, and nothing is known concerning the concentration of ruthenium, rhodium, palladium, osmium, iridium and platinum in sea-water (REVELLE et al. [1956], KRUMHOLZ, GOLDBERG et al. [1957], LOWMAN [1963a]).

Iron and cobalt are, of course, most important trace elements.

10.2. Concentration and elimination of nuclides

It has been established by the author's investigations that most species of Black Sea plants and animals concentrate iron, cobalt and ruthenium to a steady state within eight to 16 days of commencement of experiments.

A *Nitzschia* cell culture concentrated ruthenium-106 from sea-water throughout the 12 days of an experiment, although the amount per cell by the end of the experiment fell as a result of biological dilution. It is thought that dividing plankton algae are capable of concentrating considerable amounts of ruthenium-106 from the environment (BOROUGHS, CHIPMAN *et al.* [1957]).

According to JONES [1960] this radionuclide is taken up as readily by the marine diatom *Phaeodactylum tricornutum* as by fine particles of sand. The uptake of ruthenium-106 by sand is characterized by an adsorption isotherm similar to Freundlich's isotherm:

$$\frac{x}{m} = aC^b \qquad or \qquad \log \frac{x}{m} = \log a + b \log C,$$

where x is the mass of the adsorbed dissolved substance, m is the mass of the adsorbent, C is the ultimate concentration of the solution, a and b are constants. The uptake of ruthenium-106 by sand is, of course, limited by the form in which the ruthenium-106 occurs and by its valence state.

Although uptake is described in this case by the adsorption isotherm, Jones is of the opinion that the ruthenium is maintained on the particle surface by chemical forces. The amount of ruthenium-106 extracted from sea-water increased as the concentration of algal cells rose.

Laminaria digitata adsorbs more ruthenium-106 per unit of surface than do *Ulva lactuca* and *Porphyra latissima*.

Adsorbed ruthenium-106 is fairly firmly retained by particles as less than 7% of the activity passed into the solution when particles were washed with ordinary sea-water or with sea-water at pH 2, 4 or 6.

Approximately 60% of the ruthenium-106 passed into pure sea-water from contaminated *Porphyra* within an hour. In this instance the bond was not as strong as for sand, mud or diatoms. Mussel shells lost 50% of their ruthenium-106 in pure sea-water within 24 hr, but the soft tissues lost only 5%. At the end of a week 40% of the nuclide remained in the shells, and 80% in the soft tissues. When the flesh of these molluscs was cooked in boiling freshwater for 10 min 45% of the ruthenium-106 was eliminated (JONES [1960]). *Ceramium rubrum* and *Enteromorpha intestinalis* concentrated iron-59 rapidly, but gave up very little of it in pure sea-water (TAYLOR and ODUM [1960]).

10.3. Concentration factors

As stated above iron concentration in sea-water varies considerably. The concentration factors for stable iron (table 35) calculated by MAUCHLINE [1961] were based on a concentration of this element in sea-water of 5×10^{-5} g/l (GOLDBERG [1957]), and on data for its content in marine organisms (VINOGRADOV [1953], BLACK and MITCHELL [1952]).

Iron concentration factors in algae, some coelenterates, molluscs, echinoderms and arrow worms are high (thousands or ten thousands of units), but fall to hundreds of units for a number of coelenterates, molluscs, tunicates and fishes.

Iron-59 concentration factors are in general, although not invariably, lower in algae than those for stable iron. Possible explanations will be considered below. A similar phenomenon has been noted for molluscs, although a balanced accumulation was not achieved (table 35).

Table 35

Concentration factors for iron and ruthenium in marine organisms

Aquatic organisms	Concentration factors			Source
	^{54}Fe in experiments	stable Fe in nature	^{106}Ru in experiments	
Green algae:				
Enteromorpha intestinalis	300	–	–	a
Ulva rigida	730	–	95	b
U. pertusa	100*	–	360	c
Diatoms:				
Ceratium tripos	–	45000	–	d
Brown algae:				
Ascophyllum nodosum	–	5700	–	e
Pelvetia canaliculata	–	9000	–	e
Fucus spiralis	–	12500	–	e
F. vesiculosus	–	3400	–	e
F. serratus	–	1200	–	e
Laminaria sp.	–	5800	–	f
Undaria pinnatifida	–	1300	–	f
Hizikia fusiforme	–	2900	–	f
Cystoseira barbata	1060*	–	197*	b
Sargassum thumbergii	–	–	170	c
S. natans	–	–	366	g
S. fluitans	–	–	310	g

Table 35 (continued)

Aquatic organisms	Concentration factors			Source
	[54]Fe in experiments	stable Fe in nature	[106]Ru in experiments	
Red algae:				
Ceramium rubrum	100	–	–	a
Porphyra umbilicalis	–	–	800–1 000	h
P. tenera	–	2 000	–	f
Gelidium amansii	–	4 000	–	f
Phyllophora nervosa	1 650*	–	593	b
Meristotheca papulosa	–	–	60	c
Algae	–	–	660	h
	–	–	2 000	i
Flowering plants:	–	2 000	–	d
Zostera marina	435	–	181	b
Sponges:				
Halichondria sp.	–	10 000	–	d
Coelenterates:				
Cyanea capillata	–	300	–	d
Metridium dianthus	–	2 400	–	d
Crustaceans:				
Calanus	–	25 000	–	j
Decapoda				
plankton	–	–	100	k
Leander pacificus	–	–	10	c
muscles	–	–	5	c
viscera	–	–	30*	c
exoskeleton	–	–	25	c
Carcinus maenas				
exoskeleton	290	–	193	b
muscles	19	–	3	b
gills	296	–	450	b
heart	–	–	1.2	l
testis	–	–	2	l
digestive gland	–	–	3	l
digestive tract	–	–	2.5	l
Panaeus japonicus				
muscles	–	1 000	–	f
Penacopsis sp.				
muscles	–	4 000	–	f
Panulirus japonicus				
muscles	–	1 000	–	f
Paralithodes camtschatica				
muscles	–	4 000	–	f

Table 35 (continued)

Aquatic organisms	Concentration factors			Source
	^{54}Fe in experiments	stable Fe in nature	^{106}Ru in experiments	
Neptunus trituberculata				
muscles	–	2000	–	f
Molluscs:				
Limacina	–	50000	–	j
Mytilus galloprovincialis	66*	–	38*	b
	–	–	50	m
shell	49	–	49*	b
	–	–	53	m
body	99*	–	16	b
	–	–	6	m
byssus	–	–	500	m
gills	–	–	10	m
foot	–	–	3.7	m
mantle	–	–	2.5	m
muscles	–	–	2.0	m
visceral mass	–	–	18	m
Venus sp.				
shell	55*	–	–	n
body	25*	–	1	n
Ostrea gigas				
muscles	–	8000	–	f
Venerupis japonica				
muscles	–	7000	–	f
V. philippinarum				
gills	53*	–	12	c
visceral mass	43*	–	6.4	c
muscle tissues	30*	–	3.6	c
other viscera	70*	–	14	c
shell	13*	–	11	c
Meretrix meretrix lusoria				
muscles	–	13000	–	f
Haliotis diversicolor				
muscles	–	17000	–	f
H. gigantea				
muscles	–	3000	–	f
Ommatostrephes	–	10000	–	j
Turbo cornutus				
muscles	–	9000	–	f
Ostopus fangsiano				
muscles	–	6000	–	f

Table 35 (continued)

Aquatic organisms	Concentration factors			Source
	^{54}Fe in experiments	stable Fe in nature	^{106}Ru in experiments	
Echinoderms:				
Strongylocentrotus pulcherrimus				
digestive tract	30*	–	5.3	c
gonads	0.68	–	1*	c
test	10*	–	2.6	c
Stichopus japonicus				
muscles	–	78000	–	f
viscera	–	1000	–	f
Anthocidaris crassispina				
ovaries	–	10000	–	f
Arrow worms:				
Sagitta setosa	–	1800	–	d
	–	50000	–	j
Tunicates:				
Ciona intestinalis	–	300	–	d
Invertebrates	–	–	2000	k
Fishes:				
Engraulis encrasicholus ponticus				
eggs (before hatching)	–	–	12	o
larvae (just hatched)	–	–	3.6	o
larvae (24 hr old)	–	–	26.2	o
Acanthogobius flavimanus				
muscles	–	2000	–	f
	0.05	–	–	i
skeleton and scales	0.04–0.3*	–	–	i
gills and digestive tract	10*–12*	–	–	i
Chasmichthys gulosus				
(small)	–	–	2.5	c
(middle size)	5.8	–	–	c
Girella punctata	–	–	2	c
Clupea pallasii				
muscles	–	1800	–	f
Theragra chalcogramma				
muscles	–	400	–	f
Chrysophrys major				
muscles	–	400	–	f
Lateolabrax japonicus				
muscles	–	3000	–	f
Cololabis saira				
muscles	–	3000	–	f

Table 35 (continued)

Aquatic organisms	Concentration factors			Source
	^{54}Fe in experiments	stable Fe in nature	^{106}Ru in experiments	
Scomber japonicus				
muscles	–	1 800	–	f
Pleuronectes sp.				
muscles	–	600	–	f
P. platessa				
soft tissues	–	–	0.05*	h
Sardinops melanosticta				
muscles	–	2 000	–	f
Trachurus japonicus				
muscles	–	700	–	f
Odontogadus merlangus ponticus				
gills	–	–	10	l
digestive tract	–	–	20	l
muscles	–	–	0.04	l
skin and scales	–	–	0.8	l
vertebra	–	–	0.02	l
bones	–	–	0.08	l
kidneys	–	–	0.09	l
blood	–	–	0.6	l
heart	–	–	0.05	l
spleen	–	–	0.05	l
gallbladder	–	–	0.01	l
gonads	–	–	0.01	l
brain	–	–	0.005	l
eyes	–	–	0.01	l
head bones	–	–	0.4	l
liver	–	–	0.07	l
eggs (before hatching)	152	–	17.8	o
larvae (just hatched)	–	–	0.5	o
larvae (96 hr old)	–	–	15.8	o

* Stable concentration not achieved.

a TAYLOR and ODUM [1960].
b POLIKARPOV [1963].
c HIYAMA and KHAN [1964].
d MAUCHLINE [1961], calculated from VINOGRADOV's data [1953].
e BLACK and MITCHELL [1952].
f ICHIKAWA [1961], calculated from KAGAWA's data [1959].
g POLIKARPOV, ZAYTSEV et al., in the press.
h TEMPLETON [1959].
i HIYAMA [1962].
j KETCHUM and BOWEN [1958].
k TEMPLETON [1962].
l ZESENKO [1965].
m ZESENKO and POLIKARPOV [1965].
n GONG et al. [1957].
o IVANOV [1965a].

There is little information concerning ruthenium-106 concentration factors in marine organisms. Figures of up to 2000 have been given for concentration factors in algae and in certain invertebrates. Accumulation of ruthenium-106 in the soft tissues of fishes is very slight (less than one unit), although it may be significant in the muscles of crabs and in the body of mussels (table 35).

Different authors have given widely differing figures for iron-59 concentration factors in freshwater organisms. Under natural conditions in the Columbia River, the values are considerably higher than those obtained by laboratory experiments with water from Lake Bol'shoye Miassovo. Concentration of ruthenium-106 in freshwater algae is approximately the same as concentration in marine algae (taking into consideration that the data relate to dry weight) and concentration in leeches and fishes is slight (tables 21 and 22).

Cobalt concentration in sea-water varies between 10^{-6} (VINOGRADOV [1953]) and 5×10^{-7} g/l (GOLDBERG [1957]). The lower of these two quantities was taken for calculation of the concentration factors of stable cobalt (MAUCHLINE [1961]). The present author's data on cobalt concentration factors in algae (45 to 335) are in good agreement with data on stable cobalt concentration factors for algae in the Atlantic Ocean off the Canadian coast (40 to 200) (YOUNG and LANGILLE [1958]). Cobalt-60 concentration factors are far lower in sea anemones and shrimps than those of stable cobalt and the other members of the same groups of elements.

In the present author's opinion the numerical data for stable cobalt concentration factors in scallops and mussels (approximately unity) are incorrect, since according to his own information and other data in the literature there is excellent accumulation of cobalt-60 in the shells of mussels and *Venus* (152 and 500, respectively) and in the body (186 and 18) (table 36).

Cobalt-60 concentration factors in freshwater plants and animals are, in general, fairly high, and reach hundreds of units in relation to dry weight (tables 21 and 22). In leeches and some molluscs the concentration factors are tens of units.

Near the testing grounds in the Pacific Ocean iron-55, iron-59, cobalt-57, cobalt-58 and cobalt-60 have been discovered in addition to other radionuclides in various marine organisms. In these areas the liver of tunny (TOZAWA *et al.* [1957]) and of flying fishes (SEYMOUR *et al.* [1956]) contained radioisotopes of iron and cobalt. Iron-55 accounted for 21 % of the total activity of plankton in the Marshall Islands (LOWMAN [1958]).

10.4. Distribution of radionuclides by organs and tissues

The shells of *Mytilus edulis* concentrated more ruthenium-106 than the digestive gland, foot and muscles of this mollusc and 95% of the nuclide was concentrated in the shells. In plaice ruthenium-106 was concentrated mainly in gills, intestines and skin, and to a lesser extent in liver and spleen. Little or none was present in the muscles (TEMPLETON [1959]).

In the freshwater molluscs *Anodonta* 73.1% of the ruthenium-106 was contained in the shell, 23.2% in visceral mass, 2.6% in gills and 0.6% each in mantle and muscles. Taking only the soft tissues, 86.2% of the isotope was in the visceral mass, 9.7% in gills, 2.2% in mantle and 1.9% in muscles (GETSOVA *et al.* [1964]). Approximately half the ruthenium-106 was present in the shell in *Limnaea stagnalis*, *Radix auricularia*, *Calba palustris* and *Bithynia leachi*, and approximately 20% in *Radix ovata* (TIMOFEYEVA-RESOVSKAYA, POPOVA *et al.* [1958]).

Table 36

Cobalt concentration factors in marine organisms

Aquatic organisms	Concentration factors		Source
	^{60}Co in experiments	stable Co in nature	
Green algae:			
Ulva lactuca	–	440	a
U. rigida	335*	–	b
U. pertusa	380	–	c
Enteromorpha linza	116*	–	b
Bryopsis plumosa	165*	–	b
Monostroma sp.	–	15	d
Brown algae:			
Cystoseira barbata	45*	–	b
Fucus vesiculosus	–	400	e
F. serratus	–	560	e
F. spiralis	–	740	e
Ascophyllum nodosum	–	340	e
Laminaria saccharina	–	260	a
L. sp.	–	27	d
Alaria esculenta	–	160	a
Sargassum thumbergii	420	–	c
Red algae:			
Porphyra sp.	–	64	d
Algae	–	40–200	f

Table 36 (continued)

Aquatic organisms	Concentration factors		Source
	^{60}Co in experiments	stable Co in nature	
Flowering plants:			
Zostera marina	–	120	a
Sponges:			
Halichondria sp.	–	20	a
Coelenterates:			
Actinia equina	20	–	b
Metridium dianthus	–	1100	a
Cyanea capillata	–	1400	a
	–	350	g
Beröe cucumis	–	130	g
Worms:			
Myxicola infundibulum	–	20000	a
Aphrodite aculeata	–	10000	a
Arenicola marina	–	4000	a
Crustaceans:			
Centropages typicus et hamatus	–	590	g
Calanus finmarchicus	–	220	g
Leander adspersus	11*	–	b
L. pacificus	7*	–	c
L. sp.	30.6	–	m
Euphausia krohnii	–	770	g
Palinurus sp.			
muscles	–	4000	h
Hyas araneus	–	300	a
Molluscs:			
Limacina retroversa	–	9800	g
Clione limacina	–	3600	g
Ommatostrephes illicebrosa	–	210	g
O. sloani			
muscles	–	62	d
Cardium edule	–	1	a
Mytilus edulis	–	1	a
M. galloprovincialis			
shell	152	–	b
body	125–186	–	b, m
Venus sp.			
shell	500*	–	i
body	18*	–	i
Polypus sp.			
muscles	–	52	d

Table 36 (continued)

Aquatic organisms	Concentration factors		Source
	^{60}Co in experiments	stable Co in nature	
Ostrea gigas			
muscles	–	170	d
Pecten yessoensis			
muscles	–	190	d
Meretrix meretrix lusoria			
muscles	–	200	d
Venerupis philippinarum			
visceral mass	9.2	–	c
gills	17*	–	c
muscle tissues	7*	–	c
other viscera	10*	–	c
shell	36	–	c
Echinoderms:			
Asterias rubens	–	4000	a
Stichopus tremulus			
muscles	–	240	h
Strongylocentrotus pulcherrimus			
digestive tract	60*	–	c
gonads	20*	–	c
shell	5.4	–	c
Arrow worms:			
Sagitta elegans	–	10000	g
Tunicates:			
Salpa fusiformis	–	130	g
	–	60	j
Ciona intestinalis	–	400	a
Fishes:			
Squallus acanthias	–	80	a
Gadus merlangus	–	560	a
G. aeglefinus	–	200	a
G. morrhua			
adult	–	130	a
young	–	150	a
G. virens	–	140	a
G. macrocephalus			
muscles	–	36	d
Chelidonichthys kumu			
muscles	–	82	d

Table 36 (continued)

Aquatic organisms	Concentration factors		Source
	^{60}Co in experiments	stable Co in nature	
Evynnis japonica			
muscles	–	20	d
Lateolabrax japonicus			
muscles	–	30	d
Seriola quinqueradiata			
muscles	–	14	d
Germo germo			
muscles	–	28	d
Katsuwonus vagans			
muscles	–	84	d
Scomber japonicus			
muscles	–	28	d
Cololabis saira			
muscles	–	84	d
Sardinops melanosticta			
muscles	–	64	d
Clupea pallasii			
muscles	–	26	d
Chasmichthys gulosus	2.5–5.2	–	c
Girella punctata	5*	–	c
Acanthogobius flavimanus			
muscles	0.54	–	k
spleen	15	–	k
Euthynnus yaito	–	84	l
Pneumatophorus japonicus	–	28	l

* Stable concentration not achieved.

a   MAUCHLINE [1961], calculated from the data of VINOGRADOV [1953].
b   POLIKARPOV [1963].
c   HIYAMA and KHAN [1964].
d   ICHIKAWA [1961], calculated from the data of TSUCHIYA *et al.* [1935].
e   BLACK and MITCHELL [1952].
f   YOUNG and LANGILLE [1958].
g   MAUCHLINE [1961], calculated from the data of NICHOLLS *et al.* [1959].
h   ICHIKAWA [1961], calculated from the data of VINOGRADOV [1953].
i   GONG *et al.* [1957].
j   KETCHUM and BOWEN [1958].
k   HIYAMA [1962].
l   WALLAUSCHEK and LÜTZEN [1964].
m   KEČKEŠ (see footnote on page 48).

CHAPTER 11

A GENERAL REVIEW OF THE CONCENTRATION FACTORS OF RADIOACTIVE SUBSTANCES IN MARINE ORGANISMS

11.1. Fission products

Concentration factors in plants are high (of the order of hundreds or thousands of units) for the radioisotopes of yttrium, cerium, praseodymium, zirconium, niobium, iodine and ruthenium. The concentration factors of

Table 37. Concentration factors of fission products

Aquatic organisms	Cs			Sr			Y	
	^{137}Cs-^{137}Ba in experiments	in nature	stable Cs in nature	^{90}Sr-^{90}Y in experiments	in nature	stable Sr in nature	^{91}Y(and ^{90}Y) in experiments	stable Y in nature
Flagellates:								
whole	–	–	–	6	–	–	–	–
coccoliths	–	–	–	–	–	20	–	–
Green algae	1–6	–	–	1–3; 1 600	–	2	630–900	–
Diatoms	1–2	–	–	17	–	–	–	–
Brown algae	2–30*	53	16–50	6.5–40	16–60	8–90	220	–
Red algae	1–50	–	16–22	0.05–8	–	0.1–0.3	160	–
Flowering plants	2	–	–	3	–	5	210	–
Protozoa:	10*	–	–	–	–	–	–	–
Acantharia								
whole	–	–	–	–	–	110	–	–
spicules	–	–	–	–	–	60 000	–	–
Coelenterates:								
Scyphomedusae								
muscle band	–	–	2	–	–	–	–	–
whole	–	–	–	–	–	–	–	–
Coral polyps								
(sea anemones)	12	–	–	1	–	–	40	–
muscle wall	–	–	4	–	–	–	–	–
Worms:	6	–	–	–	–	–	–	–
musculo-cutaneous sac	–	–	10	–	–	–	–	–

these elements are lower in invertebrates and fishes. The strontium-90 concentration factor is striking (1 600) in the rapidly dividing cells of a green alga (*Carteria* sp.), and naturally also the strontium concentration factor in the strontium (celestine) spicules of Acantharia, for which it is 60 000 (this factor is the highest, since it was calculated for strontium sulphate). The uptake of yttrium-90 by the pelagic eggs of sea fishes is very large, and the concentration factor in the egg case may reach 10 000 (table 37).

It is noteworthy that there is good agreement between the numerical values obtained for the concentration factors of the stable and radioisotopes of cesium, strontium and the rare earths in experiments and under natural conditions. The data coincide almost completely in algae and in the soft tissues of animals. In crustaceans, and especially in the bones of fishes, and also

corresponding elements by groups of marine organisms

Ce-144Pr experiments	stable Ce in nature	^{143}Pr in experiments	^{95}Zr-^{95}Nb in experiments	^{71}Ge in experiments	^{95}Nb in experiments	stable Mo in nature	^{131}I in experiments	stable I in nature	^{106}Ru-^{106}Rh in experiments
–4 500	–	–	–	–	–	–	–	–	–
–	–	–	–	–	–	–	–	–	–
*–2 400	–	250	2 050*	16	335	–	78–300*	–	100–360
0–3 300	–	–	–	–	–	–	–	–	–
*–350	–	–	170*	66	2 000	2–15	140–300*	800–6 800	170–366
*–3 300	–	–	200–2 960*	76	470–1 100	– .	205–17 900	160	60–1 000
–	–	–	1 120*	197	1 100	–	30	–	180
–	–	–	–	–	–	–	–	–	–
–	–	–	–	–	–	–	–	–	–
–	–	–	–	–	–	–	–	–	–
–	–	–	–	–	–	–	–	–	–
–	–	–	–	–	–	–	–	–	–
–	–	130*	8*	–	–	6*	–	–	–
–	–	–	–	–	–	–	–	–	–
–	–	–	–	–	–	–	–	–	–
–	–	–	–	–	–	–	–	–	–

Table 37 (continued)

Aquatic organisms	Cs			Sr			Y	
	^{137}Cs-^{137}Ba in experiments	^{137}Cs-^{137}Ba in nature	stable Cs in nature	^{90}Sr-^{90}Y in experiments	^{90}Sr-^{90}Y in nature	stable Sr in nature	^{91}Y (and ^{90}Y) in experiments	stable Y in nature
Crustaceans:								
Entomostraca	1*–13*	–	0.1–1	0.1–1	–	0.3	–	–
Malacostraca								
whole	4–25	–	–	3*–8*	10	0.3–90	100–110	–
exoskeleton	15	–	–	15	180	180	–	–
muscles	11–18	–	17–27	0.2	–	0.1–1	–	–
gills	1	–	–	–	–	–	–	–
viscera	–	–	–	1.3	40	33	–	–
Molluscs:								
Gastropoda	–	–	–	–	–	–	–	–
muscles	–	–	15	–	–	0.1–0.2	–	–
Lamellibranchiata	–	–	–	–	–	–	–	–
mantle	–	–	–	0.75–1.9	–	–	–	–
byssus	–	–	–	–	–	–	–	–
gills	8–9.3	–	–	0.7*–1.5	–	–	–	–
shell	0.0	–	–	6*–10*	–	130	250	–
body	6*–12	–	–	0.6–1	–	–	12	–
muscles	8–10	–	3–8	0.4*–0.7	–	0.2	–	–
viscera	6.8–8.5	–	–	0.8–1.2	–	–	–	–
Echinoderms:								
holothurians								
muscles	–	–	7	–	–	0.7	–	–
viscera	–	–	–	–	–	–	–	–
sea urchins								
viscera	9.3	–	–	0.6	–	–	–	–
Fishes:	–	–	–	–				
pelagic eggs								
whole	9	–	–	0.8–4.4	–	–	10.5–233	–
casing	–	–	–	–	–	–	10000	–
attached eggs								
whole	–	–	–	–	–	–	–	–
casing	–	–	–	–	–	–	–	–
contents	–	–	–	–	–	–	–	–
whole	6–10	–	–	1–5*	–	4.2–5.5	–	–
gills	3	–	–	0.4–2.4	–	–	–	–
muscles	5–35*	–	5–23	0.03–0.15	–	0.1–1.5	–	2–10
bones	3.5	–	–	0.5*–3*	–	200	–	–
viscera	6–13*	–	–	0.45–1.7	–	–	–	–

* Balanced concentration not achieved.

Ce-144Pr ...eriments	stable Ce in nature	143Pr in experiments	95Zr-95Nb in experiments	71Ge in experiments	95Nb in experiments	stable Mo in nature	131I in experiments	stable I in nature	106Ru-106Rh in experiments
–	–	–	–	–	–	–	–	100–6000	–
0–220	–	–	–	–	26	–	20–30	–	10–100
400	–	–	350	–	60–100	–	–	–	25–190
25	–	–	2	–	1–3	–	–	20–48	3
0	–	–	460	–	90–150	–	–	–	450
200	–	–	3	–	5	–	–	–	3–30*
–	–	–	–	–	–	–	–	50000	–
–	–	–	–	–	–	–	–	–	–
0	–	–	–	–	160*	–	–	–	40*–50
–	–	–	5	–	2	–	–	–	2.5
00	–	–	600	–	3000	–	2000*	–	500
–	–	–	14	–	50	–	1.8	–	10–12
–50	–	–	4–36	16*	95–170*	–	2–60*	–	11–53
0*–350	–	–	8–13	13*	140	–	20*	–	1–16
–	–	–	2	–	7	–	0.9	40–70	2–3.6
–	–	–	20	–	30	–	0.4–1	–	6.4–18
–	–	–	–	–	–	–	–	20–60	–
–	–	–	–	–	–	–	–	490–540	–
–	–	–	–	–	–	–	100*	–	5.3
–	–	–	–	–	–	–	–	–	2.6
0–495	–	–	14.5–35	–	–	–	–	–	12–18
–	–	–	–	–	–	–	–	–	–
–	–	–	24	–	106	–	–	–	–
6	–	–	56.7	–	256	–	–	–	–
–	–	–	18.7	–	126	–	–	–	–
12	–	–	–	–	–	–	1.3–8.5	–	2–2.5
–	–	–	10	–	–	–	–	–	10
7	2–25	–	0.05	–	–	–	–	11–15	0.04–0.05*
–	–	–	0.03–0.08	–	–	–	–	–	0.02–0.08
–20	–	–	11	–	–	–	–	–	20

Note: All daughter products were in a state of equilibrium with the parent radionuclides.

Table 38

Concentration factors of induced radionuclides and the corresponding elements by groups of marine organisms

Aquatic organisms	Zn		^{32}P in experiments	stable Sb in nature	Cr	
	^{65}Zn in experiments	stable Zn in nature			^{51}Cr in experiments	stable Cr in nature
Phytoplankton	–	–	–	–	–	–
Green algae	130*–290	–	160–330*	–	70	–
Diatoms	50000*	–	–	–	–	–
Brown algae	190*	420–1400	30	–	60	100–500
Red algae	840*	80	15–257	–	80	(120000?)
Flowering plants	340	–	50	–	40	–
Sponges	–	30	–	30	–	800
Coelenterates:						
Scyphomedusae	–	1600	–	30	–	1600
coral polyps						
(sea anemones)	70*	300	30	90	20	–
ctenophores	–	–	–	–	–	–
Worms	–	6	1000	–	–	–
Crustaceans:						
Entomostraca	200*	–	–	–	–	–
Malacostraca	26*–37*	9400–15000	100–372	–	37	–
exoskeleton	150*	–	–	–	–	–
muscles	40*	1900–4400	–	–	–	–
Molluscs:	–	1400–31000	137	–	–	–
shell	10*–49*	–	27*–30*	–	2	–
body	0.25–630*	3500	29*–357	–	18	–
viscera	11*–140*	–	–	–	–	–
mantle	9*	–	–	–	–	–
muscles	10*	2600–40000	–	–	–	–
gills	422*	–	–	–	–	–
Echinoderms:	–	25–56	–	70	–	140–9000
viscera	200*	–	–	–	–	–
muscles	±	1400	–	–	–	–
Arrow worms	–	–	–	–	–	–
Tunicates	–	6600	–	30	–	–
Fishes:	3.4	280–15500	–	100–140	–	2000
muscles	1.4*	540–4400	–	–	–	–
pelagic eggs	–	–	5.5	–	–	–
gills	15*	–	–	–	–	–
bones	3*	–	–	–	–	–

* Balanced concentration not achieved.

S in periments	Mn		Fe		Co	
	^{54}Mn in nature	stable Mn in nature	^{59}Fe in experiments	stable Fe in nature	^{60}Co in experiments	stable Co in nature
	2500–6300	–	–	–	–	–
?–3.0	5500	1300–4600	100*–730	–	120*–380	15–440
	–	20–5000	–	45000	–	–
?–1.0	680	300–20000	1100*	1200–12500	45*–420	27–740
?–2.0	2000	1000–6000	100–1650*	2000–4000	–	64
	–	–	440	–	–	120
	–	3000–9500	–	10000	–	20
	–	120	–	300	–	350–1400
	–	4100	–	2400	20	1100
	–	–	–	–	–	130
	–	–	–	–	–	4000–20000
	2400	–	–	25000	–	–
	930	7500	110	–	7*–3	220–770
	–	–	–	–	–	–
	–	–	20	1000–4000	–	4000
	830	400–550000	66*	50000	–	1–9800
	–	5000–50000	13*–55*	–	36–500*	–
?–0.4	–	620–50000	25*–100*	–	18*–186	–
	–	–	43*	–	9.2–10*	–
	–	–	–	–	–	–
	–	–	30*	3000–17000	7*	52–200
	–	–	53*–70*	–	17*	–
	–	3500–33000	–	–	–	4000
	–	–	30*	1000–10000	60*	–
	–	200	–	78000	–	240
	–	–	–	1800	–	10000
	–	9500	–	300	–	60–400
	240–850	95–126000	5.8	–	2.5–5*	28–560
	–	70–320	0.05	400–3000	0.5	14–84
	–	–	152–170	–	–	–
	–	–	–	–	–	–
	–	–	0.04–0.3*	–	–	–

apparently in the shells of molluscs strontium-90 accumulation factors were considerably lower in an experiment than under natural conditions. This difference is due to a feature of the mineral metabolism of strontium, namely that this element is continuously laid down with calcium in growing mineralized tissues. Strontium-90 and ruthenium-106 uptake is least in the body and muscles of marine animals (table 37).

It therefore follows that the best concentrators, and the best biological indicators for cesium-137 are brown and red algae and the soft tissues of invertebrates and fishes; Acantharia (especially the spicules), *Cartaria* sp. from among the green algae, all species of brown algae, Coccolitophoridae, shells of crustaceans and molluscs, and bones of fishes for strontium-90; cases of pelagic fish eggs, crustaceans and shells of molluscs for the radioisotopes of yttrium; algae, sea urchins, crustaceans and the body of molluscs for cerium-144; algae for praseodymium-143, zirconium-95 and niobium-95; flowering plants for germanium-71; algae for iodine-131; and algae and crustaceans for ruthenium-106.

11.2. Neutron-induced radionuclides

In contrast to the data cited above (table 37), the concentration factors of neutron-induced radionuclides, or more accurately of the corresponding stable elements, are subject to extremely wide variation (table 38). The most important reason for this phenomenon is, apparently, the great differences between the data of various authors on the concentration of these elements in sea-water. Calculations of the concentration factors of these elements in marine plants and animals are based on separate and significantly differing concentrations in the sea. Therefore, the concentration factors may be either increased or decreased. Thus, according to the data of two authors (table 31) the concentration factors of stable chromium in six species of brown algae, including *Fucus serratus*, are hundreds of units (100 to 500), whereas MAUCHLINE [1961] gives a concentration factor of 120000 for this element in *Fucus serratus*.

It follows from the material in table 38 that variations in the magnitude of concentration factors for a number of elements within related groups of organisms may reach two, three and even four orders of magnitude (chromium, manganese and cobalt). In all instances except one the concentration factors of radioisotopes were lower than or equal to those of the stable elements. For some radionuclides in most experiments saturation was not achieved in marine organisms. The exception mentioned above relates to the concentration factors of zinc-65 in *Phyllophora* (839), and of stable zinc in

Gelidium (80). This may have been due either to differences in the concentration of stable zinc in the sea-water which reduced the concentration factor, or to a real difference in the concentration factors of zinc in these algae. The chemical elements were determined separately in marine organisms and sea-water at different times and in different waters.

The following conclusion can be reached on the basis of the fairly reliable data (table 38). Concentration factors are high for zinc, manganese, iron, cobalt and, apparently, chromium, and low for sulphur, with the remaining elements occupying an intermediate position. More detailed conclusions, such as an enumeration of indicator species, are at present impossible owing to the uncertainty of data concerning the concentration factors of these elements. There is a need for further collection of data on the concentration factors of neutron-induced radionuclides in marine organisms, and for new and convincing data on concentrations of the corresponding elements in marine organisms and in the surrounding environment at the same time and in the same water area. If this information were available it would be possible to calculate reliable concentration factors.

It must, therefore, be acknowledged that our information on the capacity of marine plants and animals to concentrate neutron-induced radionuclides, and their corresponding elements, is far less complete and far less convincing than our information on the concentration factors of fission products and their corresponding chemical elements.

11.3. Naturally occurring radioactive substances

Table 39 summarizes information on the concentration factors of naturally occurring radioactive substances.

Potassium and rubidium concentration factors in marine organisms do not exceed a few tens of units, whereas those for the other elements (especially radium) reach hundreds, thousands or tens of thousands of units. This fact is of extreme importance in calculation of the doses of ionizing radiation absorbed by marine organisms as a result of decay of natural radionuclides incorporated in them. Moreover, although potassium is most uniformly distributed in aquatic organisms, selective accumulation in structures which may be radiosensitive is a feature of many other radionuclides. For example, potassium-40 concentrations in sea-water are approximately one hundred times those of radium-226. The radiation doses absorbed by the organisms as a result of these radiations are equalized, given the moderate concentrations of radium-226 in marine organisms (with concentration factors 10 times greater than for potassium-40 (table 39), and given the factor of non-uniform distribution (narrow localization), which is also of the order of 10. If the dose is expressed in rems (the RBE = 10 for alpha-particles) the dose from radium-226 is higher than that from potassium-40. Let us take a further example: according to the data of CHERRY [1964] the doses received by *Rhizosolenia hyalina* are 20 times those received by human tissues in the same period.

Table 39

Concentration factors of naturally occurring radioactive substances (compiled from the works by POLIKARPOV [1963], CHERRY [1964] and MAUCHLINE and TEMPLETON [1964])

Aquatic organisms	Concentration factors under natural conditions							
	Be	K	Rb	Po	Ra	Th	U	Pu
Plankton	–	–	–	–	2750	1250**	50	–
phytoplankton	–	–	–	–	1000	–	–	–
zooplankton	–	–	–	2500**	100	–	–	–
Algae – macrophytes	15000	3–50	5–50	1000*	100	10	10–200*	1000*
Invertebrates								
soft tissues	–	1–15	3–50	–	50	50–300	–	–
skeleton	–	0	30	–	650	–	400	–
Fishes								
soft tissues	–	5–20	3–30	–	15–200	100	20	0.17*–13
skeleton	–	20	40	–	–	–	–	–

* In experiments.
** Balanced concentration not achieved.

It therefore follows that naturally occurring radionuclides should also form the subject of intensive radioecological investigation in seas and oceans. The general theoretical importance of establishing the part played by the natural radioecological factor in the development and functioning of marine organisms and groups of organisms is so obvious as to be indisputable. Thorough study of the artificial radioecological factor is moreover impossible in the absence of comparison with the natural radioecological factor.

CHAPTER 12

THE ROLE OF AQUATIC ORGANISMS
IN MIGRATION OF
RADIOACTIVE SUBSTANCES

12.1. The role of food chains in concentration of radionuclides by aquatic organisms

It is obvious that aquatic organisms differ from terrestrial plants and animals in that they derive many mineral substances directly by the considerable exchange that exists between them and the aquatic environment.

What is the role of nutrition in the accumulation of radionuclides by aquatic organisms? Other conditions being equal, do more radionuclides enter aquatic organisms by direct exchange with the aqueous environment or by nutrition?

KARZINKIN [1962] concluded from an extensive review of the literature on the assimilation of salts from aqueous solutions by aquatic organisms (VEL'TISHCHEVA [1951], KIRPICHNIKOV et al. [1956], MEREJKOWSKY [1879–1880], PÜTTER [1907], KROUGH [1939]), and from investigation in his own laboratory that the uptake of mineral substances by fishes occurs mainly via gills, fins and body surface, and not with food.

Marine algae are of very minor significance in the transfer of radioactive substances to animals, especially in relation to strontium and cesium (CHIPMAN [1958a]). Of the 12 species of unicellular algae investigated by Rice, only two had high strontium-90 concentration factors. Nevertheless, calculations demonstrated that filter-feeding organisms that extracted and assimilated all the strontium contained in the *Nitzschia* (strontium concentration factor 17) in one liter of water would receive as much strontium as is contained in 0.051 l of sea-water, i.e., approximately 5%. Strontium-90 concentration factors in *Carteria* are considerably higher (1616). Even here, however, it should be borne in mind that animals filter vast quantities of sea-water, and that in view of their utilization of the strontium which it contains, additional strontium derived from phytoplankton consumed as food will not make a great contribution (RICE [1956]).

Similar features have been established for zinc-65. The content of this nuclide in sea-water has been found to be $(4 \text{ to } 8) \times 10^6$ counts per minute per liter. Under equilibrium conditions each specimen of the aquatic organism investigated concentrated zinc-65 by direct uptake from solution in the following amounts expressed in relative units: 5.0×10^{-4} for the microphyte *Chlamydomonas* sp., 150 for the crustacean *Artemia salina* and 2×10^5 counts per minute for the reef fish *Chaetodon miliaris*. If it is assumed that all the zinc-65 was derived by the animals from feeding, each crustacean would have to consume 3×10^5 algal cells, and each fish 1300 crustaceans for their activity to be comparable to that accumulated directly from sea-water. As many as 3×10^6 *Chlamydomonas* cells may develop in one ml of sea-water. In this case 0.1 ml of the culture will contain the required activity, equivalent to that concentrated by the crustacean from the water. *Artemia*, however, assimilates at most not more than 5% of the zinc-65 in radioactive *Chlamydomonas* consumed. At least two ml of a culture of these active microphytes would therefore be needed for each *Artemia* to reach the same concentration as would be created by simple concentration from the water.

An even greater difference is revealed on transition of zinc-65 from crustaceans to fishes. The maximum assimilation of this nuclide from active *Artemia* by the reef fishes investigated was 10%. This indicates that, in order to achieve the stipulated level of activity each fish would require 13000 crustaceans or 3.9×10^9 *Chlamydomonas* cells, which corresponds to 1.3 l of the active microphyte culture. The amount of water swallowed by each of these fishes (weight 20 g) is, however, approximately three ml daily, and the time needed for 1.3 l is 400 days, or more than a year. In other words direct uptake from sea-water is far more effective in the concentration of zinc-65 by sea fishes (TOWNSLEY et al. [1960]). It should be noted that the density of the unicellular algal population in the investigation cited was several orders higher than that encountered under natural conditions in sea-water even during the plankton bloom. The discrepancy is therefore far more significant than has been stated.

It is evident from fig. 19 that cesium-134 concentration factors in crabs (*Carcinus maenas*) are identical in sea-water containing cesium-134 for crabs starved throughout the experiments (more than 1000 hr), and for others plentifully and systematically fed on non-radioactive worms (*Nereis diversicolor*) (BRYAN [1961]). In an environment in which cesium-137 level is constant consumption of radioactive food by dentate shrimps is apparently less significant than direct uptake from the water (TEMPLETON [1962]).

When sea fishes were fed with food containing radiostrontium, it was found that this was clearly not the main source of entry of the nuclide (TEMPLETON [1962]). It is noteworthy that sea fishes, unlike freshwater fishes, 'drink' water (KARZINKIN [1962]).

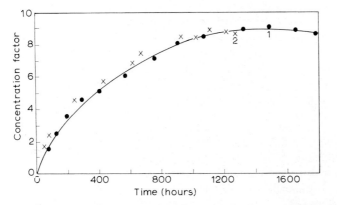

Fig. 19. Cesium-134 concentration factors in whole crabs (*Carcinus maenas*) when fed and
when fasting in radioactive sea-water (BRYAN [1961]). (1) fed, (2) fasting.

Many contradictory views have been expressed concerning freshwater organisms. Undoubtedly much depends on the concentration of organisms in freshwater, which is usually far higher than in seas and oceans. When microphytes concentrate a considerable proportion of the total radioactivity of the solution, microscopic predators and, especially, filter-feeding organisms such as lamellibranches accumulate radionuclides via the food chain. For this to happen plankton content in the water must be extremely high. The Japanese authors YOSHII *et al.* [1956] have thrown some light on this matter. They noted that the alga *Aphanocapsa koordesii*, which is plentiful at some seasons in brackish lakes, is consumed by the rotifers *Brachionus plicatilis*, which multiply rapidly, overgraze the algae, and then die for lack of food. The bodies of the dead rotifers sink to the bottom. This limnological process was experimentally reproduced. Rotifers were introduced at a rate of five specimens per ml into a vessel containing algae that had concentrated a mixture of fission products to a balanced concentration at a cell concentration of 4×10^6 cells per ml, and had extracted approximately 30% of the radioactivity from the solution. Within five days only 1% of the initial concentration of algae remained, and the rotifers had not merely concentrated the activity in the microphytes, but had also accumulated it directly from the water, to a total of 65% of the initial activity. Even when the density of the algal population was high, the rotifers accumulated approximately as much radioactivity through the food link as they derived by direct exchange with the environment.

When oysters (*Crassostrea virginica*) are fed daily with a *Carteria* suspension containing strontium-89, the accumulation of radiostrontium in the body reaches saturation within eight days from commencement of the experiment, and the concentration factor of this nuclide is only twice that in fasting oysters. The experimental and control (fasting) oysters were kept in a strontium-89 solution in sea-water (BOROUGHS, CHIPMAN et al. [1957]). The possibility that a considerable part of the strontium-89 might not be assimilated, but might be incorporated in pseudofeces, was not excluded, however. It is obvious that in migration of strontium-90 via food chain microphytes, oysters would not be significant if other species of marine algae were to be employed, since these normally have strontium-90 concentration factors that are hundreds or thousands of times less than those of Carteria, which is the outstanding concentrator of this nuclide.

RICE [1963c] has described an experiment on the uptake of cobalt-60 by Artemia from the surrounding water and from Carteria cells containing cobalt-60. A group of 50 Artemia was transferred from one vessel containing sea-water to another and then back again every 24 hr. One vessel contained a cobalt-60 solution, and the other contained 200 million Carteria in a liter without cobalt-60. Another group of 50 Artemia was also repeatedly transferred from vessel to vessel: one contained sea-water but no cobalt-60 and the other contained 200 million Carteria cells and cobalt-60. The first group of Artemia fed on radioactive Carteria and the second group on nonradioactive Carteria in a radioactive environment. It was found that the Artemia accumulated 7.5 times more cobalt-60 by consuming Carteria containing cobalt-60 than they accumulated by direct absorption from the sea-water. It follows from the experiments of Rice that the amount of a radionuclide transmitted from phytoplankton to zooplankton is a function of the phytoplankton biomass. When this biomass is many orders in excess of the natural levels, the zooplankton may concentrate more radionuclide from the phytoplankton than directly from sea-water. With the normal biomasses characteristic of seas, the quantity of radionuclides concentrated from the water is far in excess of the quantity of the same radionuclides absorbed from food.

When an alga (Caulerpa sp.) that had accumulated 2400 counts/min from a solution of ash (one g) from the deck of a Japanese fishing vessel* was fed for nine days to the crustacean Artemia salina, the radioactivity acquired by the crustacean was only 62 counts/min·g, whereas specimens kept for three days in a solution of ash with a concentration of 105 counts/min·ml concentrated activity to a level of 5770 counts/min·g (MATSUE and HIRANO

---

* Fukuru Maru No. 5.

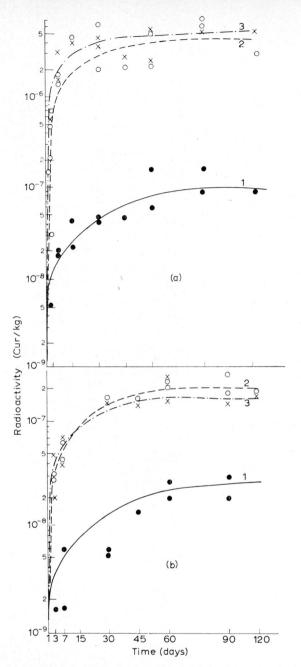

Fig. 20. Concentration of strontium-89 in bone tissue (a) and in muscle tissue (b) (activity of water 5×10^{-7} Cur/l, of food 7.6×10^{-7} Cur/kg) of yearling carp kept under different conditions: (1) aquarium in which fish received radioactive food in pure water, (2) pure chironomid larvae in radioactive water, (3) radioactive food in radioactive water (LEBEDEVA [1962]).

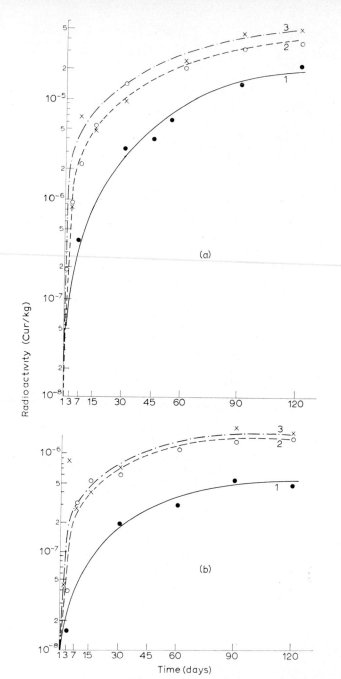

Fig. 21. Concentration of strontium-89 in the bones (a) of yearling carp and in their muscles (b) (activity of water 5×10^{-7} Cur/l, of food 4×10^{-5} Cur/kg) under different conditions: (1) aquarium in which fish received radioactive food in pure water, (2) pure chironomid larvae in radioactive water, (3) radioactive food in radioactive water (LEBEDEVA [1962]).

[1956]). The food factor was, therefore, found to be negligible small by comparison with direct uptake from water. This had also been established for calcium-45. Fishes (*Carassius auratus*) accumulate far more of this nuclide from water than from the consumption of radioactive worms (*Rhizodrilus limasus*). It may be of interest to note that whereas the fishes absorb only eight to 15% of the calcium-45 from worms consumed, white rats assimilate 79% of this nuclide from the same worms (TOMIYAMA, ISHIO et al. [1956c]).

Freshwater fishes concentrate calcium-45 and phosphorus-32 mainly from aqueous solution, and not from food. Thus, 94% of the calcium requirement is met by the surrounding solution. Even fasting fishes take up as much of this nuclide from water as those that are receiving food. The amount of calcium taken up from the water may reach three to 34 times that derived from food. Calcium is concentrated mainly via the gills, but phosphorus-32 is also concentrated via the tail fin (BOGOYAVLENSKAYA [1959], SHEKHANOVA, [1959], RUDAKOV [1958], ZHADIN et al. [1955]).

LEBEDEVA [1962] carried out an interesting investigation. When she kept yearling carp (*Cyprinus carpio*) without food in a solution of strontium-89 $(5 \times 10^{-7} \pm 0.57 \times 10^{-7}$ Cur/l) the curve depicting concentration of this nuclide in the fishes practically coincided with the curve obtained when fishes were kept in the same solution and fed daily on chironomid larvae containing strontium-89 $(7.6 \times 10^{-7} \pm 1.5 \times 10^{-7}$ Cur/kg of wet weight), as can be seen from fig. 20. The curve depicting concentration of this nuclide by fishes in pure water from chironomid larvae possessing the activity stated lies considerably below the other two curves. The pattern was the same for bone tissue as for muscles. If the level of uptake of strontium-90 by fishes when kept in active water and fed on active food is taken to be 100%, the level of uptake in the aquarium whose water was contaminated with radioactive strontium in which the fishes were not fed was 60 to 70%, whereas in the aquarium containing pure water, in which the fishes were fed with active chironomid larvae it was only 8 to 27%. Lebedeva also set up an experiment that more closely imitated natural conditions. The pattern of the experiment was the same, but the fishes were fed with more active food $(4.2 \times 10^{-5} \pm 1.3 \times 01^{-5}$ Cur/kg)*. As a result the lower curve was raised, but failed to reach the curve due solely to uptake of strontium-89 from an active solution without feeding (fig. 21). This indicates that the main source of strontium-89 contamination of fishes is not radioactive food, but the aqueous environment

---

* 8.0 times more than in water for fishes.

which contains this nuclide.

It should be noted that the radioactive chironomid larvae were obtained by keeping them for 24 hr in strontium-89 solutions of appropriate strength. The larvae were given once a day at a rate of 10 to 15% of the weight of the fish. This amount was consumed in 20 to 30 min, during which time uncontaminated chironomid larvae concentrated approximately 0.1 of the amount of strontium-89 in the radioactive food. In pure freshwater an active chironomid larva lost up to 20% of the strontium-89 in 30 min. In other words, what had taken place was not only surface contamination of the food, but mainly incorporation of strontium-89 into the tissues of the chironomid larvae.

Three months later the strontium-89 discrimination factors (the ratio of strontium units in the fishes and the food or water) in relation to the water in the muscles and in the bones was 10^{-2}, and in comparison with the food, whatever its concentration of strontium-89, was 10^{-3} (table 40).

Lebedeva justifiably concluded from this information that benthos-feeding fishes in contaminated water concentrated radioactive strontium mainly (60 to 78%) from the water, and not from food.

We have already noted that a similar conclusion has been reached in relation to uptake of calcium-45 by carp where water rather than food was the source of calcium (TOMIYAMA, ISHIO et al. [1956c].

TELITCHENKO [1962] investigated strontium-90 uptake by crucian carp from a solution with an activity of 10^{-5} Cur/l for ten days under conditions of fasting and feeding. The results obtained were as follows (strontium-90 content expressed in counts/min·g of live weight):

	Feeding		Fasting	
	Males	Females	Males	Females
Vertebral column	4000	6000	4000	3000
Skull	8000	4000	4000	4000
Scales	16000	10000	8000	8000
Gills	8000	8000	8000	5000
Intestines	4000	400	3000	1000
Muscles	200	100	100	100

These data do not reveal the vast difference noted by KRUMHOLZ and FOSTER [1957], and by DAVIS and FOSTER [1958], although these authors, unfortunately, do not provide direct evidence in support of the view that feeding is the main path by which radioactive substances enter freshwater

Table 40

Strontium-89 discrimination factor in a fish in relation to food and water (LEBEDEVA [1962])

Time of experiment (days)	Food (7×10^{-7} Cur/kg)		Water (5×10^{-7} Cur/l)		Food (5×10^{-5} Cur/kg)		Water (5×10^{-7} Cur/l)	
	bones	muscles	bones	muscles	bones	muscles	bones	muscles
1	1×10^{-4}	5×10^{-4}	1×10^{-3}	1×10^{-3}	1×10^{-5}	9×10^{-5}	2×10^{-4}	4×10^{-3}
7	1×10^{-4}	1×10^{-3}	8×10^{-3}	1×10^{-2}	2×10^{-5}	–	2×10^{-3}	–
15	3×10^{-4}	–	2×10^{-2}	–	3×10^{-4}	5×10^{-4}	6×10^{-3}	9×10^{-3}
30	5×10^{-4}	1×10^{-3}	2×10^{-2}	1×10^{-2}	5×10^{-4}	9×10^{-4}	1×10^{-2}	1×10^{-2}
60	2×10^{-3}	2×10^{-3}	3×10^{-2}	1×10^{-2}	1×10^{-3}	1×10^{-3}	2×10^{-2}	2×10^{-2}
90	2×10^{-3}	2×10^{-2}	3×10^{-2}	3×10^{-2}	2×10^{-3}	4×10^{-3}	3×10^{-2}	3×10^{-2}

animals, and especially fishes, but restrict themselves to inference. In conclusion KRUMHOLZ and FOSTER [1957] acknowledged the virtual absence of information on the efficiency of transfer of radionuclides from food organisms to aquatic predators.

A brief communication by WILLIAMS and PICKERING [1961] does not contain sufficient material for comparison of the uptake of strontium-85 and cesium-137 by *Daphnia* and by fishes directly and by feeding. Their communication fails to indicate how activity varied in the aquarium containing the *Euglena* or the concentration of the latter. They did not carry out experiments on the concentration of these nuclides with and without feeding in vessels containing radioactive water, which would have made possible unambiguous and quantitative analysis of the efficiency of both paths of uptake.

It can be concluded from analysis of table 37 that the data yielded by laboratory experiments in which the feeding of marine animals was excluded are in good agreement with measurements of the concentration factors of radioactive and stable nuclides under natural conditions. The differences in strontium concentration factors in mineralized tissues under experimental and natural conditions are due to the continuous growth of these tissues.

The pattern that emerges of the ways in which marine animals concentrate chemical elements is, therefore, as follows. The animals satisfy their requirements for most elements by direct absorption from the surrounding water. Intricate organic substances, on the other hand, are usually derived by the animals in the process of heterotrophic nutrition, i.e., at the expense of other organisms. The main source of carbon for the animals is, apparently, to be found in the food links. A certain amount of the phosphorus and trace elements incorporated in specific biologically active molecules is also apparently derived in this manner, although it is also possible that there may be direct utilization of such molecules present in sea-water as a result of their release by living marine plants. The greater part of the chemical mineral substances and, therefore, by far the greater part of the corresponding radionuclides are consequently accumulated by marine animals other than with their food.

It is obvious that direct investigations, the number of which is yet limited, must be carried on for confirmation of this general deduction or to insert amendments into it. A recent review by MAUCHLINE and TEMPLETON [1964] contains the following information on this matter. Among more than 30 works on the matter in question there are several times more scientists in favor of the direct uptake of radionuclides from water than there are those in favor of the food pathway.

Supporters of the latter prevail among the field investigators (C. Osterberg, D. G. Fleishman and others). I. L. Ophel and J. M. Judd prepared a report 'Experimental studies of radiostrontium accumulation by freshwater fish from food and water' (for the 1966 Symposium of radioecological concentration processes, Stockholm, April 1966) in which they give the ratio of these two ways of strontium-90 uptake by *Carassius auratus* under different concentration factors: when $K < 100$, water is the main source of it; when $K = 100$, significance of water and food becomes equal; when K is from 250 to 500, water contributed 30 to 8% of the strontium-90 retained by fish, respectively.

The transfer of radioactive substances along the food chains of aquatic organisms that migrate from radioactive waters to those that are less radioactive or almost pure is quite a different matter. Radionuclide concentration factors lose their meaning, and the operative factors become the elimination of radionuclides from the organism and the assimilation of radioactive food by predators. Some information has already been given at appropriate points in the narrative. To the best of the author's knowledge there have been no special studies of this matter, most probably because of the difficult and laborious nature of the task. It is, in fact, essential to investigate the relationship between the rate of elimination of the most important radionuclides from the principal food fishes and other plentiful aquatic organisms and the time spent in a radioactive aqueous environment, which may be considerable. It is also essential to study the extent of uptake of radionuclides via the alimentary canal in the varied and intricate food chains of aquatic organisms, and the features of the subsequent elimination of the radionuclides. Outlines of the elimination and assimilation of radioactive substances by aquatic predators should also be drawn up to cover the characteristic features of a wide range of variants. Unfortunately, such extremely important investigations are as yet only in the planning stage, and a number of technical and organizational problems will have to be overcome before they can be carried out.

12.2. The fate of accumulated radioactive substances after the death of aquatic organisms

What happens to the radionuclides accumulated by aquatic organisms in the course of their life when the organisms are converted to detritus after their death? As far as is known this matter has not been the subject of special investigation. It is believed that all radionuclides sink to the bottom with decomposing remains and are firmly retained by the bottom sediments until complete radioactive decay. Thus, AGRE and KOROGODIN [1960] failed to note desorption of a mixture of fission products from dead freshwater plankton (detritus) in the period required for it to settle to the bottom. Ruthenium-106 was equally concentrated by the thalli of living and dead *Porphyra*, and was adsorbed on to the surface of the cells and did not penetrate into them. No differences were noted in the accumulation of this nuclide by living and dead marine diatoms *Phiodactylum* (JONES [1960]). There are therefore no grounds for expecting any differences in the fate of ruthenium-106 when algae die that have concentrated it.

It has been demonstrated that dead algal cells and detritus concentrated more cesium-137 than live cells in the 'bloom' of freshwater algae. The definite metabolic ratio between cesium and potassium was disturbed, and cesium was accumulated by the dead algae and their organic remains in a range not dependent on potassium concentration in the environment (WILLIAMS [1960]). All that is known concerning strontium-90 is that *Carteria* cells killed by acid or by heating did not build up a concentration over a period of two days, whereas the radioactivity of live cells increased two hundred-fold over the same period (RICE [1956]). When *Fucus* containing strontium-90 were killed by heating the concentration factor fell from 39 to 30 (SPOONER [1949]). New data have been received from BARINOV [1965c, ch. 2, §2.3, p. 60].

The present author has made a special study to establish the part played by detritus formation in the further migration of cesium-137, strontium-90 and cerium-144 taking brown algae as his subject (POLIKARPOV [1961e,f]). It follows from fig. 22 that cesium-137 and especially cerium-144 were not

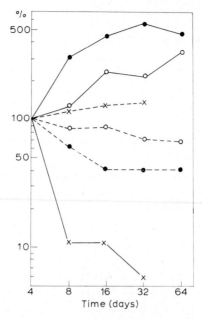

Fig. 22. Alteration of concentration factors (continuous lines) of strontium-90 (×), cerium-144 (●) and cesium-137 (○) during the decline of *Cystoseira*. Concentration of radionuclides in the sea-water shown by broken lines. Time plotted from commencement of experiment. Concentration factors and activity of sea-water on fourth day of experiment taken as 100%. Decline due to the absence of fresh sea-water commenced after four days (POLIKARPOV [1961f]).

merely retained by the detritus, but were additionally adsorbed from sea-water. This was noted over a period of two months by increasing concentration factors in dying *Cystoseira* and corresponding decrease in the level of activity in the water. Strontium-90, however, behaved in exactly the opposite manner, i.e., it was returned to the water. Its concentration factors fell to near the minimum value, and activity in the water increased. A similar phenomenon has been noted by the author in other brown algae (*Padina pavonia* and *Dictyota fasciola*). For all live brown algae strontium-90 concentration factors are between 20 and 40 (POLIKARPOV [1960c]). Loss of strontium-90 has not been noted on the death of green and red algae, since the concentration factors for these algae vary around unity.

The author carried out an experiment with the radioisotopes of strontium and calcium to establish how their ratio was affected on decomposition of *Cystoseira*. Calcium-45 concentration factors in *Cystoseira* are only a tenth those of strontium (table 41). In the course of detritus formation the ratio

Table 41

Ratio of strontium-90 and calcium-45 concentration factors in the decomposition of *Cystoseira* (decline commenced on 8th day from beginning of experiment)

Time (days)	Strontium-90 level of activity (%) in water	in *Cystoseira*	concentration factor K_{Sr}	Calcium-45 level of activity (%) in water	in *Cystoseira*	concentration factor K_{Ca}	$\dfrac{K_{Sr}}{K_{Ca}}$
2	104	90	39	100	102	4	10
4	100	100	45	100	100	4	11
8	102	110	42	111	94	2	21
16	131	70	16	118	64	1	16
32	163	37	5	122	79	1	5
64	169	28	3	131	93	1	3

varies considerably, and falls to three. Both concentration factors are reduced, but that of strontium-90 falls far more (from 45 to 3) than that of calcium-45 (from 4 to 1).

It has therefore been established that elements of the second group, and possibly also of other groups, when concentrated in excess in aquatic organisms are capable after their death of returning to the water, if they are not firmly adsorbed or bonded or incorporated in the crystalline lattices of mineralized tissues or formations, but are present in an ion-exchange state (for example in the cell sap of vacuoles).

It follows that radionuclides, including strontium-90, which is one of the most dangerous, may be returned to the water in the course of detritus formation. Further special research is needed to establish the extent of this phenomenon and the part that it plays relative to other factors in the migration of strontium-90.

12.3. Some aspects of the distribution of radioactive substances in waters and the role of aquatic organisms in their migration

Equation (3′) (chapter 2, section 2.1) can be transformed to

$$A_1 = \frac{K}{K + (1/10^{-6}B)} \times 100\%, \tag{3a}$$

where K is the concentration factor, $B = P/10^{-6}P'$, P and P' are the respective weights of the organisms and the water, B is the concentration of organisms in the aqueous medium in parts per million, i.e., the biomass (e.g., g/m^3).

Equation (3a) can be used to calculate the dependence of the percentage extraction of radionuclides A_1 by aquatic organisms from the aqueous medium on the concentration factors K and on biomass B (table 42).

Table 42

Dependence of proportion of radioactive matter extracted by aquatic organisms on biomass and on concentration factors

Biomass (g/m³)	Extraction as a percentage for the following concentration factors						
	1	10	10^2	10^3	10^4	10^5	10^6
10^{-1}	0	0	0	0	0	1	10
1	0	0	0	0	1	10	50
10	0	0	0	1	10	50	90
10^2	0	0	1	10	50	90	99
10^3	0	1	10	50	90	99	100
10^4	1	10	50	90	99	100	100
10^5	10	50	90	99	100	100	100
10^6	50	90	99	100	100	100	100

Note. Percentages given precise to whole numbers. The dashed line encloses the most realistic biomasses and concentration factors.

Mean zooplankton biomass in the open areas of the Barents Sea is 0.14 g/m³, rising at some periods and especially in the south-west to between 0.2 and 2.0 and even six to eight g/m³. The crustacean *Calanus finmarchicus* accounts for approximately 90% of the zooplankton (ZENKEVICH [1951]). The average photoplankton biomass in the open areas of the Black Sea is measured in tenths of a gram per m³, rising in bays to several grams per m³ (MOROZOVA-VODYANITSKAYA [1948, 1954, 1957], PITSYK [1950, 1951, 1954, 1960], KONDRAT'YEVA and BELOGORSKAYA [1961]). Zooplankton concentration is also expressed in tenths of a gram per m³, reaching maximum values similar to those for phytoplankton in the north-western Black Sea and along the coast of Anatolia (KUSMORSKAYA [1955], DIMOV [1960], KOVAL' [1959], NIKITIN [1945], PETIPA et al. [1963]).

At depths of two to five m layers of zooplankton of as much as 75 g/m³ are found in the Black Sea (PETIPA et al. [1963]). The plankton biomass is largest in the Caspian Sea and the Sea of Azov where normal spring and summer values are several grams per cubic meter. The maximum recorded phytoplankton biomasses in the Sea of Azov during the phytoplankton bloom have reached 200 g/m³ and even above (ZENKEVICH [1951], PITSYK [1955]).

The data in table 42 become quite clear when compared with the information on plankton biomasses in the various seas and with the concentration factors of radioactive substances. Concentration factors for cesium-137 are very low, and it is therefore impossible to detect significant concentration of this nuclide from sea-water even with the largest plankton biomasses. The same can be said of strontium-90, even including the *Acantharia*, since despite the vast concentration factor of this element in the spicules of these protozoans (up to 60000), the concentration factor for whole cells is approximately 110. Only for *Carteria* may a different pattern be expected. When its biomass is several tens of grams per cubic meter, it is capable during rapid division (when its concentration factors increase to 1600) of concentrating up to 10% of the activity from the water. It should, however, be borne in mind that biomasses of this order are exceptional, and that when the biomass is lower (a few grams per cubic meter or less) the extraction of strontium-90 from the aqueous medium is almost imperceptible. All that has been said is equally applicable to the radioisotopes of the rare earths, whose concentration factors reach 1000.

Study of the distribution of fission products between the water and aquatic organisms, taking plankton organisms as an example, has therefore revealed an interesting fact, namely that when the biomass is up to 10 g/m³ (and sometimes higher) and the concentration factors are up to 1000, the organisms account for only an insignificant part of the radioactivity present in the environment. In this connection may be mentioned the work of FEDOROV [1961] on plankton from the Barents Sea, employing almost natural biomasses in his experiments. According to his data percentages of extraction by *Calanus* after eight hours in a radioactive solution may reach 12.4% for strontium-89, 32.5% for cesium-137 and 50.0% for yttrium-91. It is a simple matter to calculate that this corresponds to the concentration factors in *Calanus*, namely, 10^6 to 10^4 units for strontium-89 and cesium-137 and 10^7 to 10^5 units for yttrium-91. The figures are even higher for a mixture of plankton, 10^7 to 10^5 for strontium-89 and cesium-137 and 10^8 to 10^6 for yttrium-91. These magnitudes are clearly unrealistic. Moreover, they contradict the data in table 37. If strontium concentration in sea-water (0.8×10^{-2} g/l is multiplied by its concentration factors in mixed plankton this also yields unrealistic values of between 0.8 and 80 kg of elementary strontium per kg of wet weight of aquatic organisms.

Despite the distribution of radioactivity between organisms and the aqueous medium that has been described (table 42), the role of aquatic organisms in the migration of chemical elements in the sea remains considerable. This role is manifested in horizontal and vertical migrations, in the processes of biological filtration and the formation of feces and pseudofeces, and also in the death of living organisms. On migration, death or consumption aquatic organisms are replaced by new individuals or generations. There are, for example, several hundred generations of plankton algae a year. The concentration of radioactive matter in a contaminated water is thereby continuously reduced. The hydrological factor undoubtedly plays a very large part in this process.

Is it not possible to make at least an approximate comparison of the significance of the two factors, biological and hydrological, in the migration of radionuclides in seas and oceans? This task was undertaken by KETCHUM and BOWEN [1958], who suggested expressions for the rate of vertical migration of a certain amount of radionuclides together with the aquatic organisms that concentrated them

$$T_B = KB(C_b \alpha_b - C_t \alpha_t)d_m/P_m \, , \tag{4}$$

and of the radionuclides present in the water mass

$$T_P = A_z(C_b \alpha_b - C_t \alpha_t)/d \, , \tag{5}$$

where K is the concentration factor, B is the biomass, C_b and C_t are the concentrations of the element in the bottom and top sections respectively of the gradient layer of water, α_b and α_t are the specific activities (ratio of the number of atoms of a radioisotope to the total number of all atoms of the given element) of a radionuclide at these two levels, d_m is the distance for which the organisms migrate, P_m is the migration time, A_z is the vertical coefficient of vertical turbulent diffusion, and d is the depth of the gradient zone. It is evident that most values should be directly determined. Hence the ratio of biological (4) and physical (5) migration is expressed as follows

$$\frac{T_B}{T_P} = B \frac{K}{A_z} \frac{dd_m}{P_m} \, . \tag{6}$$

For numerical comparison the authors adopted $d = d_m = 100$ m, a migration time of 10^5 sec and a biomass of one $g/m^3 (B = 10^{-6})$. The circulation results are given in table 43. Physical transport predominated for a concentration factor of unity. At $K = 10$ the two factors balance each other with a very low coefficient of turbulent diffusion, and when $K = 100$ and $K = 1000$ biological migration predominates over physical migration up to $A_z = 1.0$.

This biological pathway for the vertical transport of radioactive substances is characteristic of those radionuclides which accumulate rapidly to a stable

Table 43

Ratio of biological and physical transport (T_B/T_P) of elements in the sea for various mixing coefficients and concentration factors in a state of equilibrium absorption of the nuclide at each level (KETCHUM and BOWEN [1958])

Mixing coefficients A_z	T_B/T_P for the following concentration factors K					
	1	10	10^2	10^3	10^4	10^5
10^{-2}	10^{-1}	1	10	10^2	10^3	10^4
10^{-1}	10^{-2}	10^{-1}	1	10	10^2	10^3
1	10^{-3}	10^{-2}	10^{-1}	1	10	10^2
10	10^{-4}	10^{-3}	10^{-2}	10^{-1}	1	10
10^2	10^{-5}	10^{-4}	10^{-3}	10^{-2}	10^{-1}	1

level in aquatic organisms living in an active layer, and are readily eliminated in a new water layer in which the level of radioactivity is lower.

Ketchum and Bowen also suggested an expression for gravitational mixing vertically downwards of elements incorporated in homogeneous particles

$$T_g = KB'C_t v , \tag{7}$$

where B' is particle concentration and v is the settling velocity.

This process is most effective in the transport of elements firmly bonded in the organic or skeletal structures of aquatic organisms.

Formula (7) can be combined with expression (4) to give a general picture of the vertical migration of radionuclides involving aquatic organisms. It is at present impossible to find a resultant for particles in the sea, which vary greatly in size, density and shape and, therefore, in settling velocity, which is affected by the turbulent mixing of sea-water.

In addition to vertical biological transport of radioactive substances, which often exceeds physical transport, consideration must also be given to the prolonged presence in the surface layers of sea-water of radionuclides concentrated in aquatic organisms not affected by sedimentation. This applies in particular to one of the largest biocoenoses, described by ZAYTSEV [1961a,c] as the hyponeuston. The main mass of specific forms of the phyto-plankton, various forms of invertebrates and fish eggs, larvae and fry are concentrated in the hyponeuston – a surface layer of marine life only zero to five cm thick. It will therefore be understood that those nuclides which are associated with members of the hyponeuston which do not tend to settle should remain in the surface layer even when there is fairly considerable physical mixing. This phenomenon has been described (HARLEY [1956]) for cerium-144, which remains for a very long time in the surface layer of the Pacific Ocean owing to its concentration in plankton.

ZENKEVICH [1960] has summarized extensive and varied information on horizontal and vertical biocirculation in oceans. According to his data the vertical biocirculation of zooplankton in a column of water one meter square averages 10000 to 20000 km/yr, whereas horizontal circulation (fishes) under the same conditions is 2.5 to 135 km/yr. The adsorbing surface in a column of water one meter square is 10000 to 40000 m^2 for non-living seston down to a depth of 1000 m, 400 to 500 m^2 for bacteria down to 9000 m, 0.5 to 2.0 to 5.0 m^2 (reaching 31 m^2) for phytoplankton down to 100 m and 0.8 to 1.1 m^2 for zooplankton down to 4000 m.

AGRE and KOROGODIN [1960] suggested the following expression for estimates of the distribution of radioactive matter in a completely impounded body of water. These authors noted that the quantity of radionuclides in the living matter of a water was negligibly small, and was not more than 1 to 0.1 % of the total activity of all its components. The total activity of the water could therefore be expressed by the equation

$$A = CS(H + K'h),$$

where A is the total radioactive contamination of the water, C is the specific contamination of the water, H is the surface area, S is the mean depth of the water, K' is the concentration factor of radionuclides in the bottom material, h is the thickness of the adsorbing layer of bottom deposits. This formula holds for an established balance between the quantity of radioactive substances dumped and their radioactive decay.

The role of aquatic organisms, and especially of plankton forms, arises from the rapid completion of the cycles of generations of bacteria, unicellular algae and plankton animals, which, as demonstrated, have high concentration factors for many radionuclides.

Continuous transport of radioactive substances to the bottom deposits occurs as a result of the death of organisms and detritus formation. The total amount of radionuclides thus transported to the bottom of a water in the course of a single season may be hundreds of times greater than their content in the whole mass of live organisms in the water at any given moment. Therefore, the living matter of a water behaves like a pump in pumping radioactive substances from the water into the bottom deposits. There is, of course, also direct adsorption of radionuclides from the aqueous solution by the bottom deposits.

AGRE and KOROGODIN [1960] introduced the concept of the radiocapacity factor (F) of a water, where

$$F = \frac{Kh}{H + Kh},$$

to define the proportion of radioactive contamination accumulated by bottom deposits.

It is not within the scope of the present study to examine the fairly well studied problem of the biological purification of contaminated waste water, including the practical recommendations for calculation and construction of sedimentation tanks (TIMOFEYEVA-RESOVSKAYA [1956, 1957], TIMOFEYEVA-RESOVSKAYA, AGAFONOV et al. [1960], COLLINS [1960]).

SUMMARY OF PART II

The capacity of aquatic organisms to concentrate chemical elements and their isotopes is measured by the concentration factor, defined as the ratio of the concentrations of the element (isotope) in the organisms (wet weight) and in the water. The concentration factor of one element (isotope) may also be expressed as the product of the concentration factor of another element (isotope) times the discrimination factors of these elements (isotopes). Concentration factors may be calculated, given a knowledge of the percentage distribution of the element (isotope) between aquatic organisms and the aqueous medium, and also of the biomass of organisms.

When radionuclides are introduced into sea-water in amounts that do not alter the concentration of isotopic carriers in the water, the concentration factor of each radionuclide will not be dependent on its activity (Cur/l), other conditions being equal (if all the parameters of the sea-water are maintained, if the physical and chemical state remain the same, and if there is no effect on permeability or other forms of biological effect and so on). This is shown not only by special experiments, but also by the fact that the concentration factors of stable elements under natural conditions coincide with those of the corresponding radioisotopes both under natural conditions and in experiments. An obvious exception is provided by those elements (isotopes) which are exchanged slowly, or are extracted by organisms for construction of the skeleton. The concentration factors of these elements are lower in experiments than in nature.

Small changes (changes of twice or even more) in the macroscopic amounts of potassium and calcium in sea-water do not significantly affect the concentration factors of cesium-137 and strontium-90 in marine organisms.

Little is at present known concerning the role of differences in the physical and chemical state of various radionuclides (elements) in their uptake by aquatic organisms. The highly stable pH of sea-water is one factor ensuring the stability of the concentration factors of radionuclides in marine organisms.

There has been little investigation of the effect of temperature on the concentration factors of radionuclides (elements).

Radioactive substances are divided into two categories in terms of their relation to light: 1) those whose concentration in aquatic organisms is dependent on illumination (cesium-137, zinc-65, carbon-14, cobalt-60) and 2) those whose concentration is not dependent on this factor (strontium-90, cerium-144, iron-59). There has, however, been scarcely any examination of this matter. Very little is known about the significance of biological complexing agents in the concentration of radionuclides by marine organisms.

Except for the Baltic Sea, there is good agreement between the available data on the concentration factors of various groups of chemical elements in marine organisms in various oceans and seas, including the Pacific Ocean, Atlantic Ocean, Irish Sea, Black Sea, and Barents Sea. This indicates that results obtained in any sea may be universally applicable. The salinity of the Baltic Sea is only approximately a tenth of the mean salinity of the oceans.

Concentration factors of the order of hundreds and thousands of units in marine plants are characteristic for the radioisotopes of the rare earths, zirconium, niobium, iodine and ruthenium. Values are lower for marine animals. The unicellular marine alga *Carteria* sp. has an exceptionally high strontium-90 concentration factor (up to 1 600 units). The lowest strontium-90 and ruthenium-106 concentration factors have been noted in the body and muscles of marine animals. Radioactive substances are found both in the depots of the organism that correspond to their chemical properties, and to considerable extent on biological surfaces (gills, fins, surface of the shell, body, etc.). Thus, cesium-137 content is highest in the shell and muscles of crabs. There has not been sufficient investigation of the exchange of radionuclides between aquatic organisms and the environment, or of their distribution by tissues and by biochemical fractions, although significant progress has recently been made in this respect in the Soviet Union (BARINOV [1964, 1956a,b], ZESENKO [1965]).

The following may be suggested as biotic indicators in sea-water: brown and red marine algae and the soft tissues of marine animals for cesium-137; Acantharia, *Carteria* sp., brown algae, and shells and bones of marine animals for strontium-90; pelagic fish eggs, algae and many animals for the radioisotopes of the rare earths; algae for zirconium-95, niobium-95 and iodine-131; and algae and crustaceans for ruthenium-106.

In view of the considerable discrepancies between the data of different authors for the same species there is need for reexamination and considerable correction of the concentration factors of induced radionuclides, or to be more precise of their chemical elements. Our knowledge of concentration factors for radionuclides of an induced nature is confined to a very few forms of marine life. Particular importance attaches to rapid development of research on the natural radioecological factor and its comparison with the radioecological factor of artificial origin.

Analysis of the role of food chains in the sea when radioactivity is present in the water indicates that they are of secondary importance in the concentration of radioactive matter by aquatic organisms. This is the direct opposite of biological migration of radioactive matter on land. The main path for uptake of radionuclides by aquatic organisms is direct concentration from the aqueous medium. Carbon is, apparently, an exception, and the same may be true of other elements incorporated in specific biological molecules. The question must, however, be differently formulated for the transport of radioactive matter in the food chains of aquatic organisms that have migrated into 'pure' waters. Special investigation of this matter has apparently been delayed by the difficulty and effort involved.

On the death of brown algae excess strontium-90 is returned to sea-water. It is important to investigate the extent of this phenomenon in the migration of this most important radionuclide in waters.

Cesium-137 and the radioisotopes of the rare earths are not only retained by detritus, but are also additionally adsorbed, at least in the first few months after its formation. There has, however, been scarcely any investigation of these matters.

Normal and even unusually high plankton biomasses in the sea account for only an insignificant and negligibly small portion of the radioactive matter that enters sea-water. For example, a biomass of 10 g/m^3 would be required to extract as little as 1% of a radionuclide with a concentration factor of 1000, and 1000 g/m^3 would be needed to extract 50% with the same concentration factor. Nevertheless, plankton organisms, which are short-lived, reproduce rapidly, and effect migrations, function like a pump in slowly but continuously transferring a number of radionuclides to deeper levels of the sea or to the sea floor. Under certain conditions biological transport of radioactive matter may exceed physical migration in the sea. Thus, with concentration factors of 100 to 1000 in plankton and mixing coefficients of 10^{-1} to 10^{-2} the ratio of biological to physical transport in the sea is 10 to 100.

Therefore, despite inadequate study of many particular aspects, the effect of aquatic organisms on a radioactive environment has been fairly completely outlined in general terms. In all events the *main* features and characteristics of this process have been established, and the point has been reached at which those aspects that need to be studied *first* can be defined with some accuracy.

PART III

The effect of nuclear radiation on marine organisms

CHAPTER 13

THE RADIOBIOLOGICAL SIGNIFICANCE OF CONCENTRATION FACTORS

The rad is a unit of radiation dose D which is absorbed, and equal to 100 ergs per gram of irradiated matter. The dose in rads is related to the radiation dose in roentgens (r) by the expression

$$D_{rad} = fD_r ,$$

where f is the conversion factor which equals 0.88 $m\mu$ (object)/$m\mu$ (medium); $m\mu$ is the mass absorption coefficient of gamma-rays or X-rays. One roentgen corresponds to the formation in one cm^3 of air at $0°C$ and 760 mm Hg of 2.08×10^9 ion pairs carrying a charge of one electrostatic unit of the amount of electricity of each sign, or 88 ergs of absorbed energy in one gram of air. Numerical values of the conversion factor f are given in table 44. The biological equivalent of a rad (rem) is

$$D_{rem} = D_{rad} \times RBE ,$$

where RBE is the relative biological efficiency which is dependent on the type of ionizing radiation. Thus, for the same dose, the RBE is one for gamma-rays and X-rays, one for beta-particles and electrons, 10 for protons and alpha-particles, 20 for multiple-charged ions and recoil nuclei, three for thermal neutrons and 10 for fast neutrons.

The concept of RBE was introduced in relation to the dependence of biological effect on the amount of radiation energy absorbed, and on linear ionization density in cells and tissues, which is dependent in its turn on the type and energy of the quanta and particles. Large particles carrying charges and high energy are most densely ionized per unit length of path. An almost equally high linear ionization density is a feature of uncharges particles – fast neutrons, which are knocked out when recoil protons collide with atomic nuclei, i.e., operate indirectly through the latter. Electromagnetic radiation is capable of giving up its energy in whole (photoelectrons) or in part (Compton electrons) to the electrons of the atomic shell (secondary electrons), and also of forming electron-positron pairs on interaction with the atomic nucleus. The RBE of photons and beta-particles is therefore the same.

Table 44

Values of the conversion factor for conversion from radiation dose (r) of gamma-rays or X-rays to the absorbed dose (rad)*

$$f = \frac{D_{rad}}{D_r}$$

(calculated after ISAYEV [1961])

Energy of photon (MeV)	Conversion factor f		
	$\dfrac{\text{muscle tissue}}{\text{water}}$	$\dfrac{\text{bone tissue}}{\text{water}}$	$\dfrac{\text{water}}{\text{air}}$
0.010	0.91	3.48	0.92
0.015	0.91	3.92	0.90
0.020	0.92	4.17	0.89
0.030	0.92	4.43	0.88
0.040	0.93	4.18	0.89
0.050	0.91	3.50	0.90
0.060	0.90	2.93	0.91
0.080	0.89	1.81	0.94
0.10	0.88	1.35	0.96
0.15	0.87	0.95	0.97
0.10	0.87	0.88	0.98
0.20	0.87	0.85	0.98
0.30	0.87	0.84	0.98
0.40	0.87	0.84	0.97
0.50	0.87	0.84	0.97
0.60	0.87	0.84	0.98
0.80	0.87	0.84	0.97
1.0	0.87	0.84	0.97
1.5	0.87	0.84	0.97
2.0	0.87	0.84	0.97
3.0	0.87	0.84	0.97

* The rad is the unit of radiation dose which is absorbed equal to 100 erg per gram of irradiated matter, i.e., one rad = 100 erg/g = 2.39×10^{-6} cal/g = 6.24×10^7 MeV/g = 1.89×10^{12} ion pairs/g.

An alpha-particle produces several thousand ion pairs in its path in one μm of tissue, whereas a beta-particle produces only a few pairs. It therefore follows that heavy charged particles are bound to damage a cell that they pass through, whereas the effect of beta-particles on a cell will be dependent on statistical factors. This explains the difference in the *RBE* for various types of radiation.

The relation between daily dose D and radionuclide concentration C can be expressed by

$$D_{rem} = 5.13 \times 10^4 QE_{max} C \times RBE ,$$

where Q is the number of particles or quanta per disintegration, E_{max} is the energy of the quanta or particles (the mean energy of beta-particles $\bar{E}_{min} = 0.38 E_{max}$) in MeV, C is the concentration of the radionuclide in Cur/l or Cur/kg.

If the decay scheme of the nuclide is intricate, or if there are several nuclides in the mixture, QE_{max} is found separately for each type of radiation, or energy, and then summed. As already noted, RBE is unity for quanta and beta-particles. When concentration changes as a result of radioactive decay it is not difficult to insert in the formula the expression

$$C_t = C_o e^{-0.693/T} ,$$

where T is the half-life in days. The effective half-life is given by

$$T = \frac{T_b T_p}{T_b + T_p} ,$$

where T_b is the biological half-life of the nuclide and T_p is the physical half-life in days.

The radiobiological significance of the concentration factor is the ratio of the doses produced by the radiations of a given nuclide in the organism (organ) D and in the aqueous medium D', i.e.,

$$K = D/D'.$$

In other words, it is in general possible to employ the concentration factor to convert from doses in the medium to doses in the organism (organ) and vice versa as $D = KD'$ (table 45). Limitations may be imposed on the use of this ratio by the size, or more precisely by the cross-section of an aquatic organism or of its organ, which should be commensurate with (equal to or greater than) the path length of the particles, and also sometimes by the density of the aquatic organism, if it differs from that of water.

Information on doses from radionuclides and fission-product decay rates is contained in several works (BJÖRNERSTEDT [1960], CHEEK and LINNENBOM [1960], HUNTER and BALLOU [1951]).

Path lengths of alpha- and beta-particles are given in table 46. The tracks of alpha-particles are not longer than a few cell diameters, but those of beta-particles are much longer. In internal radiation a microplankton organism

Table 45

Dependence of absorbed dose on strontium-90-yttrium-90 concentration in water and on the concentration factor in an aquatic organism*

Concentration (Cur/l)	Absorbed dose (rad/day) for the following concentration factors					
	1	10	10^2	10^3	10^4	10^5
10^{-14}	5×10^{-10}	5×10^{-9}	5×10^{-8}	5×10^{-7}	$5 \times \mathit{10^{-6}}$	$\mathbf{5 \times 10^{-5}}$
10^{-13}	5×10^{-9}	5×10^{-8}	5×10^{-7}	$5 \times \mathit{10^{-6}}$	$\mathbf{5 \times 10^{-5}}$	$5 \times \mathit{10^{-4}}$
10^{-12}	5×10^{-8}	5×10^{-7}	$5 \times \mathit{10^{-6}}$	$\mathbf{5 \times 10^{-5}}$	$5 \times \mathit{10^{-4}}$	5×10^{-3}
10^{-11}	5×10^{-7}	$5 \times \mathit{10^{-6}}$	$\mathbf{5 \times 10^{-5}}$	$5 \times \mathit{10^{-4}}$	5×10^{-3}	5×10^{-2}
10^{-10}	$5 \times \mathit{10^{-6}}$	$\mathbf{5 \times 10^{-5}}$	$5 \times \mathit{10^{-4}}$	5×10^{-3}	5×10^{-2}	5×10^{-1}
10^{-9}	$\mathbf{5 \times 10^{-5}}$	$5 \times \mathit{10^{-4}}$	5×10^{-3}	5×10^{-2}	5×10^{-1}	5
10^{-8}	$5 \times \mathit{10^{-4}}$	5×10^{-3}	5×10^{-2}	5×10^{-1}	5	5×10
10^{-7}	5×10^{-3}	5×10^{-2}	5×10^{-1}	5	5×10	5×10^2
10^{-6}	5×10^{-2}	5×10^{-1}	5	5×10	5×10^2	5×10^3
10^{-5}	5×10^{-1}	5	5×10	5×10^2	5×10^3	5×10^4

* The cosmic radiation background (bold) at sea level is approximately 10^{-4} rad/day; the potassium-40 radiation dose in the sea (italic) is 10^{-5} rad/day.

Table 46

Size of beta-particle and alpha-particle tracks in biological tissue or in water (ISAYEV [1961])

Beta-particles				Alpha-particles	
energy (MeV)	thickness of layer (mm)	energy (MeV)	thickness of layer (mm)	energy (MeV)	thickness of layer (μm)
0.01	0.002	1.0	4.80	4.0	31
0.02	0.008	1.25	6.32	4.5	37
0.03	0.018	1.50	7.80	5.0	43
0.04	0.030	1.75	9.50	5.5	49
0.05	0.046	2.0	11.1	6.0	56
0.06	0.063	2.5	14.3	6.5	64
0.07	0.083	3.0	17.4	7.0	72
0.08	0.109	3.5	20.4	7.5	81
0.09	0.129	4.0	23.6	8.0	91
0.1	0.158	4.5	26.7	8.5	100
0.2	0.491	5.0	29.8	9.0	110
0.3	0.889	6.0	36.0	9.5	120
0.4	1.35	7.0	42.2	10.0	130
0.5	1.87	8.0	48.4		
0.6	2.46	9.0	54.6		
0.7	2.92	10.0	60.8		
0.8	3.63	20.0	123.0		
0.9	4.10				

receives approximately 10% of the total energy of potassium-40 beta-radiation if its diameter is 0.2 mm, and 1% if its diameter is 0.02 mm (FOLSOM and HARLEY [1957]). The energy of the beta-particles of yttrium-90 is fully expended in a tissue layer approximately four mm thick, 10% expended in a layer of approximately 0.4 mm, and 1% expended in a layer of approximately 0.04 mm. Therefore, when the objects are sufficiently small, the principal factor may be external radiation from the radioactive solution rather than internal radiation. Incorporated alpha-emitters retain their significance as internal radiation sources down to far smaller plankton diameters than are significant for beta-emitters. Thus, internal alpha-radioactivity accounted for 80% or more of the total radiation dose received by a mixture of marine plankton from various sources (CHERRY [1964]). On the other hand, macro-organisms, and especially benthic plants and animals, become sources of increased external radiation for adult aquatic organisms or for some of their development stages, for example benthic fish eggs in a given microbiocoenosis. This applies in particular to those radionuclides that are adsorbed onto biological surfaces, thus producing a 'hot' layer of beta-radiation several mm thick around macroorganisms (table 46). This layer extends for the same depth into the aquatic organism.

CHAPTER 14

RADIATION INJURY TO AQUATIC ORGANISMS FOLLOWING SINGLE DOSE EXTERNAL IRRADIATION

14.1. Some of the basic features of the biological effect of ionizing radiation

The first striking property of this form of radiation is its ready capacity for penetration, which distinguishes it from thermal, luminous and radio-wave radiation. Because of the high energy and insignificant wavelength of ionizing radiation there are no barriers to it that are effective in relation to chemical substances. Penetrating radiation affects the most fundamental and most important foundations of material structure. In a biological substrate this type of effect leads to the development of ionization of water and less frequently of organic molecules. An energy of several tens of electron-volts is needed to form a single ion pair in tissue. Ionizing radiation is usually typified by energies of hundreds of thousands or many millions of electron-volts per particle or quantum. It is therefore evident that a great many ion pairs may be produced along a single track.

As already noted, linear ionization density is dependent on the energy and size of charge of a particle. Thus, it has been stated by TIMOFEYEV-RESOVSKIY and LUCHNIK [1960] that a single alpha-particle with an energy of five MeV produces approximately 4000 ion pairs in one micron of track in a soft tissue (pea meristem), whereas a beta-particle with an energy of one MeV produces only six. Moreover, linear ionization density varies little along the track of an alpha-particle (from $1300/\mu m$ at the commencement of the track to 5200 at the end), whereas the variation for beta-particles is quite significant: from two to $1700/\mu m$, reaching an ionization density characteristic of an alpha-particle at the end of the track. These circumstances were responsible for introduction of the additional term *RBE* for irradiation of a biological substrate with equal radiation doses producing quite different mean ionization densities.

The second characteristic of the biological effect of ionizing radiation is that the insignificant amounts of energy transmitted to the biological substrate by penetrating radiation have a considerable effect. For those mammals that have been studied, a lethal dose of ionizing radiation (hundreds of roentgens) heats the tissues by only $0.002\,°C$ (the amount of thermal radiation needed to achieve a lethal effect is 10^4 to 10^5 times greater). Only an insignificant proportion of the molecules in a gram of tissue are ionized. These slight changes do not yield to analytical methods of investigation. It is known that doses that are only hundredths or thousandths of the lethal dose are biologically active. A biological effect may, therefore, be produced by absorption of a negligibly small amount of the energy of ionizing radiation. Vast doses, of the order of 10^4 to 10^6 r, are needed, however, to produce perceptible chemical change (TARUSOV [1954]). The point is vividly illustrated by the statement that radiation energy equivalent in heat terms to the energy contained in a teaspoonful of hot tea is lethal to man.

This accounts for a most characteristic feature of the biological effect of ionizing radiation, namely the considerable latent period in development of the latent reaction of radiation injury. Some time, and possibly a very long time, is in fact needed for successive realization of all the events in the long chain that intensifies the energy possibly of a single ionization an astronomical number of times to the dimensions of a biological effect. The latent period of the various forms of radiation injury may therefore vary most widely (from a few minutes to many years) in relation to the dose, and to various external factors and features of the biological substrate (KOROGODIN and POLIKARPOV [1957], BENEVOLENSKIY et al. [1957]). Leaving on one side the question of the way in which radiobiological reactions are intensified, we merely note here that the chain of events has still been very little investigated (FRITZ-NIGGLI [1961]).

It is an extremely fundamental feature of the effect of high energy radiation on biological systems that it is cumulative in nature (in relation to genetic changes and, apparently, to carcinogenic effects). It is now thought that the effect of ionizing radiation on a population is manifested in a dual manner: through genetic consequences at low dose rates, and through lethal injury at high dose rates. The problem of repair following radiation injury is an extremely intricate one, and it is only recently that there have been successful quantitative studies at the cellular level (KOROGODIN [1962]).

14.2. Radioresistance

The median lethal dose (LD_{50}), i.e., the dose at which 50 % of the organisms perish, has been introduced as a measure of radiobiological effect. In use it is most frequently related to 30 or 14 days. LD_{100} and LD_{30} express 100 % and 30 % lethality, respectively. The survival period following various doses is another measure of radiobiological effect. A curve of death rates related to dose is divided into sectors of late, early, rapid and instantaneous death.

Above a certain dose the time of rapid death is not dependent on the dose employed (EVANS [1952]).

Information concerning the radiosensitivity of aquatic organisms of various taxonomic groups from the evolutionary standpoint and at various stages in ontogeny is presented in table 47. The vast LD_{100}, LD_{50} and ionizing radiation doses tolerated by bacteria, by algae of all types and

Table 47

Biologically effective doses of ionizing radiation for external irradiation of marine and freshwater organisms

Aquatic organisms	Dose (r)				period of observation (days)	source
	LD_{100}	LD_{50}	minimum effective dose	tolerated dose		
Flagellate bacteria:						
Pseudomonas sp.	–	–	–	20000000	–	a
Euglena sp.	–	–	–	150000	–	b
Prorocentrum sp.	–	–	–	2000	–	b
Green algae:						
Chorella sp.	–	–	–	1000000	–	b
	–	50000	10000	1000	30	c, d
Eudorina sp.	–	–	–	100000	–	b
Chlamidomonas reinhardi	–	4500	–	–	–	e
Spirogira sp.	–	–	–	20000	–	b
Zygnema sp.	–	–	–	50000	–	b
Mougeotia sp.	–	–	–	70000	–	b
Cosmarium sp.	–	–	–	70000	–	b
Chaetomorpha sp.	–	–	–	50000	–	b
Ankistrodesmus sp.	–	11000	–	–	7	d
Cladophora glomerata	4500	–	–	–	120	f
Converva spirogira	8000	–	–	–	30–40	f
Diatoms:						
Nitzschia closterium	–	50000	10000	1000	30	c
Blue-green algae:						
Croococcus sp.	–	8000	–	–	7	d
Oscillaria limosa	–	–	–	8000	–	f
Protozoa:						
Paramecium sp.	600000	–	–	–	30	g
	–	–	–	30000	–	f

Table 47 (continued)

Aquatic organisms	Dose (r)				period of observation (days)	source
	LD_{100}	LD_{50}	minimum effective dose	tolerated dose		
Colpidium colpoda	–	33000	–	–	–	h
Amoeba sp.	–	120000	–	–	–	i
Coelenterates:						
Pelmatohydra oligactis	12000	–	–	8000	63	j
Hydra fusca	2000–9000	–	–	–	35–75	f
Worms:						
Planaria polychroa	5500	–	–	–	75	k
Dendrocoelum lacteum	10000	–	–	–	60	l
Crustaceans:						
Artemia salina	–	–	–	4000	–	f
moist eggs	–	20000	–	–	6	d
dry eggs	–	50000	–	–	6	d
Calliopus laeviusculus	6400	1000	–	–	35–56	d
Allorchester angustis	6400	1000	–	–	35–56	d
Daphia magna fry	8000	–	–	–	9	f
Molluscs:						
Radix japonica	–	5000	–	–	50	d
Thais lamelosa	–	1300	–	–	160	d
Physa acuta	10000	–	–	–	28–45	f
eggs	4000	–	–	–	21	f
Planorbis marginatus	10000	–	–	–	27	f
Helisoma subcrenatum at early stages						
(misotic state)	–	100	–	–	–	m
trochophore	–	500–1000	–	–	–	m
Tunicates:						
Molomla sp.	45000	–	–	–	–	n
Fishes:						
Carassius auratus	–	670	–	–	30	o
Oncorhynchus tschawytscha						
larvae	–	1000	–	–	56	p
fry	–	1250–2500	250	–	–	p

Table 47 (continued)

Aquatic organisms	Dose (r)				period of observation (days)	source
	LD_{100}	LD_{50}	minimum effective dose	tolerated dose		
O. kisutsch						
zygotes in mitotic						
state	–	16	–	–	–	q
Salmo gairdnerii						
adults	–	1 500	–	–	–	r
at 'eye' stage	2 500	1 000	38	–	30–51	r
		415–904	–	–	–	s
at germ ring stage	–	454–461	–	–	–	s
at 32 cell stage	–	313	–	–	–	s
at one cell stage	–	58	–	–	–	s
gametes in fishes	–	50–100	–	–	–	t
Pleuronectes platessa						
(from zygote till	–	–	–	492.50*	19	u
hatching)						
Amphibians:						
Rana sp.						
adults	–	700	–	–	30	v
eggs	–	–	20	–	–	v
R. catesbiana						
tadpoles	–	–	25	–	–	w
axolotl						
early gastrula	10	< 10	–	–	17	x
Triton sp.						
adults	–	3 000	–	–	30	v
zygotes	320	40	5	2	24	y

* At a dose-rate of about 25 rads/day.

a   DUBININ [1961].
b   GODWARD [1960].
c   BONHAM et al. [1947].
d   BONHAM and PALUMBO [1951].
e   JACOBSON [1957].
f   NIKITIN [1958].
g   BACK [1939].
h   GLOCKER, LANGENDORFF and REUSS [1933].
i   HARRIS et al. [1952].
j   POLIKARPOV [1957a,b].
k   NIKITIN [1938].
l   SCHMIDT [1946].
m   BONHAM [1955].
n   GROSCH and SMITH [1957].
o   ELLINGER [1939].
p   BONHAM et al. [1948].
q   DONALDSON and FOSTER [1957].
r   WELANDER et al. [1949].
s   WELANDER [1954].
t   FOSTER et al. [1949].
u   BROWN and TEMPLETON [1964].
v   BACQ and ALEXANDER [1955].
w   ALLEN et al. [1952].
x   BLINOV [1956].
y   PETERS [1960].

by invertebrates (from protozoans to tunicates) are noteworthy features. Sterilizing doses are of the order of hundreds of thousands of roentgens for all living matter, including bacteria. Extremely resistant strains of bacteria may be produced on some occasions. This was the case in the water around the submerged reactor at Los Alamos. Bacteria (*Pseudomonas*) feeding on the resin developed profusely after some time (the water became cloudy), despite the vast dose rate of 10^6 r/hr (DUBININ [1961]).

The phenomenon of radioresistance and the dependence between radio-sensitivity and hydratation were discussed by several authors (BACQ, DAMBLON *et al.* [1955], GILET and OZENDA [1961], RUGH and CLUGSTON [1955b]).

A different pattern is noted in fishes and amphibians. The LD_{50} for adult fishes, frogs and newts is the same as or no more than 1.5 or two times the LD_{50} for mammals. The earlier stages of development are the most sensitive to radiation. The LD_{50} for salmon gametes undergoing mitosis is 16r, and for axolotl embryos at the early gastrula stage is less than 10r. The dose that produces a statistically significant increase in the death rate of newts following irradiation of their ova is as little as five r. Fishes and amphibians are, therefore, readily affected in the early developmental stages even by very small doses of ionizing radiation*.

14.3. Early manifestation of radiation injury in aquatic organisms

Many investigators have established the fact of rapid development of excitation of inhibition of function in irradiated organisms, followed by return to normal or by a state of depression, sometimes leading to radiation death.

It was long ago established (CROWTHER [1926], HANCE and CLARCE [1926]) that depression disappeared rapidly in *Colpidium* and Paramecium, and that motor function returned to normal in the greater part of the cells within three hours of a radiation dose of 60000 r. JOSEPH and PROWAZEK [1902] noted that X-irradiation inhibited pulsation of the vacuoles in Paramecium. PECHENKO [1922] noted an initial stage of excitation, giving way to inhibition and subsequent normalization in the irradiated amoeba *Vahlikampfia*. Ionizing radiation intensified movement of the protoplasm in amoeba (MEDVEDEVA *et al.* [1952]). CHASOVNIKOV [1928] described activation of the

---

* In contrast to these data there is the quite unexpected recent communication of BROWN and TEMPLETON [1964] concerning the extreme radioresistance of fish embryos: a dose of 500 r failed to produce any radiation reaction. TEMPLETON [1965] has informed the author personally that there were major printer's errors in the paper and that the activity of the gamma-radiation sources was understated by 1000 times.

protoplasm of irradiated frog liver cells, followed by depression and complete immobilization, leading to vacuolation of the protoplasm and granulation of the nucleus. Similar results [intensified movement of the plasma in the first hours after irradiation, slower movement on the second day, subsequent disordered metabolism (precipitation of calcium oxalate crystals and depigmentation) and decay] were obtained on *Elodea densa* (GRECHISHKIN [1934]). It was also noted in the same paper that the effect of ionizing radiation on the photogenic bacteria *Bacterium ponticum* was to increase their luminescence.

Irradiation of newt sperm cells produced early damage to the protoplasm, and especially the chondriosome, and later damage to the nucleus (YASVOIN [1926]). The nucleus is, however, more sensitive than the protoplasm. Thus, the LD_{50} for nuclei (120000 r) is 2.3 to 2.5 times less than that for the protoplasm of amoeba (280000 r) (HARRIS *et al.* [1952]).

The median lethal dose for paramecium is between 75000 and 350000 r in relation to various conditions. It is affected by the depth of the liquid medium irradiated, the volume of the moist air space and the size of the area of contact between the medium and the air. The toxic factor, which was wholly or partly dependent on moist air in the closed vessels during irradiation, was probably oxygen or its derivatives, including H_2O_2 (WICHTERMAN and FIGGE [1954]). When the number of bacteria in the medium was large, non-genetic effects in irradiated paramecia were unaffected or only slightly intensified by hypoxia. These effects were, however, reduced by hypoxia when there were few bacteria in the medium. In other words, much of the non-genetic damage was due to H_2O_2 formed outside the cell, and intracellular hydrogen peroxide was not of basic significance in this case. In the opinion of other authors (KIMBALL and GAITHER [1952]) it appeared that genetic damage to paramecium was due to H_2O_2 or to the radical HO_2 developing locally within the nuclei. It is, however, possible to relate the effect of low oxygen content in reducing the frequency of mutations to reduced formation of H_2O_2 following irradiation, since hydrogen peroxide is not a mutagenic factor either in itself, or in combination with X-rays. Short-lived free radicals may be the factor involved. Anoxia lasting for one hour reduced the extent of damage to the hematopoietic cells of tadpoles following X-irradiation (500 r), while anoxia lasting for two hours merely delayed the development of radiation injury (ALLEN *et al.* [1954]).

According to the present author's data (POLIKARPOV [1957a,b] a hydra with a *Daphnia* in the gastral cavity contracted slightly following X-irradiation (3000 r and above), and the caudal appendage or head of the *Daphnia* protruded from the mouth.

The process of evacuation usually lasted five to 10 min, but sometimes extended to 30 min or more, when the caudal appendage entered the cavity of one of the tentacles (*Daphnia* are usually swallowed head end first). The reaction of compression of the gastral cavity was accelerated when the dose of X-radiation was increased.

It should be noted that this reaction has been produced, as in higher animals, by simple mechanical stimulation and by the effect of various chemical agents.

Extracts of unoxidized fish fat did not have this effect, but in extracts of gamma-irradiated and oxidized fish fat, and in the fatty acid fractions of this fat, the reaction of compression of the gastral cavity of hydra occurred within a period that was related to concentration of the substances. It was also demonstrated that the maximum length of hydra during locomotor cycles was reduced, and the minimum length increased, i.e., the hydra apparently lost elasticity, during and after exposure to X-irradiation (18 000 r). Although automatic locomotor reflexes ceased almost entirely after irradiation, the number of periodic 'tremors' that would under normal conditions lead to subsequent contraction of the body considerably increased, although they were manifested only as spasmodic movements of tentacles and hypostomal area. Length of the body was not reduced.

An early reaction has been noted in irradiated hydra, *Planaria* and molluscs (NIKITIN [1932, 1938]). All hydra were 'wrinkled' 10 hr after irradiation, but within two days they had recovered and were catching *Daphnia*. *Planaria* reacted within 24 hr to X-radiation by powerful contraction of the body which lasted for four days, but was followed by recovery. Following X-radiation the molluscs sank to the bottom of the aquarium and withdrew into their shells within a day of exposure. They returned to normal within 15 days.

SCOTT [1937] demonstrated the great difference in radiation doses giving rise to early and late radiation injury for certain bacteria, protozoans, worms, insects, amphibians, mammals and tissue cultures.

It can be concluded from the material presented above that the early reaction is a universal feature of organisms at various evolutionary levels.

14.4. Latent period and late radiobiological effect

A fairly long latent period for damage by comparatively small ionizing radiation doses is a characteristic feature of the whole organic kingdom. SCOTT [1937] distinguished median lethal doses for early and late death.

The median lethal dose of X-rays for algae decreases in the course of time.

A week after irradiation with soft and hard X-rays in a liquid medium, the LD_{50} was 40000 and 18000 r for *Chlorella*, 28000 and 8000 r for *Chroococcus* and 11000 r (hard rays) for *Ankistrodesmus*. When *Chlorella* was irradiated with hard X-rays in a liquid medium, and then cultivated in agar dishes the LD_{50} was 8000 r (BONHAM and PALUMBO [1951]). In RALSTON's [1939] experiments the unicellular alga *Dunaliella salina* perished shortly after irradiation or within a few weeks, depending on the X-ray dose employed.

Following a lethal dose of X-rays (600000 r) paramecia survived for a month before perishing (BACK [1939]). KOVALEVA [1948] found that irradiated paramecia did not perish or were not affected for some days or weeks. Although irradiation of the infusorian *Tillina magna* did not lead to rapid death, various morphological and behavioral disorders were noted. Nevertheless, the animals soon righted themselves and began to divide. After a radiation dose of 50000 r, *Tillina* returned to normal by the end of the first day. In 67 instances out of 88 after a radiation dose of 75000 r and in 45 out of 58 after a dose of 100000 r the first division yielded two daughter cells as against four in the controls. Considerably smaller doses led to the formation of three, five and six daughter cells. Another effect of radiation was the appearance of irregularly shaped animals. These disorders were followed by a period of normalization (the latent period) which was, however, followed by death. The death rate was very high in the period between the 4th and 14th division after irradiation. The mean number of divisions before death was slightly decreased when the dose was increased, but the percentage of lines that perished during the period increased rapidly with dose.

Unlike *Tillina*, the infusoria *Colpoda* sp. did not fully restore the mitotic rate before death 12 days later. The wave of early damage and the wave of long-term death for the two species of protozoa was therefore separated by periods of complete and incomplete normalization of two to four weeks (BRIDGMAN and KIMBALL [1954]).

A long period of radiation injury lasting for weeks or months was a feature of the coelenterates *Hydra fusca* and *Pelmatohydra oligactis* (NIKITIN [1938], SCHMIDT [1946], POLIKARPOV [1957a,b]). The radiation sickness of hydra was revealed in a general reduction of the body extending from apex to base, reduction and disappearance of interstitial cells in the tentacles and then of the nematocysts, swelling of the myoepithelial cells and weakening and loss of reaction to mechanical stimuli. The most indicative features in this respect are the changes that occur in the maximum length of irradiated hydra in course of time (POLIKARPOV [1957b]) and the dynamics of luminescence and photoresistance in hydra treated with fluorochrome in experiments and in controls (BIRUKOV *et al.* [1958]).

Following irradiation of the worm *Dendrocoelum lacteum*, radiation injury appeared within a period that was not related to dose. The only effect of the dose was that the injury developed more rapidly and was more profound. The latent period was 13 to 15 days for a dose range of 1 000 to 10 000 r. The latter dose was absolutely lethal and the damage that developed within two weeks of irradiation led to death of all the worms within 1.5 to two months (SCHMIDT [1946]). A dose of 5 500 r was lethal to the worm *Planaria polychroa* within 2.5 months (NIKITIN [1938]).

First investigation in radiobiology of *Artemia salina* was made by GAJEWS-KAJA [1923] at Sevastopol biological station.

The LD_{50} for adult crustaceans (Amphipoda) was 10 000 r over a period of five weeks. The LD_{50} for adult molluscs (*Radix* and *Thais*) that led to death within a week was 20 000 r, that for death within 40 days was 8 000 r (*Radix*) and within 160 days was 13 000 r (*Thais*) (BONHAM and PALUMBO [1951]). A dose of 4 000 r was lethal for *Physa acuta* within one to 1.5 months after irradiation. A three-month latent period was needed for development of the reactions of injury that led to the death of *Planorbis marginatus* following a dose of 10 000 r (NIKITIN [1938]).

A dose of 2 600 r delayed the first cleavage of the sea urchin *Arbacia*, but only a considerably larger dose (50 000 r) prevented fertilization and cleavage of the eggs (SCOTT [1937]). The first cleavage of irradiated sea urchin eggs is delayed more, the higher the radiation dose (HENSHAW [1932, 1940]). The effect of ionizing radiation upon developing sea urchin was studied by HSIAO and DANIEL [1960].

Within two weeks of X-irradiation (3 500 to 5 000 r) adult fishes (*Lebistes*) lost weight and refused food, and subsequently developed bruises and sores on the body. The irradiated fish perished within 18 to 50 days (SAMOKH-VALOVA [1935]). After golden carp (*Carassius carassius*) had received an X-irradiation dose of 4 000 r their behavior became abnormal by the 10th day, and death ensued by the 13th day. The liver, intestines and gonads were damaged (SAMOKHVALOVA [1938]).

The LD_{50} of X-rays for goldfish (*Carassius auratus*) is approximately 1 000 r. Following radiation doses of 500 to 10 000 r atrophy of the lymphoid tissue, lymphocytic pyknosis and damage to brain cells were noted (ELLINGER [1939]). Approximately the same median lethal dose has been established for the larvae (1 000 r) and fry (1 250 r) of the king salmon *Oncorhynchus tscha-wytscha* within two months of irradiation. The lowest effective doses after 48 days were 250 r to produce death, 500 r to reduce the weight of doomed

fish and 100 r to reduce their length. With radiation doses of 750 and 1250 r
the elements of the blood were damaged within 14 to 21 days of exposure.
The concentration of hematopoietic cells in the fish spleen fell within seven to
14 days after a radiation dose of 750 r (BONHAM *et al.* [1948]).

Carp irradiated with gamma-rays (50 or 300 r) between the age of three
days and three months did not differ from the controls in growth rate and
viability. A dose of 1000 r inhibited growth and activity only at an early age
(six days). The LD_{100} was approximately 3500 r for all groups. The irradiated
carp died rapidly (by the 15th to 25th day) after appearance of signs of injury
on the 13th to 18th day. Doses of 10000 to 40000 r produced an early wave of
radiation injury lasting for between a few hr and one to two days, followed
by a period of apparent well-being, and finally by radiation death within 10
to 25 days (GOLOVINSKAYA and ROMASHOV [1958]).

Following X-irradiation of male and female trout (50 r) the survival rate of
larvae hatched from the eggs of these fishes was reduced. With a dose of
100 r or above there was also some reduction in the weight of the year-old
progeny of the irradiated trout. The absolutely lethal dose was 1000 r.
Surviving embryos from fishes that had received radiation doses of 500 or
750 r revealed a whole range of abnormalities (fig. 23).

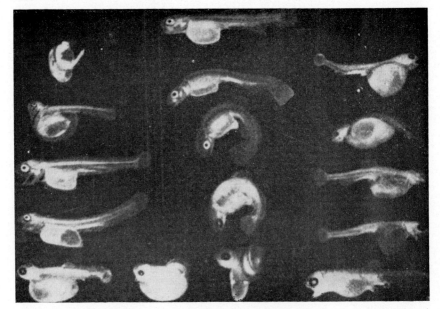

Fig. 23. Abnormal rainbow trout larvae from X-irradiated parents. Upper larva normal
(FOSTER *et al.* [1949]).

The effects of embryonic development of goldfish from exposing the parent fish to X-rays were studied by WONG and WANG [1960].

There have been a number of studies of effect of doses of ionizing radiation (usually very large) on sex organs of fishes, on spermatogenesis and oogenesis, fertility and the 'Hertwig effect' (NATALI [1940, 1942a,b], NEYFAKH [1956a,b, 1960, 1961a,b], SAMOKHVALOVA [1935, 1938], OPPERMANN [1913a,b], RUGH and CLUGSTON [1955a], SOLBERG [1938]), and detailed reviews have also appeared (GOLOVINSKAYA and ROMASHOV [1958], ROMASHOV and GOLOVINS-KAYA [1960]).

NEYFAKH [1956a,b,1959,1962] carried out very interesting work on changes in radiosensitivity during fertilization of loach eggs. The eggs were irradiated, usually with a dose of 400 r, at intervals of five to 10 min between fertilization and commencement of cleavage. This procedure led to the establishment of two periods of low radioresistance, the first (55 min) was the time of fusion of the male and female nuclei and the second was at the first division of the zygote. At these periods, doses of 100 or 200 r gave rise to developmental abnormalities of the loach embryos.

Irradiation at a certain stage in development produces characteristic abnormalities in all organs of the fish *Macropodus opercularis*. There have been fairly detailed studies of the harmful effect of X-rays on various stages of embryogenesis in relation to formation of blood and hematopoietic organs, eyes, central nervous system, sexual cells, muscles, digestive organs, heart, ears and organs of the lateral line, pronephros, olfactory organs, notochord and chromatophores (ALLEN and MULKAY [1960]).

There was a considerable latent period after irradiation of the eggs of *Oncorhynchus tschawytscha*, even when the dose employed was 2500 r. Signs of radiation injury were not noted until a month had elapsed, but led within 51 days to death of all fry from paralysis, hemorrhage and other causes (WELANDER *et al.* [1948]).

Gamma-irradiation (800 r or above) of goldfish (*Carassius auratus*) in the early stages of development delayed hatching of the larvae from the eggs, increased the number of abnormal larvae, and reduced the general length of the fry by comparison with the controls (OKADA *et al.* [1956]).

A similar pattern of radiation injury is also characteristic of amphibians. Late changes in irradiated cells of frog corneal epithelium were noted 180 days after irradiation, the damage being revealed in the appearance of polynuclearity and increased size of the cell elements as a result of swelling (STRELIN [1934]). A radiation dose of 200 r led to death before development of the larva in irradiated axolotl eggs, a dose of 300000 r to death before the first cleavage (GLOCKER *et al.* [1933]) and a dose of 600000 r to rapid death of the eggs (SCOTT [1937]).

When one blastomere of a frog embryo is irradiated with X-rays, the irradiated part does not function on gastrulation (VINTEMBERGER [1931]). SHEKHTMAN and KLYUPFEL' [1930] demonstrated that although an X-ray dose of 20 r was fatal to frog eggs, even a dose of 72000 r did not halt division

Table 48

Death of newt larvae on 24th day of development after X-irradiation of ova (PETERS [1960])

Dose (r)	Number of specimens	Number of deaths (%)	Probability of difference, P	Death from radiation (%)
2	700	18	0.7	3
Control	150	16		
5	650	18	<0.05	7
Control	160	11		
10	552	23	<0.001	19
Control	125	6		
16	200	34	0.001	24
Control	133	14		
20	167	27	<0.001	22
Control	80	6		
20	490	31	<0.001	27
Control	160	6		
40	255	53	<0.001	46
Control	133	14		
50	200	53	<0.001	46
Control	150	13		
50	322	70	<0.001	65
Control	100	13		
160	200	82	<0.001	80
Control	120	8		
200	285	87	<0.001	86
Control	200	5		
240	400	95	<0.001	94
Control	120	8		
320	400	100	–	100
Control	305	13		
6400	100	100	–	100
Control	80	6		

before formation of the mouth of the blastopore. A dose of 80 r led to death of the developing embryos within four days, but one of 400 r was lethal within 24 hr of irradiation. Irradiation of axolotl eggs in the meta-telophase of the first segmentation division led to death of the embryo within 20 days (GLOCKER et al. [1933]).

In an extensive study on the ova of the newt *Triton alpestris*, PETERS [1960] employed a range of X-ray doses between the absolutely lethal dose and the minimum dose that produced a statistically significant effect. His statistical analysis of differences between the experimental and control data (table 48) demonstrated a real effect from a dose of 5 r ($P < 0.05$) and above ($P < 0.001$). The LD_{50} was approximately 40 r, and the LD_{100} was 320 r. The probability of a difference between the experimental data (table 49) was high for two and

Table 49

Comparison of the effect of similar doses on the death of newt larvae (on 24th day of development) (PETERS [1960])

Dose (r)	Number of experimental larvae (natural deaths excluded)	Radiation death (%)	Probability of difference, P
2	588	3	
5	578	7	< 0.001
5	578	7	
10	519	19	< 0.001
10	519	19	
16	172	24	0.5
16	172	24	
20	618	26	0.3
10	519	19	
20	618	26	< 0.01
20	618	26	
40	219	46	< 0.001
40	219	46	
50	454	58	< 0.01
50	454	58	
160	184	80	< 0.001
160	184	80	
200	271	86	0.1
200	271	86	
240	368	94.5	< 0.001

five, five and 10, 10 and 20, 20 and 40, 40 and 50, 50 and 160 and 200 and 240 r. No significant difference was noted between the effect of doses of 10 and 16, 16 and 20 and 160 and 200 r, although the general pattern was maintained.

When larvae that had survived irradiation (10 or 20 r), and were normal in appearance, were exposed to a high temperature (35.8°C) for 6.5 hr it was found that there was a statistically significant increase in their death rate by comparison with control larvae. This may indicate that the viability of irradiated organisms is reduced owing to the latent development of radiation injury.

When frog sperm was irradiated with a dose of 1 000 r developmental abnormalities increased to 100%, but above 50 000 r most (91%) of the embryos were normal, but had a haploid set of chromosomes. It was therefore concluded that disturbances of the genetic apparatus were possible up to 1 000 r (RUGH [1960]).

When *Ambystoma* larvae were irradiated at doses of 100 to 10 000 r (at the 26th, 30th, and 34th stages of Harrison) it was found that the central nervous system and the eyes were the most radiosensitive. A high proportion of the embryos irradiated at early stages of development were microcephalic. Cardiovascular abnormalities developed in all embryos that received radiation doses of 250 r or more at the 26th stage of Harrison, and 500 r or more at the 30th and 34th stages (COPENHAVER *et al.* [1960]).

BLINOV [1956] made a detailed study of differences in the sensitivity to X-rays of amphibian embryos at various stages in their development. Irradiation of fertilized axolotl ova before cleavage (480 to 3 200 r) was lethal at the late blastula stage, whereas doses of 10 to 160 r were lethal mainly at the neurula stage.

Following radiation doses of 480, 1 600 and 3 200 r the cleavage period was the same as for the control. Neither a fatal outcome nor delayed development were, therefore, noted at the cleavage stage. This latent period was not affected by an increased dose. The irradiated embryos reached the neurula stage three to five days later than the controls.

Irradiation of axolotl embryos at the early blastula stage was similar in general results to irradiation of the ova prior to cleavage.

When axolotl embryos were subjected to doses of 1 600 or 3 200 r in the early gastrula stage, their development was immediately halted, and death occurred during the same stage. A dose of 450 r was lethal mainly in the neurula stage. There were no apparent morphological changes in some of the larvae that died at this stage. The various types of damage noted in the embryos included suppression of development of the tail bud and considerable disturbance in brain and sense organs (absence of rudiments of one or both eyes, destruction of the crystalline lens, underdevelopment of auditory vesicles, formation of detritus in an area of the brain). Tables 50 and 51

Table 50

Dependence of time of death of axolotl embryos on radiation dose for irradiation at early gastrula stage (BLINOV [1956])

Time of death (days)	Number of dead embryos as a percentage for the following doses (r)							
	3 200	1 600	480	160	80	40	20	10
2	–	–	–	–	–	–	–	–
3	–	–	–	–	–	–	–	–
4	–	–	–	–	–	–	–	–
5	100	84.6	20.0	71.4	–	–	–	–
6	–	15.4	80.0	21.4	46.1	30.0	27.2	12.5
7	–	–	–	–	23.0	30.0	–	12.5
8	–	–	–	–	7.6	10.0	9.9	–
10	–	–	–	7.1	15.3	10.0	–	–
11	–	–	–	–	7.6	–	–	–
12	–	–	–	–	7.6	10.0	18.1	12.5
13	–	–	–	–	–	–	–	–
14	–	–	–	–	–	10.0	18.1	50.0
15	–	–	–	–	–	–	27.2	12.5
16	–	–	–	–	–	–	–	–

Table 51

Dependence of time of death of axolotl embryos by development stages on radiation dose for irradiation at early gastrula stage (BLINOV [1956])

Developmental stage	Number of embryos, as a percentage for the following doses (r)							
	3 200	1 600	480	160	80	40	20	10
Gastrula	100	69.2	20.0	–	–	–	–	–
Neurula	–	30.8	80.7	71.4	46.1	40.0	18.1	25.0
Early tailbud	–	–	–	14.2	–	30.0	18.1	–
Middle tailbud	–	–	–	7.1	23.0	–	9.9	–
Early formation of gillbuds	–	–	–	–	–	10.0	–	–
Early formation of gill bundles	–	–	–	7.1	30.7	–	–	12.0
Appearance of lateral gill branches	–	–	–	–	–	–	36.3	37.5
Commencement of hatching	–	–	–	–	–	20.0	17.6	25.0

depict the distribution of axolotl embryos with respect to time and by stages
of radiation death following irradiation at the early gastrula stage. A dose of
10 r was absolutely lethal.

Axolotl embryos were far less sensitive when irradiated at the neurula and
tailbud stage. Doses of 10 and 20 r were ineffective, and doses of 40 and 80 r
were lethal for only 20% of the embryos. In this case the LD_{50} was between
160 and 480 r. Whatever the dose employed, however, none of the larvae
were fully normal on hatching. Growth of the experimental embryos was
clearly less than that of the controls, even when the radiation dose was 10 r
(table 52).

Table 52

Length of embryos under normal conditions and following X-irradiation (various doses)
at the neurula and tailbud stage (BLINOV [1956])

Dose (r)	Number of embryos	Length of embryo (mm) after irradiation				
		14th day	25th day	37th day	44th day	65th day
Control	4	12.3	15.6	20.5	24.5	36.0
10	5	12.1	15.2	18.8	21.1	29.0
20	5	11.5	15.2	18.7	19.3	28.0
40	4	12.1	14.6	17.6	19.1	27.3
80	4	11.2	13.9	14.3	Perished	
160	4	11.2	13.5	13.8	Perished	

Development of the gills, muscles and notochord of the irradiated axolotl
embryos was almost normal. The spinal cord was less damaged than the
brain. Chromatophores formed less rapidly in experimental subjects than in
controls. Development of the excretory system was not noticeably affected,
but formation of blood was greatly reduced. The characteristic radiation abnormalities were in disproportional development of the body, particularly
microcephalism, whereas damage at the early stages of development was
located in the tail region.

It may therefore be concluded from information reviewed here that
radiation injury generally follows the same kinetic pattern for aquatic
organisms of various evolutionary levels from unicellular organisms to
vertebrates. If the organisms survive, early effects are replaced by the latent
period that may often be very lengthy, thereafter by a wave of long-term
injury, terminating in recovery or more often in death. The doses that produce long-term radiation effects are far lower than those which produce early

signs of injury. It may therefore happen that late death occurs without there having been early symptoms.

Many aquatic organisms are very suitable subjects for radiobiological research. Unicellular algae, paramecia, hydra, sea urchins, fishes and amphibians have served as the subjects for many investigations of the biological effect of ionizing radiation. Examination and assessment of these matters do not fall within the scope of the present book. There are several monographs that deal with various aspects of the effect of ionizing radiation on a biological substrate (GORODETSKIY et al. [1961], GRAYEVSKIY and SHAPIRO [1957], TARUSOV [1954], FRITZ-NIGGLI [1961], ZIMMER [1961], BACQ and ALEXANDER [1955], HOLLAENDER [1954], LEA [1955], TIMOFEEFF-RESSOVSKY and ZIMMER [1947] and other works).

CHAPTER 15

THE EFFECT OF CHRONIC IONIZING IRRADIATION ON AQUATIC ORGANISMS

15.1. Biological role of the natural background of ionizing radiation

There are known to be three permanent sources of ionizing radiation: cosmic rays, environmental radioactivity and radioactive substances *in vivo* (BARANOV [1955], PERTSOV [1964a,b]). The relative significance of each of these factors for aquatic organisms varies in relation to the depth at which the organism dwells, radioactive properties of water and bottom, and size of the aquatic organism.

Cosmic rays produce a dose rate of 35.0 mrad/yr at the surface of water. The dose rate decreases with depth to 28.8% at a depth of 10 m, and to 1.35% at 100 m (table 53) (FOLSOM and HARLEY [1957]).

Table 53

Absorption of cosmic rays in the sea (FOLSOM and HARLEY [1957])

Depth	Dose rate	
(m)	(mrad/yr)	in relation to dose rate at surface (%)
0	35	100
10	10.1	28.8
20	4.86	13.9
50	1.40	4.0
100	0.47	1.35
200	0.15	0.42
300	0.074	0.21
500	0.030	0.087
1000	0.009	0.025
4000	0.007	0.002

Almost all the natural radioactivity of sea-water is due to the potassium-40 concentration of 3×10^{-10} Cur/l (table 54). Of the total of 728 potassium-40 dis/min·l of sea-water, beta-decay accounts for 660 and gamma-decay for 68.

The beta-particles produce a dose rate of 2.7 mrad/yr in seawater and the gamma-quanta of potassium-40 0.9 mrad/yr. The radioactivity of marine organisms is also due in the main to the natural radioisotope of potassium. The absorbed doses from incorporated natural alpha-radioelements are, however, often far higher than those from potassium-40.

Table 54

Natural radioactivity of sea-water (REVELLE et al. [1956])

Isotope	g/l	dis/min·l	Total amount in ocean ($\times 10^6$ metric tons)	Total activity in ocean ($\times 10^6$ Cur)	Gamma-energy (MeV)
^{40}K	4.5×10^{-5}	720	63 000	460 000	1.5*
^{87}Rb	8.4×10^{-5}	13.2	118 000	8 400	absent
^{238}U	2.0×10^{-6}	6**	2 800	3 800	0.05–0.82
^{235}U	1.5×10^{-8}	0.18**	21	110	0.06–0.18
^{232}Th	1×10^{-8}	0.012**	14	8	0.03–0.08
^{226}Ra	3.0×10^{-13}	1.8**	4.2×10^{-4}	1 100	0.18–0.60
^{14}C	4×10^{-14}	0.42	5.6×10^{-5}	270	–
^{3}He***	8×10^{-17}	1.5	1.5×10^{-9}	12	–

* $\gamma/\beta = 0.1$.
** Activity of nuclide and daughter products.
*** Only in the upper 50 to 100 m of water.

In a large fish the beta-activity of potassium-40 is 5 800 dis/min·kg, which corresponds to a tissue dose rate of 24 mrad/yr, and gamma-activity is 300 dis/min·kg, or 3.7 mrad/yr. Doses from potassium-40 in marine microscopic organisms (radius 0.01 mm or less) can be disregarded, i.e., the radiation dose rate to unicellular organisms consists of cosmic-ray dose (35 mrad/yr at the surface of the sea) and potassium-40 in sea-water (3.6 mrad/yr). Abyssal marine deposits may create a radiation dose rate of 40 to 620 mrad/yr in microorganisms (fig. 24).

CHERRY [1964] has demonstrated in a recent paper that marine phyto-plankton receive 23 mrad/yr, or 230 mrem/yr, from natural alpha-emitters. This is a minimum figure. The marine diatoms Rhizosolenia hyalina, whose content of natural alpha-emitters is approximately 96 pCur/g of dry matter, are exposed to the effect of doses amounting to 2 800 mrem/yr.

For quite obvious reasons the most important task facing radiobiologists is the study of biological effect of small and extremely small doses of ionizing

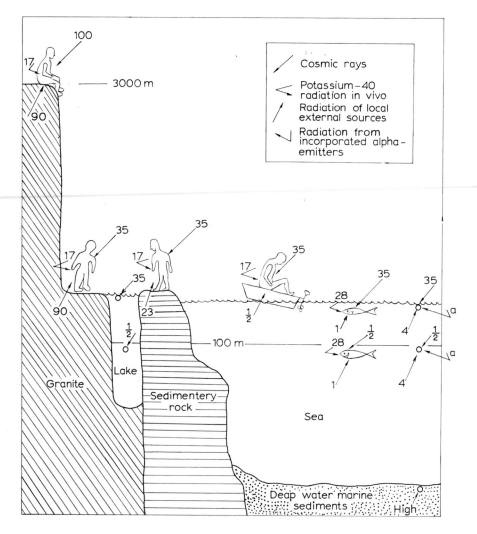

Fig. 24. Radiation dose rate from natural sources of ionizing radiation under various conditions (mrad/yr). Total annual dose for man from all beta-gamma-sources 207 mrad/yr on granite cliffs at an altitude of 3 000 m, 142 mrad/yr at sea level, 75 mrad/yr on sedimentary rocks, 52 mrad/yr on the sea. A large fish receives 64 mrad/yr at the surface and 30 mrad/yr at a depth of 100 m. Microorganisms receive 39 mrad/yr at the surface and 5 mrad/yr at a depth of 100 m (FOLSOM and HARLEY [1957]). (a) 23 to 280 mrad/yr (230 to 2800 mrem/yr) from incorporated alpha-emitters (CHERRY [1964]).

radiation. In view of the great experimental difficulties, there is still no final direct proof of the presence or absence of a direct proportionality between biological effect and dose, although it is improbable that there should be a threshold for the effect of ionizing radiation on the organism, in view of the specific nature of the physical agent. If no such threshold exists, then even the background radiation of man by cosmic rays and the radiations of natural radioactive substances may, according to the calculations of BRUES and SACHER [1952], reduce the 'ideal' life span by approximately a year. When studying the lysogenic activity of bacteriophages, MARCOVICH [1956] established that a dose of 0.3 r had a statistically significant biological effect. VINOGRADOV [1956, 1957a] demonstrated that exclusion of potassium-40 from fungi and from frog muscles had no effect on their functions when the preparations were still irradiated by cosmic rays and by the radiations of natural radioactive substances.

Experimentally produced fluctuations of the intensity of cosmic radiation affected the physiological activity of crabs *Uca pugnax* (BROWN *et al.* [1955]). These authors intensified the cosmic-ray dose (*bremsstrahlung*) by using lead screens between 0.3 and 2.1 cm thick at a height of approximately 13 cm above the animals. In controlled experiments they used planks of wood as screens. It follows from table 55 that the pigment system of the crabs reacted quite differently to that of the control. The diurnal rhythm of the reaction to intensification of cosmic radiation showers was concentration of the pigments (the body became lighter in color) at the beginning of the daytime phase of the endogenous daily cycle and dispersion of the pigment (the body became darker) for most of the rest of the day.

A regular relationship between fluctuations in cosmic radiation over a period of years and physiological functions (oxygen demand or spontaneous motor activity) in certain marine plants and animals (*Uca pugnax, Venus mercenaria, Fucus* sp.) and freshwater and terrestrial plants and animals has been described by BROWN *et al.* [1958].

Possible biological significance of natural background of ionizing radiation was discussed by SCHAEFER [1955].

Maximum permissible radiation doses have at present been worked out only for man. It is highly indicative that these levels have been repeatedly lowered, and according to the most recent health regulations (Russian list: Anonymous [1960]) the permissible irradiation of the whole population of any locality is no more than twice the natural background, i.e., 0.01 mr/hr. In this connection it is of interest that aquatic organisms are subjected to the greatest radiation effect from natural sources, both at the surface of

Table 55

Effect of altering cosmic ray dose rate on the black chromatophores of *Uca pugnax* (BROWN et al. [1955])

Series of observations	Time	State of melanophores (mean figures)		Experiment minus control	Number of observations**	Number of groups of crabs**
		experiment (lead)	control (wood)			
I	May 25–30	3.95	3.83	+0.12	860	43
II	29–31	3.58	3.49	+0.09	480	24
III	June 2–6	3.86	3.74	+0.12	500	20
IV	18–21	3.59	3.51	+0.08	3625	145
V	28	2.57	–	+0.23	260	13
	–	2.56*	2.34	+0.22	–	–
VI	29	4.26*	4.02	+0.24	180	9

Mean for all experiments plus 0.110 ± 0.0195

* Whereas in other instances the lead screens employed were of the same thickness (1.8 cm), the lead screens in these instances were varied in thickness (from 0 to 1.8 cm). The lead was placed in position or removed at a rate of approximately 0.3 cm/hr.
** Experimental and control animals.

the water, i.e., the biocoenosis of the hyponeuston, and on the bottom of deep water (marine sediments, the benthos). The radiation dose rates of natural radioactive substances on the sea floor may be 10 times or 100 times the cosmic radiation background at sea level.

15.2. Effect of various levels of radioactivity in an aqueous medium on the vital processes of aquatic organisms

It follows from tables 53 and 54 that the cosmic-ray background at the surface of the water (10^{-4} rad/day) is approximately 10 times the potassium-40 radiation background (10^{-10} Cur/l) in sea-water (10^{-5} rad/day). The potassium concentration factor in sea fishes and their eggs is approximately 10 (VINOGRADOV [1953]), i.e., the dose rate due to potassium-40 radiations in the organisms is approximately the same as the dose rate of the cosmic-ray background. It would therefore appear that potassium-40 concentrations might be taken as the maximum permissible concentration of any radioactive substance for aquatic organisms. This assumption cannot, however, be accepted for two reasons: 1) because of irregular distribution of radioisotopes of the other elements by organs and tissues, which contrasts with the distribution of potassium and 2) because of the high concentration factors of a number of radionuclides. By virtue of both factors, critical tissues are subject to incomparably larger radiation doses than the natural background, even though the concentration of the radionuclides in the solution may be only 10^{-10} Cur/l. This effect is dependent on the radioresistance of individual tissues and of the organism as a whole. An extremely important part is played both by the total dose absorbed by the tissue, and, in all probability, by the 'microgeometry' of the distribution of doses in cells, which is conditioned, in the first instance, by the physicochemical nature of the radioisotopes of the various elements and by the type and energy of the radiation.

As already noted (table 47) low radiosensitivity to a single external radiation dose is somatically a feature of plants and invertebrates. This pattern was established by BLINKS [1952] in the area of the Bikini Atoll, where the radiation level a year after nuclear testing was two to three times background, and approximately 50 times background on the coral reefs near the site of the explosion. Dose rate in the substrate was between 1×10^{-3} and 1×10^{-2} and even reached 0.6×10^{-1} r/day, i.e., approximately 600 times greater than that of cosmic rays. The major portion of this radiation was absorbed by algae found on rocks. Blinks investigated green, brown and red algae (*Valonia, Dictyosphaeria, Bryopsis, Halimeda, Udotea* and *Cladophoropsis*) for permeability, uptake of salts, bioelectric potentials, pigment content, photo-

synthesis, respiration and also for enzymatic activity and calcification. The
only abnormality was in catalase content. Oxygen formation by catalase was
extremely high in *Halimeda*, *Udotea*, and *Cladophoropsis*. Marine plants are
therefore resistant to increased doses of ionizing radiation, at least for a year.
If radiogenetic changes had occurred they were still in a recessive state.

The present author established fairly high radioresistance in experiments
on the effect of sulphur-35 on the rate of cell division in the marine unicellular
alga *Prorocentrum micans* (POLIKARPOV and LANSKAYA [1961]) (fig. 25). A
temporary stimulation effect was noted in a solution with a concentration of
0.8×10^{-5} Cur/l, and an inhibitory effect, commencing with the third generation, was noted at concentrations of 0.8×10^{-4} Cur/l and above. It should be
borne in mind that this alga has been found to be relatively more sensitive to

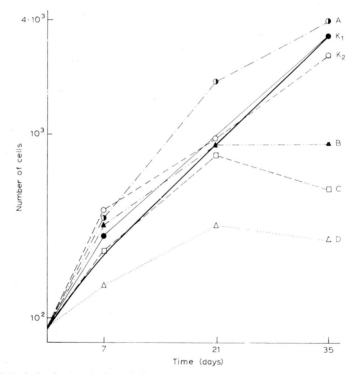

Fig. 25. Variation in the number of algal cells in the course of time. K_1 and K_2 = control;
A, B, C, D = experimental series with concentrations of the emitter of 8×10^{-6}, 8×10^{-5},
8×10^{-4} and 8×10^{-3} Cur/l, respectively; heavy line denotes theoretical curve (POLI-
KARPOV and LANSKAYA [1961]).

Table 56

Effect of various strontium-90 and strontium-89 concentrations on breeding of *Daphnia* (TELITCHENKO [1958])

Substance	Concentration	Amount	Number of progeny from 10 females in 80 days	Data for last generation		
				period between moults (days)	period taken to reach sexual maturity (days)	litter of a single female
^{90}Sr	10^{-5} Cur/l	7	917	6	13	4–8
^{90}Sr	10^{-6} Cur/l	9	1963	4	9	6–18
^{90}Sr	10^{-7} Cur/l	9	2511	4	8–9	11–39
^{89}Sr	10^{-8} Cur/l	9	2617	3–4	6–7	9–29
^{89}Sr	10^{-9} Cur/l	9	2711	3	8	12–28
Stable Sr	5 mg/l	6	854	7	14–15	2–11
Stable Sr	1 mg/l	9	2117	5	10	5–31
Water	–	9	2661	4	8	11–28

Note. The chlorides of ^{90}Sr, ^{89}Sr and stable strontium were employed.

external radiation than other microphytes owing to its large chromosomes (GODWARD [1960]). It has been shown that freshwater algae and lower crustaceans are resistant to the beta-radiation of phosphorus-32 at an activity of up to 10^{-5} to 10^{-4} Cur/l (VINBERG and GAPONENKO [1958]), and of strontium-90 and strontium-89 at an activity of up to 10^{-5} Cur/l (TELITCHENKO [1958], TIMOFEYEVA-RESOVSKAYA [1958b]).

The information in table 56 demonstrates that a significantly harmful effect on the multiplication of *Daphnia* was established over a period of 80 days by strontium-90 concentrations of 10^{-5} Cur/l and above. The numbers of *Daphnia* were 34% of the control, the number of progeny of a single female was far less and the period between moults and the maturation time were far longer than in the control. According to VOYNAR's data [1960] a stable strontium concentration of one mg/l, which does not affect the multiplication of *Daphnia* (TELITCHENKO [1958]), is 10 times its concentration in natural freshwater (0.1 mg/l). It therefore follows that the effect that has been established is not due to the chemical properties of strontium radioisotopes, but to their radioactivity. In fact, 10^{-5} Cur/l of strontium-90 is only 10^{-7} mg/l in terms of weight, which is millions of times less than the natural concentration of strontium in freshwater (TELITCHENKO [1958]).

It has been demonstrated that fishes that survive external radiation and are normal in appearance produce inferior progeny with radiation abnormalities.

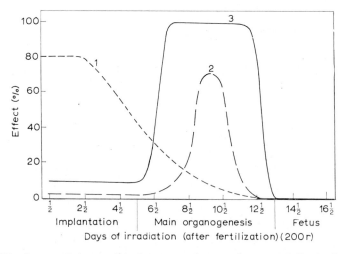

Fig. 26. Death rate and abnormalities in mouse embryos and neonates following irradiation at various stages in development (RUSSELL and RUSSELL [1954]). (1) death rate in embryogenesis, (2) death rate at birth, (3) abnormalities.

On the other hand, there is extensive evidence in mammalian radiation embryology (RUSSELL and RUSSELL [1954]) that radiation in the very early stages of embryogenesis is followed primarily by death of embryos, whereas radiation at later stages leads to abnormalities (fig. 26).

It is highly probable that radioactive substances incorporated in a biological substrate have a biological effect at far lower doses than are effective in external radiation. Academician SPITSIN [1960] has established that addition of radionuclides to catalysts considerably increases their efficiency. Nevertheless, when a catalyst was externally irradiated by beta-particles or gamma-quanta in a dose equal to the radiation dose of an incorporated radionuclide the same results were not noted (the results were weaker than in the action of external radiation). It may therefore be concluded that the effect of external radiation cannot be compared with that of a radionuclide introduced into an organic system.

There have, unfortunately, been very few studies of the biological effects of solutions of radionuclides on the development of fishes. Results should therefore be treated as extremely tentative. Japanese scientists have investigated the effect of various solutions of radioactive ash, collected from the deck of the fishing vessel Fukuru Maru No. 5 after the nuclear weapon tests in 1954 at Bikini Atoll, on development of eggs of freshwater fishes. It was found that an aqueous extract of ash in a concentration of approximately 10^{-6} Cur/l had a 100% lethal effect on survival of *Pseudorasbora parva parva* larvae when the eggs placed in the solution were at the stage of the early anlage of the annular gastrula. Later, from commencement of the formation of the optic vesicles, the effect of this concentration was noticeably weakened (table 57). The eggs were kept in Petri dishes containing 30 ml of solution. Damage which developed in course of the experiment consisted of slight mobility of the embryos, curvature of the body, reduction of body length and of the number of myotomes, underdevelopment of the fins, delayed hatching and abnormality of the chromatophores. The most characteristic form of damage was the curved body, which prevented complete emergence of some of the larvae. Fertility and development of the eggs were not, however, affected when female *Oryzias latipes* were kept in two liters of a solution of ash of approximately the same concentration for five days without food before being placed in ordinary water, where they were crossed with normal males. No abnormalities appeared in the nine months after hatching (HIBIYA and YAGI [1956]).

This result can be explained on the basis of TELITCHENKO's [1961, 1962] data, according to which the strontium-90 concentration factor for eggs

Table 57

Effect of an ash extract on development of the freshwater fish *Pseudorasbora parva parva* (HIBIYA and YAGI [1956])

Stage of development	Activity of solution (counts/min · ml)	Number of eggs	Number of dead eggs and larvae	Number of larvae		
				still within the egg case	hatched	surviving
Early anlage of annular gastrula	265	16	8*	3	5	0
	urban water supply (control)	16	3	0	13	10
Commencement of formation of optic vesicles	265	40	7*	10	23	4
	urban water supply (control)	40	4	0	36	36

* All perished at time of hatching.

within the female is less than unity (there was almost no penetration of yttrium-90 into the eggs). Eggs in the ovaries are, therefore, largely 'protected' from the penetration of at least these two fission products.

MIKAMI et al. [1956] have described experiments with the eggs of the tropical aquarium fish Zebra danio (table 58). A radioactive residue (92 mg) was obtained from rainwater with an activity of 4.2×10^{-10} Cur/l. Concentrations of between 2.1×10^{-10} and 2.1×10^{-8} Cur/l were obtained by dissolving this residue in water. The control eggs were kept in pure water. The eggs were placed in the solutions within one hr of fertilization. It was found that concentrations of more than 2.1×10^{-9} Cur/l had a noticeably lethal effect. The formation of black and yellow pigments was retarded in concentrations of 2.1×10^{-10} Cur/l, and retarded absorption of the vitelline mass and inhibition of the growth and differentiation of the embryos were noted at concentrations of 4.2×10^{-10} Cur/l and above. Hatching of the larvae was delayed when the concentration was 4.2×10^{-10} Cur/l. The abnormalities included distorted vertebral column, underdevelopment of the edge of the iris and tail, microcephalism, cyclops and acephalia. Histological examination revealed that development of the central nervous system was particularly affected, as was development of sense organs, alimentary canal and digestive gland (figs. 27–30). Morphogenesis and differentiation of the alimentary canal and the digestive gland ceased at concentrations of more than 0.6×10^{-8} Cur/l. The authors concluded that radioactive rainwater affected the embryonic development of Zebra danio eggs.

A radionuclide mixture of unknown composition was employed in the studies that have been considered. The extremely damaging effect on development of fish eggs may possibly have been largely due to the presence of alpha-emitters in the solutions. The possible chemical effect of impurities was not excluded, but the general dependence of biological effect on dose (concentration of radioactive substances) remained that specific to the effect of ionizing radiation.

Radiobiological experiments with the eggs of sea fishes have been commenced in the author's laboratory (POLIKARPOV and IVANOV [1961, 1962a, b,c], ZAYTSEV and POLIKARPOV [1964], IVANOV [1965b]). The effect of strontium-90-yttrium-90 with an activity of from 10^{-14} and 10^{-12} to 10^{-4} Cur/l was investigated in experiments conducted between 1960 and 1964 mainly on pelagic eggs of Black Sea fishes (Scorpaena porcus, Engraulis encrasicolus, Trachurus mediterraneus ponticus, Mullus barbatus ponticus, Serranus scriba etc.) (tables 59 and 60; figs. 31–33).

The controls included sea-water, sea-water plus stable strontium chloride

Table 58

Effect of a mixture of radionuclides isolated from radioactive rainwater on development of the eggs of the aquarium fish *Zebra danio* (MIKAMI *et al.* [1956])

Serial number of experiments	Characteristics of development of eggs in relation to time		
	14 IX (beginning of experiment)	15 IX	16 IX
I	Pure water (control)	Eyes formed in 13 out of 14. Perceptible development of black pigment. One egg perished.	7 hatched out of 13
II	2.1×10^{-10} Cur/l	Eyes formed in 11 out of 12. Slight formation of pigment. One egg perished.	8 hatched out of 11
III	4.2×10^{-10} Cur/l	Eyes formed in 25 out of 27. Very slight formation of pigment. Two eggs perished.	14 hatched out of 25
IV	2.1×10^{-9} Cur/l	Eyes formed in 23 out of 30. Very little pigment. Seven eggs perished.	6 hatched out of 23
V	4.2×10^{-8} Cur/l	Seven eggs out 24 perished. Eyes failed to form in remaining 17.	None of 17 hatch Very little pigment
VI	2.1×10^{-8} Cur/l	All 20 eggs perished.	–

* The vertebral curvature was mainly spinal or dorsal, but left or right in some instances.

IX (orning)	17 IX (evening)	18 IX	20 IX	22 IX
atched, low ment dent.	1 hatched	All hatched.	Shape and size normal in all cases.	Fixed.
atched.	1 hatched.	All hatched.	Vertebral curvature* and other abnormalities in 3 out of 11. Apparent abnormalities 3/11 (27%).	Fixed.
hatched, ow pigment formed.	1 hatched, yellow pigment just evident.	All hatched.	Vertebral curvature* and other abnormalities in 6 out of 25. Apparent abnormalities 6/25 (24%).	Fixed.
atched, mation of ck pigment rded and ow pigment evident.	8 hatched, yellow pigment just evident.	All hatched, but resorption of yolk retarded.	Vertebral curvature* and other abnormalities in 10 out of 23. Apparent abnormalities 10/23 (43%).	Fixed. Damage to edge of iris noted.
atched, slight elopment of ck pigment.	6 hatched, yellow pigment very slightly apparent and resorption of yolk almost completely halted.	3 still not hatched.	Vertebral* and other abnormalities in 14 out of 17. Apparent abnormalities 14/17 (82%).	Fixed. Damage to edge of iris noted.
	—	—	—	—

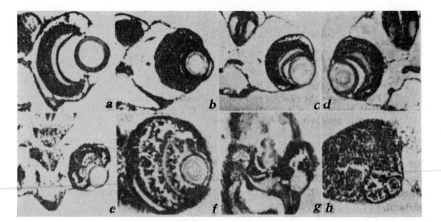

Fig. 27. Effect of radioactive rainwater on development of the eye of *Zebra danio*. (a) tap water (control), (b) 2.1×10^{-10}, (c) 4.2×10^{-9}, (d) 2.1×10^{-9}, (e) 4.2×10^{-9}, (f) 0.6×10^{-8}, (g) 0.8×10^{-8} and (h) 1×10^{-8} Cur/l (MIKAMI *et al.* [1956]).

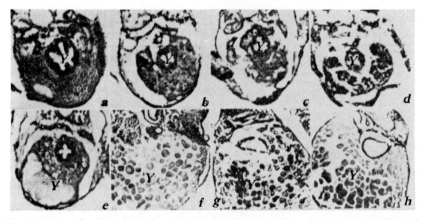

Fig. 28. Effect of radioactive rainwater on development of the duodenum and liver of *Zebra danio*. Radioactivity levels of water (a–h) as in fig. 27; Y is yolk mass (MIKAMI *et al.* [1956]).

and sea-water plus a small amount of distilled water. All these media were produced to verify the effect of corresponding concentrations of the accompanying substances (distilled water, stable strontium) when the initial radioactive solution was added to sea-water in the experimental aquaria. In no case was a statistically significant difference noted between controls.

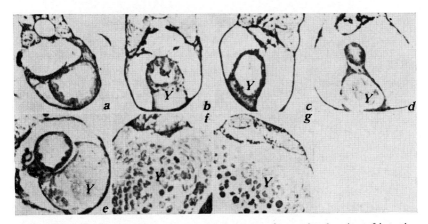

Fig. 29. Effect of radioactive rainwater on development of constricted region of intestines of *Zebra danio*. Same notations as in fig. 27; Y is yolk mass (MIKAMI *et al.* [1956]).

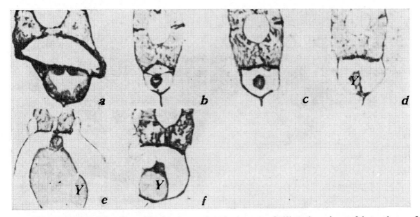

Fig. 30. Effect of radioactive rainwater on development of dilated region of intestines of *Zebra danio*. Same notations as in fig. 27; Y is yolk mass (MIKAMI *et al.* [1956]).

The fish eggs were netted from the sea or obtained from laboratory spawning fishes and were introduced into experimental vessels shortly after fertilization. The anchovy and stone perch larvae hatched within 20 to 24 hr, the mullet and horse-mackerel larvae within 40 to 48 hr, the scorpion fish larvae in 2.5 days, and the *Trachurus* larvae in five days (DEKHNIK [1961]).

Table 59

Probability of difference between experimental and control data at various concentrations of strontium-90-yttrium-90 (POLIKARPOV and IVANOV [1961])

Subject, stage of development and effects compared	2×10^{-12}		2×10^{-10}		2×10^{-8}		2×10^{-6}		2×10^{-4}	
	t*	P**	t	P	t	P	t	P	t	P
Anchovy eggs, gastrulation										
Death at early stages	–	–	1.288	0.2	2.335	0.05	–	–	–	–
Hatching delayed	–	–	0.638	0.5	1.257	0.2	–	–	–	–
Total number of abnormal larvae	1.220	0.2	2.703	0.02	5.374	<0.01	–	–	–	–
Number of larvae with vertebral abnormalities	1.220	0.2	3.726	<0.01	5.917	<0.01	–	–	–	–
Number of dorsoventrally curved larvae	1.220	0.2	1.927	0.1	3.057	0.02	–	–	–	–
Number of laterally curved larvae	–	–	–	–	3.749	<0.01	–	–	–	–
Total number of larvae with other abnormalities	–	–	1.288	0.2	2.171	0.07	–	–	–	–
Stone perch eggs, cleavage										
Death at early stages	–	–	–	–	3.622	0.08	7.890	0.03	11.08	<0.01
Hatching delayed	–	–	–	–	5.414	0.03	11.17	<0.01	14.00	<0.01

* t, student's criterion.
** P, probability of difference.

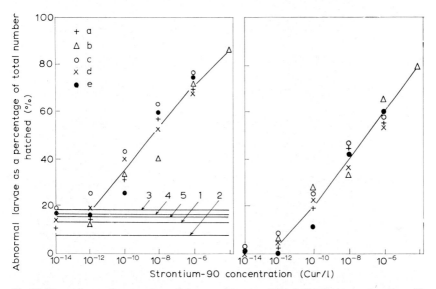

Fig. 31. Dependence of proportion of abnormal larvae of Black Sea fishes on strontium-90 concentration in surrounding sea-water. (1) and (a) = abnormal mullet larvae, (2) and (b) = green wrasse (Labridae), (3) and (c) = horse-mackerel, (4) and (d) = anchovies, (5) and (e) = from a mixture of pelagic eggs (the figures denote the proportion of abnormalities in controls of the various species of fishes). Left-hand graph without allowance for control, right hand graph with allowance for control (Polikarpov and Ivanov [1962a]).

It is evident from table 59 and fig. 31 that the most sensitive parameter of the effect of strontium-90-yttrium-90 solutions on development of pelagic eggs of the anchovy and other rapidly developing sea fishes is the proportion of abnormal larvae (at 10^{-10} Cur/l, $P = 0.02$) relative to the total number of larvae to hatch. Hatching of the larvae was delayed and eggs perished in early stages at concentrations of 10^{-8} Cur/l and above (Polikarpov and Ivanov [1961]). Like the author, Brown [1962] did not note the death of freshwater fish eggs (*Salmo trutta* and *S. salar*) in a strontium-90-yttrium-90 solution with an activity of 10^{-9} Cur/l.

The author's investigations have revealed various radiation abnormalities in the larvae of sea fishes, and especially abnormalities of the vertebral column (fig. 33).

The effect of strontium-90-yttrium-90 and other radionuclides on the eggs of freshwater and sea fishes have recently been the subject of investigations in a number of Soviet and non-Soviet institutes. Particular mention may be made of contradictory data on the eggs of the plaice *Pleuronectes platessa*.

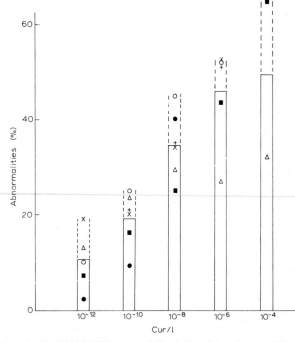

Fig. 32. Damage to the developing eggs of Black Sea fishes in strontium-90-yttrium-90 solutions. ○ *Trachurus mediterraneus ponticus*, ■ *Engraulis encrasicholus ponticus*, △ *Scorpaena porcus*, × *Serranus scriba*, + *Mullus barbatus ponticus*, ● a mixture of pelagic eggs. Abnormal prelarvae calculated as a proportion of the number hatching (ZAYTSEV and POLIKARPOV [1964]).

Table 60

Probability of difference between experimental data with strontium-90-yttrium-90 and controls (IVANOV [1965c])

Concentration	Anchovy eggs				Black Sea turbot eggs			
	death of eggs		hatching of abnormal prelarvae		death of eggs		hatching of abnormal prelarvae	
(Cur/l)	t^*	P^{**}	t	P	t	P	t	P
10^{-5}	18.06	0.01	7.21	0.01	8.40	0.01	–	–
10^{-6}	6.02	0.01	5.94	0.01	6.30	0.02	4.88	0.01
10^{-7}	2.19	0.1	3.23	0.01	2.92	0.1	–	–
10^{-8}	–	–	2.84	0.02	–	–	4.83	0.01
10^{-9}	–	–	2.29	0.05	–	–	–	–
10^{-10}	–	–	2.26	0.05	–	–	2.46	0.05
10^{-11}	–	–	–	–	–	–	–	–
10^{-12}	–	–	–	–	–	–	1.40	0.2

* t, student's criterion.   ** P, probability of difference.

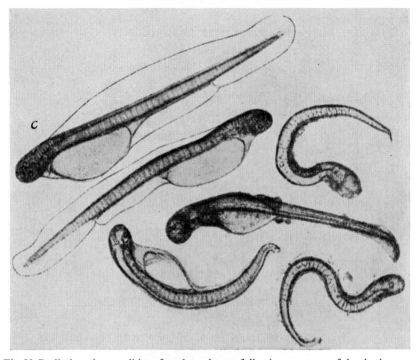

Fig. 33. Radiation abnormalities of anchovy larvae following exposure of developing eggs to effect of strontium-90-yttrium-90. C is normal larvae from control (POLIKARPOV and IVANOV [1962a]).

FEDOROV, PODYMAKHIN, SHCHITENKO *et al.* [1964] noted that eggs of these fishes from the Barents Sea were highly sensitive to slight radiocontamination of sea-water by strontium-90-yttrium-90, whereas BROWN and TEMPLETON [1964] did not detect any effect on the eggs of this species from the British coast in their preliminary investigation, even when strontium-90-yttrium-90 concentrations in the solution were 10^{-4} Cur/l. It may be recalled that these authors failed to note any signs of damage to *P. platessa* embryos from an absorbed dose of 500 rad. The striking and exceptional nature of these results calls for further, and if possible joint, investigation.

N.V. Gamezo experimented in the author's department in 1964 on the effect of strontium-90-yttrium-90 on the developing pelagic eggs (outside the mucous sac) of *Scorpaena porcus* and *Sargus annularis*. He observed that the detectable 'threshold' of the statistically significant biological effect of solutions of this radionuclide varied between 10^{-11} and 10^{-9} Cur/l (with a mean of approximately 10^{-10} Cur/l) in relation to certain features of the experiments. Gamezo also found that the mucous sac of the *S. porcus* prevented penetration of the radioisotopes of yttrium and strontium to the eggs. It should be noted that the mucous sac soon dissolves and the eggs liberated. This factor must be considered when analysing the effect of radionuclides on the development of freshwater fishes, many of whose eggs are protected by mucous until hatching of the larvae.

Therefore, fishes in early stages of development are most sensitive to solutions of radionuclides. Invertebrates and especially algae are far more resistant in this respect.

15.3. Effect of radioactive substances on populations and biocoenoses

Some information has already been given on the considerable radio-resistance of plants and invertebrates. These observations are supported by existing studies of the effect of radionuclides on the formation of biocoenoses.

TIMOFEYEVA-RESOVSKAYA [1958b] investigated the rate of formation of freshwater periphyton in aquaria containing various concentrations of a mixture of uranium fission products (between 3×10^{-6} and 0.6×10^{-3} Cur/l). Perceptible stimulation of formation of the periphyton biomass, amounting to several hundred % of the control, was noted at all concentrations. Activity in the water fell rapidly by 50% within 30 days and by 95% by the 140th day as a result of absorption by foulings.

In addition to stimulation of periphyton biomass, the structure of the association was affected by radiations of the radionuclides. Ten forms were abundant, namely, the infusoria *Colpidium* and *Vorticella*, the green algae *Cosmarium* and *Staurastrum* and the diatoms *Asterionella*, *Tabellaria*, *Synedra*, *Fragilaria*, *Navicula* and *Melosira*.

No differences were noted in any experimental level (control, 0.5×10^{-4}, 1×10^{-4} and 2×10^{-4} Cur/l) on first examination in March (two months after commencement of the experiment). At that time *Asterionella* and *Tabellaria* were most plentiful and *Vorticella* and *Colpidium* slightly less so. In the control aquaria, composition remained the same in April, but by end of the period in June the proportion of *Asterionella* and *Vorticella* had even further increased, and the leading role of *Tabellaria* and *Colpidium* had been taken over by *Synedra* and *Fragilaria*. All the originally dominant forms had disappeared from the concentrations of 0.5×10^{-4} and 1×10^{-4} Cur/l by June, with *Cosmarium* and *Staurastrum* now most plentiful and *Fragilaria* and *Synedra* slightly less so. At a concentration of 2×10^{-4} Cur/l *Synedra* was already one of the dominant forms in April, and by the end of the experiment *Synedra, Cosmarium* and *Navicula* predominated, with *Asterionella* and *Melosira* slightly less plentiful (fig. 34).

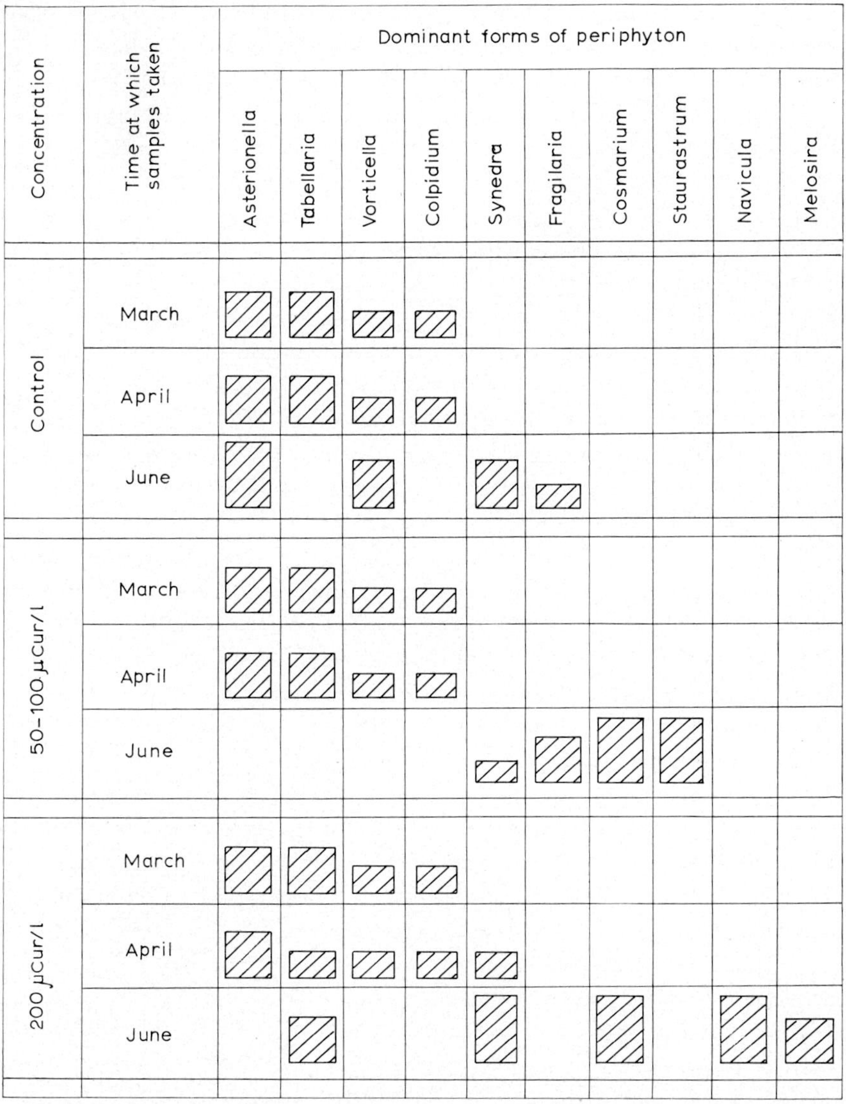

Concentration	Time at which samples taken	Dominant forms of periphyton									
		Asterionella	Tabellaria	Vorticella	Colpidium	Synedra	Fragilaria	Cosmarium	Staurastrum	Navicula	Melosira
Control	March	high	high	medium	medium						
Control	April	high	high	low	medium						
Control	June	high		high		high	low				
50–100 μCur/l	March	high	high	medium	low						
50–100 μCur/l	April	high	high	low	low						
50–100 μCur/l	June					low	medium	high	high		
200 μCur/l	March	high	high	medium	low						
200 μCur/l	April	high	low	low	low	low					
200 μCur/l	June			medium		high		high		high	low

Fig. 34. Sequence of dominant forms in periphyton biocoenoses for various fission product concentrations. Determinations approximately two, three and five months after commencement of experiments. Frequency of occurrence of dominant forms indicated by height of columns (high, medium and low) (TIMOFEYEVA-RESOVSKAYA [1958b]).

It is of interest to calculate the doses possibly received by the periphyton in these experiments. Bearing in mind that the radioactivity of the water fell to 5% as a result of accumulation by foulings, we find that, at an activity of 0.6×10^{-3} Cur/l, 2×10^{-6} Cur/cm^2 were adsorbed from the three liters on to the periphyton on walls of vessels (assuming that the fouling surface of these aquaria was 1000 cm^2), and that 10^{-8} Cur/cm^2 were adsorbed at an activity of 3×10^{-6} Cur/l. The first concentration produced a mean absorbed dose in the path of the maximum penetration of beta-particles to one cm of 40 rad/day and the second a dose of 0.2 rad/day. Doses ranging from a few thousand to a few tens of rads therefore stimulated development and reconstruction of the periphyton in the course of the experiment (five months).

DOLGOPOLSKAYA et al. [1959] investigated the fouling by marine organisms of fouling glasses with an yttrium-90 activity of 0.88×10^{-3}, 1.95×10^{-4}, 1.46×10^{-4} and 3.3×10^{-6} Cur/cm^2 placed in the sea. Between 6 August and 16 November approximately a third of the initial amount of the nuclide decayed. Within 10 days of commencement of the experiment, barnacles were discovered on the glasses. There were, however, only three small freshly settled specimens (up to one mm) in an active zone of 0.88×10^{-3} Cur/cm^2. There were more barnacles on the glass with an activity of 1.95×10^{-4} Cur/cm^2, but they were considerably smaller than those on the neutral surfaces of the glasses. Within two months the total thickness of the fouling layer on the control glasses and on inactive surfaces was 25 cm, consisting of two or three layers of barnacles surmounted by areas of bryozoans, Botryllus, calcareous worms and a scattering of small mussels. Surfaces with an activity of 0.88×10^{-3}, 1.95×10^{-4} and 1.46×10^{-4} Cur/cm^2 were quite free of fouling organisms. The pattern was the same after 102 days. The barnacles were now arranged in several layers with a base diameter of 10 to 15 mm and considerable portions of the glasses were overgrown by bryozoans, colonial ascidians and mussel colonies. A surface activity of 3.3×10^{-3} Cur/cm^2 stimulated development of organisms that settled on the surface.

Calculation of the doses demonstrates that mean absorbed doses on the active surface of the glasses at a distance of up to one cm were of the order of 1000 to 10000 rad/day for 1.46×10^{-4} and 0.88×10^{-3} Cur/cm^2, and 40 rad/day for 3.3×10^{-6} Cur/cm^2. During the experiment (98 days), the absolutely lethal doses were hundreds and tens of thousands of rads and the stimulating doses were tens of rads. It is evident that doses that accelerate and intensify the process of development are of similar order for freshwater and marine fouling biocoenoses.

The effect of X-rays on the rate of attachment of *Bugula* larvae was studied by LYNCH [1958].

KRUMHOLZ [1956] established by observations of biological processes over a number of years that there was a degeneration of fishes in White Oak Lake, into which chemical and radioactive waste from the nuclear plants at Oak Ridge were discharged. Unfortunately, the chemical contamination of the lake water makes it difficult to estimate the effect of the radioactive substances. It is, however, quite possible that radioactive substances were playing a key role in degeneration of the fishes of this lake, since the external radiation dose was 57 r/yr and the internal dose far higher. Krumholz is of the opinion that the effect of this radiation was to shorten the life span and growth rates, and probably to reduce fertility.

The lake fishes are capable of accumulating very significant amounts of a number of radionuclides. Thus, the bone of *Pomoxis nigromaculatus* and *Lepomis m. macrochirus* accumulate radiostrontium in amounts that exceed its concentration in the surrounding water 20000 to 30000 times. Radio-phosphorus is also concentrated in their bones. Cesium and the rare earths are concentrated in soft tissues of these fishes. In the summer of 1952, *P. nigromaculatus* (17.5 cm long) contained 10^{-6} Cur on the average and the level of activity was slightly lower in *L. m. macrochirus* (15 cm long).

In very detailed investigations by MARSHALL [1963] groups of 50 hatched *Daphnia* were continuously irradiated with gamma-rays at various dose rates (between 22.8 and 75.9 r/hr). It was found that a biological effect was manifested at a dose rate of 20 to 50 r/hr and above. The control (non-irradiated) population consisted predominantly of young *Daphnia*. There was a shift in favor of old individuals in the irradiated populations, especially at a dose rate of 75.9 r/hr. The life span was shortened as the dose increased. It was noteworthy that the growth of the individuals was increased as the dose rate rose. This is explained by the failure of eggs to develop in irradiated *Daphnia* and by the resulting economy of material and energy. Marshall justifiably concludes that ionizing radiation affected the population mainly by its effect on the reproductive function.

It may, therefore, be concluded from laboratory experiments and investigations under natural conditions that have been cited, that radioactive substances produce structural modifications in biocoenoses and affect their biomass, as a result of radiation effect on the most radiosensitive components of the biocoenoses, which subsequently affects the existing interrelationships. Radioresistant species that are in a 'recessive position' in a biocoenosis may thereby become dominant species. The overall effect of radioactive substances on populations and biocoenoses is, undoubtedly, steadily to impoverish natural groupings of living creatures. The existing information on this matter is, however, extremely inadequate. Considerable difficulties of method and of technique must be overcome before it will be possible to carry out extensive research on radiation biocoenology. Such research will be of great theoretical and practical significance.

CHAPTER 16

THREAT OF RADIOACTIVE CONTAMINATION
TO LIFE IN THE SEA
AND TO BIOLOGICAL PRODUCTIVITY

One of the most important tasks facing science in the nuclear age is to predict the extent of risk to aquatic organisms from radiocontamination of seas and oceans. The fate of his marine food resources is far from being a matter of indifference to man. It is already clear that mankind will make ever increasing use of the 'untouched pastures' of seas and oceans as a source from which to derive a vast quantity of foodstuffs. The biological productivity of the sea is, therefore, an aspect of general economics.

The following conclusions can be drawn from table 61, which summarizes information on maximum permissible concentrations of radionuclides in freshwater and sea-water and in marine food organisms. It is particularly noteworthy that the maximum permissible concentrations of all radionuclides adopted in the United States for the population (0.1 of the maximum permissible concentrations for nuclear energy workers) are larger by approximately an order of magnitude than those adopted in the USSR. It is also evident from the table that there are quite significant discrepancies from document to document in the maximum permissible concentrations of radionuclides (for man) in sea-water. Thus, for strontium-90 it is 2×10^{-11} Cur/l in some cases and 1×10^{-12} Cur/l in general (PRITCHARD [1960]), while according to the report of the US Committee on Oceanography and Fisheries (Anonymous [1960b]) it is 2.5×10^{-8} Cur/l (adopted in the US) and 3.3×10^{-9} Cur/l (tentative in the US), or according to a United States report on industrial radioactive waste disposal (Joint Committee on Atomic Energy, Congress of the US [1959a]) it is 5×10^{-11} Cur/l. Therefore, the differences for strontium-90 amount to four orders of magnitude. Considerable differences are also to be noted for other radionuclides: two orders of magnitude for chromium-51, zinc-65, niobium-95, and ruthenium-106, three for cesium-137 and four for iodine-131.

The following pattern is revealed by comparison of tables 2 and 61. According to Pritchard's data the maximum permissible strontium-90 concentration in sea-water is 10^{-12} Cur/l. At present this concentration has been reached in the Pacific Ocean and in the Irish Sea. In some parts of the Pacific

Table 61

Maximum permissible concentrations (MPC) for man of radionuclides in water and in marine organisms

| Nuclide | MPC in drinking water (Cur/l) | | MPC in sea-water (Cur/l) | | | | |
| | Soviet standard* | US standard** | tentative standards in US (a) | | adopted in US (b) | tentative standards in US (a) | (c) |
			local situation	general situation			
^{3}H	3×10^{-7}	3×10^{-6}	–	–	–	–	–
^{14}C	2×10^{-7}	8×10^{-7}	–	–	–	–	–
^{24}Na	8×10^{-9}	3×10^{-8}	–	–	–	–	–
^{32}P	5×10^{-9}	2×10^{-8}	–	–	9.6×10^{-11}	4.5×10^{-12}	5×10^{-1}
^{35}S	7×10^{-9}	6×10^{-8}	–	–	1×10^{-5}	1.1×10^{-8}	1.2×10
^{42}K	6×10^{-9}	1×10^{-7}	–	–	–	–	–
^{45}Ca	3×10^{-9}	9×10^{-9}	–	–	2×10^{-7}	1.2×10^{-7}	9×10^{-9}
^{51}Cr	5×10^{-7}	2×10^{-6}	2×10^{-8}	7×10^{-10}	5.4×10^{-8}	2×10^{-8}	2×10^{-8}
^{55}Fe	3×10^{-8}	8×10^{-7}	1×10^{-9}	3×10^{-11}	3×10^{-9}	1.4×10^{-9}	8×10^{-1}
^{59}Fe	1×10^{-8}	6×10^{-8}	7×10^{-11}	2×10^{-12}	6×10^{-11}	6×10^{-12}	6×10^{-1}
^{60}Co	1×10^{-8}	5×10^{-8}	3×10^{-11}	1×10^{-12}	8×10^{-10}	5×10^{-11}	5×10^{-1}
^{64}Cu	6×10^{-8}	2×10^{-7}	3×10^{-10}	1×10^{-11}	–	–	–
^{65}Zn	1×10^{-8}	1×10^{-7}	1×10^{-10}	3×10^{-12}	4×10^{-10}	7×10^{-12}	2×10^{-1}
^{90}Sr	3×10^{-11}	1×10^{-10}	2×10^{-11}	1×10^{-12}	2.5×10^{-8}	3.3×10^{-9}	5×10^{-1}
^{95}Zr	2×10^{-8}	6×10^{-7}	4×10^{-10}	1×10^{-11}	–	–	–
^{95}Nb	3×10^{-8}	1×10^{-7}	3×10^{-9}	1×10^{-10}	3×10^{-8}	5×10^{-9}	5×10^{-8}
^{106}Ru	3×10^{-9}	1×10^{-8}	1×10^{-10}	3×10^{-12}	1.6×10^{-10}	1×10^{-10}	1×10^{-1}
^{131}I	6×10^{-10}	2×10^{-9}	3×10^{-10}	1×10^{-11}	1×10^{-7}	1.6×10^{-9}	2×10^{-2}
^{137}Cs	1×10^{-9}	2×10^{-8}	4×10^{-9}	1×10^{-10}	1.6×10^{-7}	1.3×10^{-10}	4×10^{-8}
^{144}Ce	3×10^{-9}	1×10^{-8}	–	–	1.5×10^{-11}	1×10^{-11}	1×10^{-2}
^{182}Ta	1×10^{-8}	4×10^{-8}	3×10^{-10}	1×10^{-11}	–	–	–
^{192}Ir	1×10^{-8}	4×10^{-8}	–	–	–	–	–
Mixture of beta and gamma-emitters	5×10^{-11}	10^{-11}	–	–	–	–	–
Mixture of alpha-emitters	5×10^{-11}	10^{-11}	–	–	–	–	–

* Health regulations for 'Work with radioactive substances and sources of ionizing radiations' (Russian list: Anonymous [1960]).
** 0.1 of the MPC for nuclear energy workers (National Committee on Radiation Protection [1959].

	in India (e)	Concentration factors in marine organisms		MPC in edible marine organisms (Cur/kg of wet weight)				
		(b)	(a)	tentative standards in US (a)		adopted for the Irish Sea (f)	(d)	in India (e)
				local situation	general situation			
–	–	–	–	–	–	–	2×10^{-3}	–
	2×10^{-9}	–	–	–	–	–	3×10^{-5}	1×10^{-5}
$\times 10^{-4}$	–	–	–	–	–	–	8×10^{-5}	–
10^{-11}	2×10^{-12}	2×10^{5}	–	–	–	–	2×10^{-6}	3×10^{-7}
10^{-5}	9×10^{-8}	5	–	–	–	–	5×10^{-5}	5×10^{-7}
10^{-5}	–	–	–	–	–	–	1×10^{-4}	–
10^{-7}	4×10^{-8}	20	–	–	–	–	5×10^{-6}	1×10^{-7}
	4×10^{-8}	10^{3}	3×10^{2}	7×10^{-6}	2×10^{-7}	–	5×10^{-4}	3×10^{-5}
	7×10^{-10}	10^{4}	3×10^{3}	3×10^{-6}	8×10^{-8}	–	–	1×10^{-5}
10^{-9}	5×10^{-11}	10^{4}	3×10^{3}	2×10^{-7}	5×10^{-9}	–	4×10^{-5}	1×10^{-6}
10^{-8}	2×10^{-9}	10^{4}	3×10^{3}	1×10^{-7}	3×10^{-9}	–	2×10^{-4}	8×10^{-7}
$\times 10^{-7}$	–	5×10^{3}	2×10^{3}	7×10^{-7}	2×10^{-8}	–	8×10^{-4}	–
$\times 10^{-7}$	1×10^{-10}	5×10^{3}	3×10^{3}	3×10^{-7}	1×10^{-8}	–	6×10^{-4}	8×10^{-7}
10^{-10}	3×10^{-10}	20	13	3×10^{-10}	1×10^{-11}	$(1–5) \times 10^{-8}$	8×10^{-9}	2×10^{-9}
	1×10^{-8}	–	5×10^{2}	2×10^{-7}	6×10^{-9}	–	–	1×10^{-6}
	2×10^{-8}	2×10^{2}	1.2×10^{2}	3×10^{-7}	1×10^{-8}	–	–	2×10^{-6}
	2×10^{-10}	10^{3}	3×10^{2}	3×10^{-8}	1×10^{-9}	$(1–3) \times 10^{-6}$	–	2×10^{-7}
10^{-10}	4×10^{-9}	10^{2}	28	1×10^{-8}	3×10^{-10}	–	3×10^{-7}	2×10^{-7}
10^{-7}	6×10^{-9}	50	18	7×10^{-8}	2×10^{-9}	–	1.5×10^{-5}	5×10^{-8}
	1×10^{-8}	8×10^{3}	–	–	–	–	–	2×10^{-7}
	–	3×10^{2}	–	1×10^{-7}	4×10^{-9}	–	–	–
	–	–	–	–	–	–	9×10^{-6}	–
	–	–	–	–	–	$5 \times 10^{-5}–1 \times 10^{-6}$	–	–
	–	–	–	–	–	$3 \times 10^{-7}–1 \times 10^{-8}$	–	–

(a) PRITCHARD [1960].
(b) Anonymous [1960b].
(c) Committee on the Effects of Atomic Radiation on Oceanography and Fisheries [1959].
(d) Committee on Oceanography [1959b].
(e) PILLAI and GANGULY [1961].
(f) TEMPLETON [1962].

Ocean and in the Irish Sea it even reached 10^{-11} Cur/l in 1954, 1956, 1959 and 1960.

The eggs of sea fishes have been found to be most sensitive to the effect of low concentrations of strontium-90-yttrium-90 within the range 10^{-12} to 10^{-10} Cur/l. In other words, a strontium-90 concentration of 10^{-12} Cur/l is critical for the early development stages of sea fishes as well as for man. It is important to bear in mind that most of the eggs of sea fishes are of the pelagic type, and are to be found mainly at the surface, where they form a part of the hyponeuston, which is the largest biocoenosis in the world, concentrating between 40 and 90% of all phytoplankton and zooplankton, and fish eggs and larvae in the zero to five cm layer of upper one-meter horizon of the sea-water (ZAYTSEV [1961a]).

The hyponeuston occurs under conditions of the best illumination and aeration, and it may possibly be of the greatest importance that it occurs under conditions in which it is enriched by various products of mineral and organic origin carried mainly in dust form from the land by air currents. Since radioactive fallout produced by nuclear weapon tests is found initially in this biocoenosis, in which it creates locally increased concentrations of radioactive substances rather than 'mean' concentrations, and since the biocoenosis is inhabited by the most radiosensitive biological objects, it is quite evident that the hyponeuston is the most readily damaged and 'critical' biocoenosis in the world. There is, in fact, nothing to protect pelagic fish eggs, which have only a very thin case, and whose small size is combined with a relatively large surface area. These features of pelagic fish eggs make them an excellent adsorbent of the rare earths and of other rare trace elements and their radioisotopes (including yttrium-90, cerium-144 and ruthenium-106). The fish eggs are therefore subjected to increased doses of ionizing radiation, which give rise to abnormal larvae, doomed to die or be consumed.

If we are to form some impression of the possible, although as yet little examined, forms of interaction between the various components of the hyponeuston with radioactive matter in the surface layer of the sea, and of the possible biological consequences, we must first consider some of the basic features of this highly distinctive biocoenosis (fig. 35). The following information on the hyponeuston of the Black Sea has been kindly supplied by Yu. P. Zaytsev.

The biomass of the hyponeuston is, as already noted, considerably higher than that of the plankton levels in the Black Sea, although there are far more species in the entire inhabited water column of the sea than in the zero to five cm layer (table 62). At the same time many species (obligates of the hypo-

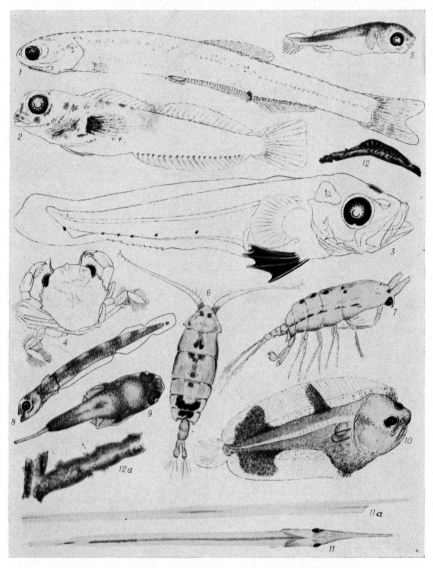

Fig. 35. Some specimens of the Black Sea hyponeuston (ZAYTSEV [1964]).

Slightly pigmental organisms with the glassy-transparent body: (1) anchovy larva (length 27 mm), (2) *Blennius* larva (length 21 mm), (3) *Trachinus* larva (length 9 mm), (4) young crab *Portunus* (length of cephalothorax 3 mm).

Organisms are colored in blue-green: (5) mullet larva *Mugil saliens* (length 5 mm), (6) *Pontella mediterranea* (length 3 mm), (7) *Anomalocera patersoni* (length 3 mm).

Organisms representing different floating objects on the sea surface by their form, color of body and by their behavior: (8) larva of *Belone belone euxini* (length 13 mm), (9) fry of *Callionymus* (length 6 mm), (10) fry of *Solea* (length 10 mm), (11) fry of *Belone belone euxini* (length 130 mm), (11a) a scrap of *Zostera marina*, (12) Isopoda *Idotea stephenseni* on a piece of a floating object, (12a) a scrap of the brown alga *Cystoseira barbata*.

Table 62

Abundance of the hyponeuston and the plankton population of the water column (ZAYTSEV [1960, 1961a, b, c, 1962])

Subject	Average for Black Sea				
	specimens/m³	mg/m³	%		
	0–5 cm	0–5 cm	0–5 cm	5–25 cm	25–45 cm
Pontellidae	61.1	–	100	1.8	1.6
Larvae:					
Decapoda	17.7	–	100	32	19
Isopoda	0.2	–	100	5	5
Anchovy eggs*	53.09	–	100	38	32
	–	–	–	–	–
Horse-mackerel, red mullet, grey mullet and bluefish eggs**	0.7 –	– –	100 –	43 –	23 –
Anchovy prelarvae and larvae	3.34	–	100	50	42
Red mullet, grey mullet and bluefish prelarvae, larvae and fry	0.2	–	100	30	20
Total hyponeuston	–	1425	100	30	36
	–	3435	100	29	28
	–	2110	100	19	12
	–	155	100	17	12

* Mean volume of egg 0.514 mm³, mean weight 0.5181 mg.
** Mean volume of egg 0.26 to 0.28 mm³, mean weight 0.2702 mg.

areas of abundant development of hyponeuston in Black Sea					Comments
▸ecimens/m³	mg/m³	%			
-5 cm	0–5 cm	0–5 cm	5–25 cm	25–45 cm	
▸200	7892	100	0.2	0.0	Almost absent below 50 cm
583	1678	100	0.0	0.0	
▸2	–	100	25	8	Almost absent below 50 cm
	–	–	–	–	
▸8	455	100	24	6	Between 1 and 6 eggs per m³
5	163	100	10	10	in each 20 cm layer between 50 cm and 20 to 25 m
▸2	64	100	25	14	0.001 to 0.002 on average in
▸	12	100	33	6	each 20 cm layer between 50 cm and 10 to 20 m; maximum 1 to 3 eggs per m³
▸1	–	100	37	17	0.1 to 1 on average in each 20 cm layer between 50 cm and 20 m; maximum 1 to 3 specimens per m³
	–	100	17	4	0.01 to 0.02 specimens per m³ on average in each 20 cm layer between 50 cm and 20 m
–	–	–	–	–	
–	–	–	–	–	
–	–	–	–	–	
–	–	–	–	–	

neuston) are found almost exclusively in this near-surface layer. These components include the early developmental stages of many plentiful and valuable food fishes (fig. 36). A period of development in the hyponeuston is a feature of a great many species of pelagic and benthic animals. Their larvae remain for several weeks or months in the uppermost zero to five cm of water. A variety of organic and mineral particles derived from the atmosphere, including insects, pollen, spores and dust particles remain long on the surface of the sea. The animals of the hyponeuston actively consume the organic fraction of these precipitates.

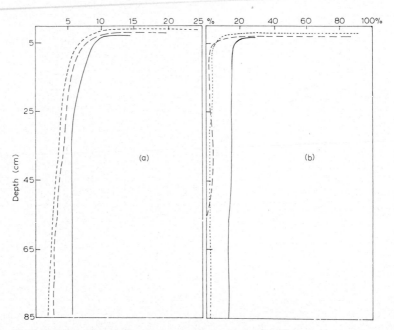

Fig. 36. Vertical distribution of organisms at the surface of the Black Sea as percentages of their total numbers in the pelagic zone. (a) eggs of mullet (dotted line), horse-mackerel (dashed line) and anchovy (continuous line); (b) Pontellidae (dotted line), megalopa stages of crabs (dashed line), and zoea stage and mysid larvae of Decapoda (continuous line) (Zaytsev and Polikarpov [1965]).

The basic structure of the biocoenosis is unaffected by wave heights of up to 2 to 2.5 m. In the summer the hyponeuston is retained in the zero to five cm layer for the period of a few weeks between storms and is immediately restored after a storm abates.

Nectobenthic and benthic animals (including shrimps, sandhoppers and polychaetes) ascend to the uppermost zero to five cm layer of water at night (1900 to 2100 hr) and descend in the morning (0500 to 0800 hr) in the shallows (out to depths of 80 to 100 m). Their appearance in this layer increases the biomass of the hyponeuston between 10 and 20 times. During the hours of darkness these migrants feed on the hyponeuston and are, in their turn, to some extent consumed by pelagic fishes. Benthic migrants are temporary, but not very characteristic components of the hyponeuston of the shelf.

Pelagic eggs and the larvae and fry of many fishes (including grey mullets, anchovy, horse-mackerel, red mullets and bluefish) form distinct aggregations in the hyponeuston.

Various organisms spend different periods of time in the hyponeuston. The larvae of bottom-dwelling animals (molluscs and crustaceans) may spend one or more weeks there, whereas Pontellidae and some Isopoda spend their whole life (many months), and for fishes (eggs, larvae, fry) the periods may be two months (anchovy, horse-mackerel, bluefish), two to three months (red mullet), or seven to eight months (mullets, including the golden and striped mullet). Nocturnal migrants from the benthos spend between eight and 10 hr daily in the hyponeuston during the warm months of the year.

Zaytsev estimates the overall mean biomass of the hyponeuston in the Black Sea at about 100 000 to 300 000 net registered tonnage. The hyponeuston is, of course, not uniformly distributed in the sea, but occurs as large patches, mainly where currents converge. Qualitative indices are not the same in these patches as the mean for the water (0.2 to 3.4 g/m^3, table 62), but may reach several tens of grams per cubic meter (ZAYTSEV [1960, 1961a,b,c,1962]). The first point to emerge from the information that has been given is an explanation for the distinctive 'barrier' role of the hyponeuston in the sea in relation to radioactive substances with high concentration factors. This barrier role has previously been noted. In accordance with the data in table 42 the aquatic organisms of the hyponeuston extract several % of the radioactivity when the biomass of the hyponeuston is 10 g/m^3 or above and the concentration factors of radionuclides in the inhabitants of the surface layer of water are of the order of 1000 or above. This proportion of radioactivity extracted by the organisms may remain in the hyponeuston for a long time.

To make quantitative estimates one clearly needs to determine the concentration factors of the most important radioactive substances in the most plentiful forms of the hyponeuston.

It should also be borne in mind that transfer of radionuclides from the hyponeuston to the benthos and *vice versa* is accelerated on the shelf by the daily migrations of benthic organisms temporarily found in the hyponeuston. This circumstance should be specially considered in relation to the dumping of nuclear waste in coastal waters. It applies to all instances of discharge of radioactive effluents in shallows (out to depths of 80 to 100 m). Almost the whole of the Baltic Sea is affected, as depths do not in the main fall below 50 or 100 m.

Nothing is as yet known concerning the biological effect of radioactive substances on structure and biomass of the hyponeuston. Nevertheless, the fact that the hyponeuston invariably contains the eggs, larvae and fry of many valuable food fishes, whose early development stages are highly radiosensitive, is indication enough. If the proportion of abnormal fish larvae is increased as the concentration of radioactive substances in sea-water rises, reproduction of fish stocks may be adversely affected and the structure and biomass of the hyponeuston may be modified. Marine radioecologists have the greatest possible interest in the most rapid development of a wide-ranging research program on all aspects of the hyponeuston.

It may be envisaged in a most general form that a whole series of fishes and other radiosensitive animals may die out, and that there may be abundant proliferation of bacteria, microphytes and radioresistant lower invertebrates under conditions of lowered consumption by fishes and as a result of intricate biocoenotic reconstructions in the hyponeuston.

The general pattern of the far reaching consequences that may ensue from modification of the hyponeuston by radiation may be depicted as follows: on the one hand, inhibition and degeneration of a number of food fishes and other radiosensitive organisms and, on the other hand, rapid development and multiplication of bacteria, microphytes and radioresistant forms of invertebrates.

It is far more difficult to foresee the specific forms and quantitative aspects of these consequences, the more so because 1) we still do not know the 'maximum permissible' concentrations of the various radionuclides in sea-water for the normal activity and reproduction of marine organisms, and especially fishes, under natural conditions and 2) there is still a lack of information on the range of radiation injury to fishes, i.e., on the proportions of eggs, larvae and fry that die and that survive to reach maturity (sterile or capable of reproduction), and on the nature and rate of radiogenetic degeneration of sea fishes.

Nevertheless, the available information indicates, on the one hand, that the developing eggs of sea fishes are extremely radiosensitive and, on the other hand, that the strontium-90-yttrium-90 concentrations needed to increase the rate of occurrence of abnormal fish larvae are very close to the concentrations of this radionuclide in the Pacific Ocean and Irish Sea. It is this circumstance, which indicates a possibly critical situation in the hydrosphere, that prompted us to start work on the possible consequences of radiation damage to the hyponeuston (ZAYTSEV and POLIKARPOV [1964]).

It is selfevident that a reduction of fish stocks, other conditions being equal, should be the result of the chronic effect of some harmful factor operative throughout a sea or a large expanse of an ocean, i.e., practically throughout the hydrosphere. Global radioactive contamination is a factor of this nature.

The reduction of stocks of a given species of fish under the influence of such a factor on eggs in the hyponeuston may, in a first approximation, be represented by an exponential function.

Table 63 gives information on the periods needed to halve the stocks of various species of Black Sea fishes in relation to the proportion of the eggs in the hyponeuston affected by radiation damage (between five and 50%).

Table 63

Dependence of the time needed for halving of fish stocks (T) on the proportion of damaged eggs in the hyponeuston (figures adjusted to the nearest whole number) (ZAYTSEV and POLIKARPOV [1964])

Percentage of eggs damaged in hyponeuston	T (years)		
	mullet	horse-mackerel	anchovy
5	55	68	100
10	29	34	50
20	15	18	25
30	11	13	17
40	8	10	13
50	7	8	10

There are grounds for the assumption that fisheries will be adversely affected and may in many instances cease to function if fish stocks are halved or more than halved. It is evident that, from the commercial standpoint, radiation damage must not affect more than 10% of the eggs in the hyponeuston, however catches will be perceptibly reduced (T is between 29 and 50 yr).

The features described relate to a chronic radiation effect of sustained intensity. Changes in the content of radioactive substances in sea-water should also affect the proportion of damaged eggs in the hyponeuston. In other words, fluctuations of fish stocks due to a radioecological factor of variable intensity may be asynchronously superimposed on the ordinary fluctuations.

It emerges from all that has been said that the hyponeuston is not only the most extensive biocoenosis in the world, but that it is also, probably, the one most sensitive to radioactive contamination.

The considerable discrepancies (up to 10000 times) between the permissible concentrations given by various authors for radionuclides for sea-water and the coincidence (or near coincidence) between the lowest of these concentrations and those already existing in sea-water prompt the conclusion that it would be very risky to further increase the radiocontamination of the seas, especially bearing in mind that the level at which strontium-90-yttrium-90 begin to have a perceptible harmful effect on development of the free eggs of sea fishes is within the range 10^{-12} to 10^{-10} Cur/l.

When working in conjunction with L. P. Salekhova, the author established that, like pelagic eggs, demersal eggs, which are very small and have a very thin case, are also radiosensitive. Concentration of radioactive substances by the aquatic organisms of benthic biocoenoses increases radiation dose rates on their surfaces, which often serve as the substrate for attachment of demersal eggs. Specific conditions therefore create additional doses for the developing stages of animals in benthic communities. Such situations should be the subject of special investigations, since it is possible that microphytes and other organisms may contribute a significant part of the total dose received by fish eggs. The direct connection between the hyponeuston and benthos on the shelf, which arises from the mutual periodic migrations of their inhabitants, and increases the transport of radioactive substances from one biocoenosis to the other, should also be taken into consideration as one way in which radiation doses particularly affecting demersal fish eggs may be increased.

In other words, a knowledge of the 'mean' concentrations of radionuclides in sea-water is quite inadequate for evaluation of their biological effects. Whatever the mean concentration, the situation may, in relation to the source of radioactive contamination, be quite different in the hyponeuston and benthos and actual concentrations may be many orders of magnitude higher than the mean concentration.

It should, moreover, be emphasized that the maximum permissible concentrations of radionuclides in sea-water given in table 61 were calculated in relation to transfer to the human organism. Investigations on maximum permissible concentrations for marine organisms as such are as yet in their infancy, but must be expanded as a matter of urgency.

Local contamination of marine food resources may reach quite significant levels as a result of nuclear weapon tests, incidents at nuclear plants on the coast and in vessels and discharge of highly active waste products into the sea. It is not, however, a part of our present task to consider specific instances of radioactive contamination in relation to local or global causes. The significance of such investigations is such that they should form the subject of a special treatise.

We shall therefore merely refer to the publication of HIYAMA [1960], which contains calculations of the amounts of certain radionuclides consumed by man in sea foods. Radionuclides may also find their way indirectly into the human organism from the sea when cattle and poultry are fed on crushed radioactive molluscs, bone meal and algae, and when fields are fertilized with the processed remains of marine organisms.

Considerable interest attaches to comparison of the hazard to man occasioned by feeding on radiocontaminated marine organisms and the hazard to the vital activity of the marine organisms themselves. This can be defined by the expression

$$x = \frac{C_{a_2}}{C_{b_2}} \frac{K}{M},$$

where $M = M_a/M_b$, $K = K_a/K_b$, C_{a_2} and C_{b_2} are the concentrations of radionuclides a and b in the sea-water, M_a and M_b are the maximum permissible concentrations of radionuclides a and b in the water (National Committee on Radiation Protection [1959]), K_a and K_b are the concentration factors of the nuclides a and b.

The greatest hazard to marine organisms will clearly be created by the radionuclides with the highest concentration factors, including cerium-144, yttrium-91 and yttrium-90 (when there is a constant source of supply). Radioactive substances with moderate and low concentration factors, such as strontium-90, will be less hazardous for them. Strontium-90 is, however, one of the most hazardous radionuclides for man. Unlike the radioisotopes of the rare earths and certain others, it is readily and effectively absorbed in the human intestines (SEMENOV and TREGUBENKO [1957], ZAKUTINSKIY [1959]). Thus, the hazard of strontium-90 relative to that of cerium-144 in edible aquatic organisms is 1:10 for man when C_{a_2} and C_{b_2} are equal. Therefore, despite the low concentration factor of strontium-90, it is no less hazardous to the human organism when contaminated food of marine origin is regularly consumed than is cerium-144, if there are equal concentrations of both radionuclides in the water. Under these conditions strontium-90 is also more hazardous to man than, for example, cesium-137 (between three and 30 times more hazardous).

Intensive biological migration of radionuclides from water to land is shown, for example, by the very high concentration factors of phosphorus-32, which is an important component of neutron-induced radioactive substances, in gulls, swallows, ducks and geese, for which the concentration factors are between 25000 and 500000, and in the yolk of duck and geese eggs (up to 1500000 (table 26)), and the considerable concentration factors of strontium-90, a most important fission product of heavy nuclei, in the bones of domestic ducks (420), and in their muscles and internal organs (50 and 85, respectively) (table 24). When duck carcasses were used in the preparation of various types of soup approximately 10% of the strontium-90 passed into the stock, but the figure was 50% when fish bones were used and the pH of the medium was the same (less than five). The strontium-90 content of the liquid stock was therefore increased between three-fold and twelve-fold owing to the higher strontium-90 content in bones than in muscles (approximately 50 times higher).

All the strontium-90 in fish bones is consumed in tinned fish. In this instance the fish bones are a far greater hazard than the soft edible tissues (MAREY et al. [1958]).

To estimate the radiation hazard of consuming radioactively contaminated sea food one must make a study of 1) the uptake of various radionuclides by the human organism from all food sources, 2) their proportions and systematic occurrence in the food and 3) their assimilability. A comparison must then be made with the maximum permissible concentrations of radionuclides in food products. These matters pass outside the sphere of radioecology into that of radiation hygiene.

SUMMARY OF PART III

The radiobiological importance of the concentration factor is that it can, in general, be used for conversion from doses in the environment to doses in the aquatic organism (or organs and tissues) and *vice versa*. It is evident that, owing to the high absorption of nuclear particles in water, radioactive substances incorporated in the tissues and organs of aquatic organisms are a source of ionizing radiation. In calculating the doses absorbed by microscopic creatures consideration should be given to their diameter, the track length of the particles and the concentration factors of the radionuclide. Benthic microorganisms and demersal fish eggs experience additional irradiation in the 'hot zone' on the surface of macrophytes which concentrate radioactive substances.

The most radioresistant forms are unicellular organisms and macrophytes, and to a lesser extent invertebrates. Fishes and amphibians are far more radiosensitive. The early developmental stages of the eggs of fishes and amphibians are the least radioresistant. Thus, the LD_{50} for the zygotes of salmon during mitosis is 16 r, and for the early gastrula of the axolotl is 10 r. A dose of five r was found to have a statistically significant damaging effect on newt zygotes.

Analysis of available material on early and long-term reactions of radiation injury in aquatic organisms following single external irradiation with penetrating radiation demonstrates that, in general, these reactions follow the same kinetic pattern whatever the evolutionary position of the organisms, i.e., between unicellular organisms and vertebrates.

Despite several detailed special studies, the biological role of the cosmic background remains a matter of dispute. The largest cosmic radiation doses are received by the hyponeuston, and the largest doses from natural radioactive substances are received by the benthos on the sea floor (sometimes between 10 and 100 times greater than the cosmic background). Marine plankton organisms may receive absorbed doses of up to 280 rad/yr (2800 rem/yr) from incorporated natural alpha-emitters.

The radiosensitivity of aquatic organisms in solutions of radioactive substances follows the sequence established for single external irradiation. It is, however, highly probable that biologically effective radiation doses from radionuclides incorporated in a biological substrate are far lower than those for external irradiation of short duration. Academician Spitsyn is of the opinion that it is impossible to compare the effect of external and internal irradiation.

Studies of the effect of incorporated radioactive substances on development of eggs of sea fishes are as yet in their infancy. It would appear from existing data that the frequency of occurrence of abnormalities in larvae and fry is increased when developing eggs are exposed by strontium-90-yttrium-90 concentrations of 10^{-12} to 10^{-10} Cur/l or above. Other radionuclides, or their daughter products, with high concentration factors in eggs have a similar harmful effect.

Owing to the great technical difficulties involved, we have almost no information at all concerning the reactions of communities of aquatic organisms to the effect of radioactive substances.

Differences between the maximum permissible concentrations of radionuclides in sea-water laid down in various books and manuals are between two and four orders of magnitude. For example, the maximum permissible strontium-90 concentration in sea-water (for man) appears as between 10^{-12} and 10^{-8} Cur/l. Such significant differences introduce a great deal of uncertainty.

The maximum permissible concentrations for marine organisms remain unknown, and only a few initial attempts have been made to determine them. Thus, strontium-90-yttrium-90 has a harmful effect on the developing eggs of sea fishes at concentrations of 10^{-12} to 10^{-10} Cur/l and above.

There is an urgent need for extensive research on this matter because of the concentration of life (up to 40 to 90% of all plankton, and fish eggs and larvae) in the uppermost zero to five cm of sea-water. In other words, the hyponeuston, which is the largest biocoenosis in the world, is most probably the most vulnerable to radioactive fallout, which probably produces concentrations of radioactive matter in the surface layer far higher than those throughout the water column. Damage to the hyponeuston by radionuclides threatens to disrupt reproduction of stocks of sea fishes, i.e., to disrupt the commercial productivity of seas and oceans.

It may be concluded from our estimate of the comparative hazard of various radionuclides by human consumption of radioactive marine organisms, and the hazard to the organisms themselves, that, for example, when strontium-90, the radioisotopes of the rare earths and cesium-137 are present in equal concentrations in sea-water, strontium-90 presents the greatest risk to man despite its low concentration factors in aquatic organisms, but that the radioisotopes of the rare earths present the greatest hazard to aquatic organisms, because of their high concentration factors and, therefore, their high radioactivity.

Therefore, although we are still in the early stages of investigation of the effects of radioactive substances on marine organisms, populations and biocoenoses, we are already in a position to draw the important conclusion that *further radioactive contamination of sea-water is inadmissible.*

PART IV

Conclusions

CHAPTER 17

MARINE RADIOECOLOGY;
ACHIEVEMENTS AND FUTURE PERSPECTIVES

We have attempted to establish some of the basic features of the interaction between a radioactive environment and marine organisms, without considering matters concerned with specific situations in given waters for a given source of radioactive contamination. It is the author's opinion that such specific problems may only be solved from a knowledge of general radioecological features. It has been shown that, unless we have this knowledge, quite unfounded conclusions may be drawn (concerning, for example, the distribution of radionuclides between sea-water and the plankton), or that quite different matters (for example, the role of food links in radiocontaminated and relatively pure water as a source of radioactive substances for aquatic organisms) may be confused.

If the simple empirical approach is replaced by rigorous analysis, it should be possible to isolate improbable data conditioned by extraneous factors, and to avoid incorrect conclusions and impractical recommendations, as well as to anticipate and predict developments under certain circumstances. Thus, as already noted, a contradiction has developed between the theoretically established and verified facts and the highly empirical view advanced by some investigators of the existence of an inverse relationship between the concentration factors of radioactive substances in aquatic organisms and their concentration (in terms of radioactivity) in the water, without allowance for the concentration of stable carriers. For example V. M. Zhogova (ZHOGOVA [1962]) obtained a concentration factor of 600 million for strontium-90 in freshwater bacteria. Considering that freshwater contains approximately 0.1 mg/l of stable strontium (VOYNAR [1960], VINOGRADOV [1957b]), we find that one kg of bacteria should contain 10^{-7} kg/l $\times 0.6 \times 10^9$ or approximately 60 kg of stable strontium. Even if stable strontium concentration was far lower in the water in which the experiments were carried out (ZHOGOVA [1962]) than the mean content of this element*, a strontium concentration factor of this order is not realistic.

---

* It is possible that the freshwater in which the experiments were conducted contained almost no stable strontium, but even so Zhogova's strontium-90 concentration factor is unrealistic. (Russian Editor)

Development of theoretical principles for investigations of the radio-ecology of aquatic organisms and radiation hygiene of waters is therefore a task of some urgency.

To what extent may it be considered that this task has been completed? The conditions that influence or fail perceptibly to influence concentration factors have now been established in general form. Given ecological and hydrological information, it is possible to use the concentration factors of radioactive substances in marine organisms that have been established to estimate the part played by aquatic organisms in the migration of certain substances in the sea and to calculate the doses of ionizing radiation absorbed by organisms on the basis of radionuclide concentration in the water, or to find the concentrations in sea-water from the concentrations in marine organisms. There has been sufficient research on the biological effect of ionizing radiation to permit a rough prediction of the reaction of various groups of marine organisms to the chronic effect of radioactive substances of a given concentration in water.

Nevertheless, some of the achievements in marine radioecology merely serve to underline the problems that still have to be solved. It is obvious that there is still inadequate information concerning the concentration factors of a number of radionuclides (or stable chemical elements) in marine organisms. This applies in particular to the radioisotopes of ruthenium, zirconium, niobium and plutonium. Existing concentration factors must also be verified and where necessary corrected. More precise information is needed on the significance of detritus formation in the general migration of radioactive substances. Little has been published concerning the radiosensitivity of marine plants and, especially, of marine animals. There is hardly any information concerning the effect of radionuclides on marine populations and biocoenoses.

In addition to these tasks, which are the province of radioecology, radioecologists also have considerable interest in rapid and extensive development of hydrobiological studies of the hyponeuston, benthos and schools of fishes, since the lack of this information holds back progress in a number of branches of radioecology.

It is obvious that there is an intimate connection between marine radioecology and hydrologic and oceanographic data in relation to the disposal of radioactive wastes, especially on the sea floor. Is there, in fact, any likelihood that radioactive waste that has not completely decayed may ascend to the surface, where it will be immediately involved in intensive biological circulation and begin to exert a biological effect? It was previously held that the period of complete vertical mixing in seas and oceans was extremely prolonged (of the order of many millennia). Thus artificial radioactive substances should decay completely within this period. Recent studies of hydrological conditions in the area of the deepest oceanic trenches around the Philippines (BOGOROV and KREPS [1958], BOGOROV and TAREYEV [1960], KREPS [1959]) and of the hydrology of the Black Sea (VODYANITSKIY [1958], BOGDANOVA [1959], SKOPINTSEV [1959]) have, however, yielded new information of rapid vertical exchange between bottom and surface waters in the Pacific Ocean and in the Black Sea. Thus, the time taken by bottom waters in the Black Sea to ascend to the surface is not 2500 to 5000 yr as previously assumed, but only 60 to 130 yr. During this time the radioactivity of cesium-137 and strontium-90 is reduced only to one-fifth to one-thirtieth.

Attention should also be paid to currents which may bring radioactive substances to the surface in a far shorter period. It has been demonstrated in chapter 2 that strontium is not adsorbed by marine sediments, especially at a $pH < 7$ characteristic of the depths of the Black Sea (owing to the presence of hydrogen sulfide). It follows therefore that strontium-90 may remain in sea-water for a very long period, until complete decay, and be gradually concentrated by marine organisms, including those of food value.

It is also significant that, in addition to their transport by water masses, radioactive substances may migrate apparently in an adsorbed state on marine microorganisms. This, of course, is applicable to those radionuclides which have high concentration factors in microorganisms.

In relation to the increasing practice of dumping radioactive effluents into the surface layers of the Irish Sea, the Mediterranean and other seas, and in the Pacific Ocean, it has already been noted in the reports of the United States National Academy of Sciences (Anonymous [1960b]) that many specialists from the USSR and some European countries have protested against any such dumping (YEMEL'YANOV [1962], KOLYCHEV [1961], SPITSYN and KOLYCHEV [1960], FONTAINE [1956,1959], ALIVERTI [1960], FAGE [1960]). Nevertheless, increasing amounts of radioactive material are being dumped in this manner by the nuclear enterprises of a number of Western European countries and by the United States, but not by the Soviet Union (YEMEL'-YANOV [1962], SEREDA [1962]).

The importance attached by the nuclear powers to the problem of the effect of ionizing radiations on biological processes in the hydrosphere may be illustrated by the examination of this problem in the United States. Great attention is paid in the United States program to biological aspects of artificial radioactivity in sea-water (Committee on Oceanography [1959a]).

The planned cost of the research is \$1 138 000 a year, which is approximately 20% of total annual expenditure of the program which includes research on safety precautions and physical and chemical aspects.

Marine radioecological research will, therefore, undoubtedly increase in importance.

Radioecology is, in truth, a child of and an integral part of the nuclear age. Development of all aspects of radioecology should not lag behind progress in the production and use of this form of energy, but should outstrip it, if it is to fulfil its tasks of creating the scientific foundation for the predictions and recommendations that will be needed. Its scope extends to all sources of radioactive contamination of the biosphere as a whole and of its constituent parts, i.e., sources that arise from the 'normal' functioning of nuclear plants and nuclear powered ships, as well as from accidents to them and, of course, from nuclear weapon tests.

Radioecologists are now being trained rapidly in many countries and scientific institutions concerned with radioecology are being established. ODUM [1959] has correctly stated that no nuclear installation or project will be able to manage without the services of a radioecologist in the not-too-distant future, since such a specialist will be the only person capable of predicting the radiological consequences of radioactive contamination of the environment.

As much information as possible concerning real and potential biological changes in the biosphere, as a result of the escape of radioactive substances from man's conscious control, must be accumulated as rapidly as possible. Much, and perhaps everything may depend upon the rapidity with which conclusions are reached and on their correctness. For, only one mistake can be made when the safety of life on earth is at stake.

Decontamination is possible only in small areas of the biosphere and not in the biosphere as a whole. It is also impossible to saturate the organic kingdom with protective substances.

Quickening of the process of mutation and intensification of carcinogenesis, reconstruction of existing biocoenoses and disturbance to secular biogeochemical cycles may prove to be irreversible especially for highly organized creatures which are also the most radiosensitive.

The situation with regard to the radiation hazard to life in the sea is made all the worse by the fact that marine organisms have received far smaller doses of ionizing radiation from cosmic rays and natural radioactive substances throughout geological history than have terrestrial organisms. It is therefore possible that the evolutionary process of the inhabitants of the sea

may be modified by the effect of additional doses of radiation. It is of fundamental significance that the ontogeny of many marine forms begins with a
free, unprotected ovum, readily capable of adsorbing considerable quantities
of a number of radionuclides, and therefore readily affected by increased
biologically effective doses of ionizing radiation.

The information set out in the present book demonstrates unequivocally
that marine radioecology is, as yet, in its infancy. Many problems must be
solved rapidly, not merely out of idle scientific curiosity, but because their
solution is imperatively demanded by the nuclear age. If radioecology is to
live up to its motto of 'ensuring the radiation safety of the biosphere' we
must create specialized radioecological laboratories, departments, stations
and institutes, including those specializing in marine radioecology, and must
draw on young scientific workers both from the new sciences of radiobiology,
biophysics, radiation chemistry, nuclear physics and radioelectronics, and
from the classical disciplines of ecology, embryology, biochemistry, analytical
chemistry, spectroscopy and mathematics.

CHAPTER 18

GENERAL CONCLUSIONS

The following general conclusions may be drawn from our analysis of the state of research on marine radioecology.

1. Appearance in the hydrosphere of artificial radioactive substances, a new and biologically unusually effective environmental factor, necessitated research on the effect of radionuclides on marine organisms and on the part played by aquatic organisms in the migration of radionuclides in seas and oceans. These problems are the province of marine radioecology, a branch of radioecology (ecological radiobiology).

2. It is the task of marine radioecologists to study the features of the interaction between a radioactive environment and marine organisms and thereby to evolve the scientific principles from which to predict the nature, level, rate of radiocontamination of marine organisms, extent of the hazard to life in the sea and to its biological productivity, and to man's consumption of radioactive sea food whose levels of radioactivity derived from various concentrations of radioactive substances in sea-water. A further task is to develop measures to prevent the harmful effects of radiation.

3. The capacity of aquatic organisms to concentrate radionuclides measured by the concentration factor, defined as the ratio of the concentration of a radionuclide (or stable chemical element) in the organism (live weight) and in the aqueous medium. Other conditions being equal, the concentration factors of a radionuclide are not dependent on the radioactivity level in water (in particular, when the concentration of the corresponding stable carrier is not affected, radioactive and stable isotopes of the same element are found in the same chemical form, there is isotopic exchange between them and no radiation injury to aquatic organisms) and remain constant in the same or related species from different seas (with the exception of the Baltic Sea, which has less salinity). The stability of concentration factors is favored by the stable pH of sea-water. Concentration is affected by physical and chemical states of radioactive substances in solution, water temperature, illumination and certain other factors.

4. Almost every radionuclide (or stable chemical element) is associated

258

with several species of aquatic organisms that function as biological concentrators (biological indicators). All radioactive substances can be divided
into groups on the basis of concentration factors that have been established
for marine organisms, namely, very high concentration factors (tens of
thousands), high (hundreds or thousands), medium (tens), low (units) and
very low (less than unity). Similarly, various groups of marine organisms
exhibit differing capacities to concentrate a given radionuclide.

5. We know that on the land radioactive substances enter organisms
mainly via food chain. By contrast, when radionuclides are present in an
aqueous medium the main way in which most are accumulated by aquatic
organisms is by direct absorption from the water. The transfer of radioactive
substances along food chains may be of importance in waters only if these
substances are absent from the environment after, for example, the migration
of radioactive aquatic organisms into 'pure' waters.

6. At any given moment aquatic organisms account for only an extremely
insignificant portion of radioactive substances in the sea when the biomass is
normal, and even when it is high. The maximum possible biomass of oceanic
plankton (10 g/m^3) could not extract more than 1% of a radionuclide (with a
concentration factor of $1\,000$) from the water. Nevertheless, in view of the
rapid and continuous succession of generations, and of intensive reproduction and migrations, plankton organisms act as a pump, and continually
transfer many radioactive substances to deeper layers of water or to the sea
floor. When the concentration factor of a radionuclide is $1\,000$ and the
mixing coefficient is 0.1, biological transfer of radioactivity downwards
vertically is 10 times hydrological transfer.

7. Excess strontium-90 accumulated during life is returned to the environment from dead marine plants in the process of detritus formation. In this
respect strontium-90 differs from the radioisotopes of the rare earths, and
apparently from other nuclides. Strontium-90 is, therefore, not only more
hazardous to man than other radionuclides in food, but is also more mobile
in the sea, because it remains in the water for a long time and repeatedly con-
taminates organisms.

8. The features of radiation injury to aquatic organisms exhibit the general
radiobiological pattern, whatever the taxonomic position of the organism.
This applies to the whole range of marine life from unicellular organisms to
vertebrates.

9. The fishes are the most radiosensitive forms of marine life, and the algae
are the most radioresistant. Aquatic invertebrates occupy an intermediate
position. The developing eggs of fishes are sensitive particularly to the effect

of external irradiation and incorporated nuclides. In the early stages of development fish eggs are affected by an external radiation dose of approximately 16 r. The frequency of occurrence of abnormalities in the larvae of sea fishes has been shown to be increased at strontium-90-yttrium-90 concentrations of 10^{-12} to 10^{-10} Cur/l and above, i.e., at levels that have already been reached in the Pacific Ocean and in the Irish Sea.

10. There are considerable discrepancies between the data given in the literature for the maximum permissible concentrations of various radioactive substances in sea-water for man feeding on radioactive marine organisms. The difference may be as much as from 100 to 10000 times for the same radionuclides. Thus, the maximum permissible strontium-90 concentrations in sea-water that have been proposed range from 10^{-12} to 10^{-8} Cur/l. There is hardly any information concerning maximum permissible concentrations of radioactive substances in sea-water for individual aquatic organisms and associations. It has already been noted that, allowing for a safety factor, the maximum permissible strontium-90 concentration for developing fish eggs should be less than or around 10^{-12} to 10^{-10} Cur/l.

11. The greatest concentration of marine life, including the pelagic eggs of sea fishes, is to be found in the uppermost layer (zero to five cm) of the ocean (the hyponeuston); therefore, organisms in this layer are exposed to the harmful effects of high levels of radioactive fallout. Radioactive damage to the hyponeuston might, above all, disrupt natural replacement of fish stocks.

12. The comparative hazard of strontium-90, cesium-137, and radioisotopes of the rare earths is different for man's consumption of sea food and for sea organisms themselves because of the different ways in which these nuclides are concentrated and, in the last analysis, because of the different absorbed tissue doses of radiation. At equal concentrations in sea-water, and allowing for the concentration factors, strontium-90 represents the greatest hazard to man, and the radioisotopes of the rare earths the greatest hazard to aquatic organisms.

13. It has been shown that further radioactive contamination of the seas and oceans is inadmissible, because it entails great risk of a) producing irreversible changes in the hydrobiosphere, b) disrupting the resources upon which the fisheries depend and c) producing dangerous levels of contamination in the marine organisms consumed by man. To avoid these radiation consequences, it is essential to end all nuclear weapon tests and the dumping of liquid and solid radioactive waste into the seas and oceans.

14. In the nuclear age there is a need for rapid and thorough development of marine radioecological research, since available information is inadequate

for establishing policy recommendations or framing of measures to avert and remove local radiation hazards which may arise in the operation of nuclear-powered ships and nuclear reactors on the coast, or when considerable quantities of radionuclides are employed in large-scale experiments at sea. While the risk of nuclear war employing rockets remains, there is also the task of foreseeing the radioecological consequences in the hydrosphere in case of the military use of nuclear weapons.

As a product of the nuclear age, marine radioecology, therefore, takes a direct part in the noble cause of ensuring the radiation safety of the hydro-biosphere.

Appendices

SUPPLEMENTARY TABLES

Table A.1

Mean composition of the natural radioactivity of sea-water (Popov [1964])

adionuclide	Element	Half-life	Concentration (g/l)	Activity (dis/min · l)		Disintegration energy (cal/day · l)
)K	K*	1.36×10^9 yr	4.5×10^{-5}	660β	90γ	5.6×10^{-8}
⁷Rb	Rb*	6.6×10^{10} yr	5.6×10^{-5}	10β		1.5×10^{-10}
⁵In	In*	6.0×10^{14} yr	2×10^{-5}	$3 \times 10^{-4} \beta$		8×10^{-16}
³Ca	Ca*	2×10^{16} yr	9×10^{-4}	$7 \times 10^{-4} \beta$		–
³La	La*	7×10^{16} yr	3×10^{-10}	$10^{-6} \beta$	$2 \times 10^{-5} \gamma$	2×10^{-16}
⁴Sn	Sn*	1.5×10^{17} yr	2×10^{-7}	$10^{-11} \beta$		–

Radioactive families

a) *uranium*

³UI	U*	4.5×10^9 yr	3×10^{-6}	2α	0.4γ	⎫
↓						
⁴UX₁	Th	24.3 days	4×10^{-17}	2β	0.3γ	
↓						
⁴UX₂	Pa	1.15 min	1×10^{-21}	2β	0.1γ	1×10^{-9}
↓						
⁴UZ	Pa	6.7 hr	1×10^{-20}	0.06β	0.006γ	
↓						
⁴UII	U	2.5×10^5 yr	2×10^{-10}	2α	0.5γ	⎭
↓						
)Io*	Th	8.1×10^4 yr	1×10^{-12}	0.04α	0.001γ	1×10^{-11}
↓						
³Ra*	Ra	1620 yr	8×10^{-14}	0.2α		⎫
↓						
²Rn*	Em*	3.825 days	5×10^{-19}	0.2α		4×10^{-20}
↓						
³RaA	Po	3.05 min	3×10^{-22}	0.2α	$6 \times 10^{-5} \beta$	
↓↘						
²¹⁸At	At	2 sec	1×10^{-25}	$5 \times 10^{-5} \alpha$		⎭

Table A.1 (continued)

Radionuclide	Element	Half-life	Concentration (g/l)	Activity (dis/min · l)	Disintegration energy (cal/day · l)
^{214}RaB ↓	Pb	26.8 min	3×10^{-21}	$0.2\,\beta$ $0.2\,\gamma$	
^{214}RaC ↓↘	Bi	19.7 min	2×10^{-21}	$0.2\,\beta$ $0.2\,\gamma$ $7 \times 10^{-5}\,\alpha$	
^{210}RaC″	Tl	1.4 min	5×10^{-26}	$7 \times 10^{-5}\,\beta$	
^{214}RaC′ ↓	Po	1.64×10^{-4} sec	3×10^{-28}	$0.2\,\alpha$	4×10^{-20}
^{210}RaD ↓	Pb	22 yr	1×10^{-15}	$0.2\,\beta$ $0.2\,\gamma$	
^{210}RaE ↓	Bi	5 days	6×10^{-19}	$0.2\,\beta$	
^{210}Po	Po	139 days	2×10^{-17}	$0.2\,\alpha$ $0.2\,\gamma$	
^{129}I	I	1.7×10^7 yr	2×10^{-18}	$10^{-9}\,\beta$ $10^{-9}\,\gamma$	10^{-20}

b) *actinium*

^{235}AcU ↓	U*	7.1×10^8 yr	2×10^{-8}	$0.1\,\alpha$ $0.1\,\gamma$	2×10^{-11}
^{231}UY ↓	Th	25.5 hr	8×10^{-20}	$0.1\,\beta$ $0.1\,\gamma$	
^{231}Pa* ↓	Pa	3.3×10^4 yr	5×10^{-14}	$0.005\,\alpha$ $0.005\,\gamma$	
^{227}Ac ↓↘	Ac	22 yr	4×10^{-17}	$0.005\,\beta$ $5 \times 10^{-5}\,\alpha$	
^{223}AcK	Fr	21 min	6×10^{-25}	$5 \times 10^{-5}\,\beta$	
^{227}RdAc* ↓	Th	18.7 days	8×10^{-20}	$0.005\,\alpha$ $0.005\,\gamma$	
^{223}AcX ↓	Ra	11.2 days	5×10^{-20}	$0.005\,\alpha$ $0.005\,\gamma$	
^{219}An ↓	Em*	3.92 sec	2×10^{-25}	$0.005\,\alpha$ $0.002\,\gamma$	1×10^{-11}
^{215}AcA ↓	Po	1.83×10^{-3} sec	1×10^{-27}	$0.005\,\alpha$	
^{211}AcB ↓	Pb	36 min	1×10^{-22}	$0.005\,\beta$ $0.002\,\gamma$	
^{211}AcC ↓↘	Bi	2.16 min	6×10^{-24}	$0.005\,\alpha$ $2 \times 10^{-5}\,\beta$	
^{211}AcC′	Po	0.52 sec	8×10^{-29}	$2 \times 10^{-5}\,\alpha$	
^{207}AcC″	Tl	4.79 min	1×10^{-23}	$0.005\,\beta$ $0.005\,\gamma$	

Table A.1 (continued)

Radionuclide	Element	Half-life	Concentration (g/l)	Activity (dis/min · l)	Disintegration energy (cal/day · l)
			c) *thorium*		
^{232}Th*	Th	1.4×10^{10} yr	5×10^{-8}	$0.01\,\alpha$ $0.003\,\gamma$	
↓					
^{228}MsTh$_1$	Ra	6.7 yr	2×10^{-17}	$0.01\,\beta$	
↓					
^{228}MsTh$_2$	Ac	6.13 yr	2×10^{-21}	$0.01\,\beta$ $0.01\,\gamma$	
↓					
^{228}RdTh*	Th	1.9 yr	7×10^{-18}	$0.01\,\alpha$ $0.004\,\gamma$	
↓					
^{224}ThX	Ra	3.64 days	4×10^{-20}	$0.01\,\alpha$ $5 \times 10^{-4}\,\gamma$	
↓					
^{220}Tn	Em*	54.5 sec	7×10^{-24}	$0.01\,\alpha$	3×10^{-11}
↓					
^{216}ThA	Po	0.158 sec	2×10^{-26}	$0.01\,\alpha$	
↓					
^{212}ThB	Pb	10.6 hr	4×10^{-21}	$0.01\,\beta$ $0.01\,\alpha$	
↓					
^{212}ThC	Bi	60.5 min	4×10^{-22}	$0.01\,\beta$ $0.005\,\gamma$	
↓↘					
^{208}ThC	Tl	3.1 min	7×10^{-23}	$0.005\,\beta$ $0.005\,\alpha$	
^{212}ThC′	Po	3.03×10^{-7} sec	2×10^{-32}	$0.01\,\alpha$	

Products of the interaction of cosmic radiation and atmospheric elements

^{3}H*	H	12.2 yr	3×10^{-16}–6×10^{-18}	6–$0.2\,\beta$	1×10^{-13}
^{14}C*	C	5568 yr	4×10^{-14}	$0.4\,\beta$	3×10^{-12}
^{10}Be*	Be	2.7×10^6 yr	10^{-13}	$10^{-3}\,\beta$	3×10^{-14}
^{32}Si*	Si	710 yr	9×10^{-19}	$3 \times 10^{-5}\,\beta$	3×10^{-15}
↓					
^{32}P	P	14.3 days	5×10^{-23}	$3 \times 10^{-5}\,\beta$	3×10^{-15}
			Total	780	6×10^{-8}

* The asterisk denotes elements and nuclides experimentally determined in the ocean.

Table A.2

Elementary chemical composition of sea-water (KRUMHOLZ, GOLDBERG and BOROUGHS [1957])

Element	Content (g/l)	Element	Content (g/l)	Element	Content (g/l)
H	108	Ti	1×10^{-6}	Cd	5.5×10^{-8}
He	5×10^{-9}	V	2×10^{-6}	In	$<2 \times 10^{-5}$
Li	2×10^{-4}	Cr	5×10^{-8}	Sn	3×10^{-6}
B	4.8×10^{-3}	Mn	2×10^{-6}	Sb	$<5 \times 10^{-7}$
C	2.8×10^{-2}	Fe	1×10^{-5}	I	5×10^{-5}
N	5×10^{-4}	Co	5×10^{-7}	Xe	1×10^{-7}
O	857	Ni	5×10^{-7}	Cs*	5×10^{-7}
F	1.3×10^{-3}	Cu	3×10^{-6}	Ba	6.2×10^{-6}
Ne	3×10^{-7}	Zn	1×10^{-5}	La	3×10^{-7}
Na	10.5	Ga	5×10^{-7}	Ce	4×10^{-7}
Mg	1.3	Ge	$<1 \times 10^{-7}$	W	1×10^{-7}
Al	1×10^{-5}	As	3×10^{-6}	Au	4×10^{-9}
Si	3×10^{-3}	Se	4×10^{-6}	Hg	3×10^{-8}
P	7×10^{-5}	Br	6.5×10^{-2}	Tl	$<1 \times 10^{-8}$
S	0.9	Kr	3×10^{-7}	Pb	3×10^{-6}
Cl	19	Rb	1.2×10^{-4}	Bi	2×10^{-7}
Ar	6×10^{-4}	Sr	8×10^{-3}	Rn	9×10^{-18}
K	0.38	Y	3×10^{-7}	Ra	3×10^{-14}
Ca	0.4	Mo	1×10^{-5}	Th	7×10^{-7}
Sc	4×10^{-8}	Ag	3×10^{-7}	Pa	3×10^{-6}

* 3×10^{-7} g/l (BOLTER et al. [1964]).

Table A.3

Physical characteristics of the beta-radiation of the most important radionuclides (KODOCHIGOV [1962])

Isotope	Half-life	Type of transformation	Total number of beta-particles of a given energy to one disintegration	Mean number of particles of a given energy to one disintegration	Energy of the upper spectral limit (MeV)	Mean energy of beta-particles (MeV)	Thickness of Al half-value layer (g/cm²)	Thickness of full value Al layer (g/cm²)
^{3}H	12.46 yr	β^-	1	1	0.01795	0.00569	0.00028	0.00062
^{14}C	5568 yr	β^-	1	1	0.155	0.053	0.0026	0.026
^{22}Na	2.60 yr	β_1^+,γ	1	0.0006	~1.8	0.73	0.082	0.850
		β_2^+,γ	–	~1		0.225	0.016	0.183
^{24}Na	15.06 hr	β^-,γ	1	1	1.390	0.55	0.057	0.620
^{31}Si	2.62 hr	β^-,γ	1	1	1.471	0.58	0.062	0.65
^{32}P	14.3 days	β^-	1	1	1.701	0.68	0.077	0.790
^{35}S	87.1 days	β^-	1	1	0.169	0.056	0.003	0.032
^{36}Cl	4.4 × 10⁵ yr	β^-	1	1	0.714	0.24	0.021	0.254
^{40}K	1.31 × 10⁹ yr	β^-,γ	0.89*	0.89*	1.32*	0.51**	0.080	0.583
^{42}K	12.44 hr	β_1^-,γ	1	0.75	3.58	1.61	0.400	1.807
		β_2^-	–	0.25	2.04	0.85	0.105	0.970
^{45}Ca	152 days	β^-	1	1	0.245	0.075	0.0055	0.057
^{52}Mn	6.0 days	K, β^+,γ	0.35	0.35	0.58	0.21	0.016	0.20
^{59}Fe	45.1 days	β_1^-,γ	1	~0.50	0.46	0.14	0.012	0.145
		β_2^-	–	~0.50	0.257	0.080	0.005	0.060
^{57}Co	270 days	β^+,γ	–	–	0.26	0.082	0.0055	0.061
^{58}Co	72 days	K, β^+,γ	0.15	0.15	0.47	0.145	0.013	0.150
^{60}Co	5.27 yr	β^-,γ	1	1	0.306	0.098	0.0065	0.080
^{63}Ni	85 yr	β^-	1	1	0.067	0.015	–	0.0075

Table A.3 (continued)

Isotope	Half-life	Type of transformation	Total number of beta-particles of a given energy to one disintegration	Mean number of particles of a given energy to one disintegration	Energy of the upper spectral limit (MeV)	Mean energy of beta-particles (MeV)	Thickness of Al half-value layer (g/cm^2)	Thickness of full value Al layer (g/cm^2)
^{64}Cu	12.8 hr	K, β^-, γ	0.58	0.39	0.571	0.19	0.015	0.190
		β^+	–	0.19	0.657	0.22	0.019	0.235
^{65}Zn	250 days	K, β^+, γ	0.025	0.025	0.325	0.1	0.007	0.085
^{72}Ga	14.3 hr	β^-_1	1	0.09	3.15	1.36	0.235	1.60
		β^-_2	–	0.08	2.52	1.06	0.151	1.250
		β^-_3	–	0.11	1.5	0.59	0.063	0.675
		β^-_4	–	0.32	0.9	0.325	0.030	0.350
		β^-_5	–	0.40	0.6	0.2	0.017	0.205
^{74}As	17.5 days	β^-_1, γ	1	0.51	1.36	0.525	0.055	0.604
		β^-_2	–	0.49	0.69	0.24	0.020	0.255
		β^+_1	–	0.11	1.53	0.62	0.064	0.690
		β^+_2	–	0.89	0.92	0.33	0.030	0.340
^{76}As	26.8 hr	β^-_1, γ	1	0.60	3.04	1.3	0.220	1.520
		β^-_2	–	0.25	2.49	1.04	0.150	1.217
		β^-_3	–	0.15	1.29	0.49	0.051	0.566
^{82}Br	35.87 hr	β^-, γ	1	1	0.455	0.14	0.012	0.145
^{86}Rb	19.5 days	β^-_1, γ	1	0.80	1.82	0.74	0.083	0.853
		β^-_2	–	0.20	0.72	0.25	0.022	0.257
^{89}Sr	53 days	β^-	1	1	1.463	0.57	0.060	0.650
^{90}Sr	19.9*** yr	β^-	1	1	0.61	0.27	0.017	0.210
^{90}Y	61 hr	β^-	1	1	2.18	0.91	0.115	1.049
^{91}Y	61 days	β^-	1	1	1.537	0.61	0.064	0.700
^{95}Zr	65 days	β^-_1, γ	1	0.99	0.371	0.12	0.0076	0.107
		β^-_2	–	0.01	0.84	0.3	0.026	0.320
^{95}Nb	25 days	β^-, γ	1	1	0.160	0.046	0.0026	0.027

Nuclide	Half-life	Transition					
^{99}Mo	67 hr	β_1^-, γ	1	0.80	1.23	0.465	0.047	0.525
		β_2^-	—	0.20	0.45	0.145	0.0115	0.135
		β_3^-	—	weak	0.08	0.02	0.0014	0.0095
^{103}Ru	39.8 days	β_1^-, γ	1	~0.99	0.217	0.065	0.0041	0.047
		β_2^-	—	~0.01	0.698	0.022	0.020	0.250
^{106}Ru	1.0 yr	β^-	1	1	0.0392	0.011	—	0.0028
^{106}Rh	30 sec	β_1^-, γ	1	0.82	3.55	1.55	0.300	1.800
		β_2^-	—	0.18	2.30	0.95	0.130	1.20
^{110m}Ag	270 days	β_1^-, γ	1	~0.03	2.86	1.22	0.197	1.430
		β_2^-	—	~0.03	2.12	0.88	0.110	1.000
		β_3^-	—	~0.35	0.530	0.18	0.012	0.165
		β_4^-	—	~0.58	0.087	0.023	0.0045	0.011
^{115m}Cd	43 days	β_1^-, γ	1	~0.98	1.61	0.64	0.070	0.725
		β_2^-	—	~0.02	0.7	0.24	0.020	0.250
		β_3^-	—	weak	0.3	0.09	0.0061	0.078
^{114m}In	49 days	K, γ	1.84	—	—	—	—	—
^{114}In	72 sec	K, β_1^-	—	≤0.97	1.984	0.82	0.100	0.950
		β_2^-	—	~0.0001	0.65	0.225	0.018	0.225
		e^-****	—	0.84	0.164	0.08	0.0028	0.028
^{123}Sn	136 days	β^-	1	1	1.42	0.50	0.055	0.625
^{124}Sb	60 days	β_1^-, γ	1	0.21	2.291	0.96	0.125	1.120
		β_2^-	—	0.07	1.69	0.68	0.075	0.770
		β_3^-	—	0.07	0.95	0.34	0.031	0.350
		β_4^-	—	0.26	0.68	0.23	0.019	0.235
		β_5^-	—	0.39	0.50	0.17	0.0115	0.155
^{131}I	8.14 days	β_1^-, γ	1	0.007	0.815	0.28	0.0245	0.300
		β_2^-	—	0.872	0.608	0.20	0.016	0.200
		β_3^-	—	0.093	0.335	0.10	0.007	0.085
		β_4^-	—	0.028	0.250	0.07	0.005	0.058
		e^-	—	0.028	—	—	—	—
^{134}Cs	2.3 yr	β_1^-, γ	1	0.75	0.648	0.22	0.018	0.225
		β_2^-	—	0.25	0.09	0.024	0.001	0.012

Table A.3 (continued)

Isotope	Half-life	Type of trans-formation	Total number of beta-particles of a given energy to one disintegration	Mean number of particles of a given energy to one disintegration	Energy of the upper spectral limit (MeV)	Mean energy of beta-particles (MeV)	Thickness of Al half-value layer (g/cm²)	Thickness of full value Al layer (g/cm²)
^{137}Cs	33 yr	β_1^-	1	0.92	0.523	0.17	0.013	0.160
		β_2^-	–	0.08	1.17	0.40	0.045	0.500
^{137m}Ba	2.6 min	γ	–	–	–	–	–	–
^{140}Ba	12.8 days	β_1^-, γ	1	0.60	1.022	0.36	0.035	0.400
		β_2^-		0.40	0.480	0.15	0.012	0.150
^{140}La	40 hr	β_1^-	1	~0.10	2.26	0.88	0.120	1.100
		β_2^-		~0.20	1.67	0.62	0.074	0.750
		β_3^-		~0.70	1.32	0.46	0.052	0.570
^{141}Ce	33.1 days	β_1^-, γ	1.16	0.33	0.581	0.18	0.015	0.200
		β_2^-		0.67	0.442	0.13	0.011	0.130
		e^-		0.16	–			–
^{144}Ce	282 days	β_1^-, γ	1	0.70	0.300	0.09	0.0061	0.078
		β_2^-		0.30	0.170	0.046	0.003	0.032
^{144}Pr	17.5 days	β^-, γ	1	>0.99	2.97	1.22	0.210	1.500
^{143}Pr	13.7 days	β^-	1	1	0.932	0.34	0.031	0.350
^{147}Nd	11.3 days	β_1^-, γ	1	~0.60	0.83	0.27	0.025	0.317
		β_2^-		~0.15	0.60	0.19	0.017	0.200
		β_3^-		~0.25	0.38	0.042	0.008	0.110
^{147}Pm	2.6 yr	β^-	1	1	0.223	0.07	0.0042	0.045
^{152}Eu	13 yr	β^-, γ	1	1	1.58	0.63	0.070	0.72
^{152}Eu	16 yr	β_1^-, γ	1	~0.10	1.9	0.78	0.090	0.900
		β_2^-		~0.40	0.7	0.24	0.020	0.250
		β_3^-		~0.50	0.3	0.09	0.0061	0.078
^{155}Eu	1.7 yr	β_1^-, γ	1	0.20	0.243	0.07	0.0048	0.051
		β_2^-	–	0.80	0.154	0.04	0.0026	0.026

Isotope	Half-life	Radiation						
^{166}Ho	27.3 hr	β_1^-,γ	1	~0.89	1.84	0.75	0.080	0.85
		β_2^-	—	~0.11	0.55	0.18	0.015	0.175
^{169}Er	9.4 days	β^-	1	1	0.33	0.10	0.007	0.086
^{175}Yb	102 hr	β_1^-,γ	1	—	0.50	0.165	0.012	0.151
		β_2^-	—	—	0.13	0.035	0.002	0.020
^{177}Lu	6.8 days	β_1^-,γ	1	0.65	0.495	0.16	0.0115	0.150
		β_2^-	—	0.17	0.17	0.12	0.0076	0.107
		β_3^-	—	0.18	0.37	0.05	0.003	0.032
^{181}Hf	45 days	β^-,γ	1	1	0.408	0.13	0.0095	0.115
^{182}Ta	111 days	β^-,γ	1	1	0.525	0.17	0.014	0.163
^{185}W	73.2 days	β^-,γ	1	1	0.428	0.14	0.010	0.130
^{186}Re	92.8 hr	β_1^-,γ	1	0.80	1.07	0.37	0.036	0.420
		β_2^-	—	0.20	0.93	0.335	0.030	0.352
^{190}Os	16.0 days	β^-,γ	1	1	0.143	0.04	0.0024	0.024
^{192}Ir	74.37 days	β^-,γ	1	1	0.66	0.225	0.0185	0.225
^{197}Pt	18 hr	β^-,γ	1	1	0.670	0.23	0.019	0.230
^{198}Au	2.69 days	β^-,γ	1	1	0.963	0.35	0.033	0.380
^{203}Hg	47.9 days	β^-,γ	1	1	0.208	0.058	0.004	0.042
^{204}Tl	3.5 yr	K, β^-	~0.98	~0.98	0.765	0.27	0.022	0.280

* From STEHN [1960].
** Calculated by D. S. Parchevskaya.
*** 28 yr (STEHN [1960]).
**** Conversion electron.

Table A.4

Physical characteristics of the gamma-radiations of the most important radionuclides
(KODOCHGOV [1962])

Isotope	Half-life	Mode of transformation	Gamma-radiation energy (MeV)	Number of quanta of a given energy to a single disintegration	Gamma constant	Total gamma constant	Activity of one μCur of the gamma-radiation source (μg equivalent Ra)
^{22}Na	2.6 yr	β^+, γ_1	1.277	~1	6.60	12.56	1.49
		γ_2	0.511	~2	5.96	–	–
^{24}Na	15.1 hr	β^-, γ_1	4.14	~0.0004	0.0065	19.06	2.27
		γ_2	2.76	~1	11.9	–	–
		γ_3	1.38	~1	7.15	–	–
^{40}K	1.31×10^9 yr	β^-, γ	1.46*	0.11*	8.75**	8.75**	1.042**
^{42}K	12.44 hr	β^-, γ	1.5	0.17	0.16	0.16	0.019
^{51}Cr	27.8 days	K, γ	0.32	0.08	0.15	0.15	0.018
^{52}Mn	6.0 days	K, β^+, γ_1	1.46	0.35	2.59	8.01	0.95
		γ_2	0.94	0.35	1.85	–	–
		γ_3	0.73	0.35	1.49	–	–
		γ_4	0.51	0.7	2.08	–	–
^{54}Mn	310 days	γ	0.84	1	4.83	4.83	0.58
^{55}Fe	2.94 yr	K, roentgen	0.07	0.00002	8×10^{-6}	8×10^{-6}	1×10^{-6}
^{59}Fe	45.1 days	β^-, γ_1	1.29	0.43	2.88	6.25	0.74
		γ_2	1.097	0.567	3.34	–	–
		γ_3	0.195	0.028	0.03	–	–
^{58}Co	72 days	K, β^+, γ_1	0.81	~1	4.68	5.57	0.66
		γ_2	0.51	0.3	0.89	–	–
^{60}Co	5.27 yr	β^-, γ_1	1.33	1	6.9	13.2	1.57
		γ_2	1.17	1	6.3	–	–
^{64}Cu	12.8 hr	K, $\beta^-, \beta^+, \gamma_1$	1.34	~0.005	0.04	1.2	0.14
		γ_2	0.51	0.38	1.16	–	–

Isotope	Half-life	Radiation					
^{65}Zn	250 days	K, β^+, γ_1	1.12	45	2.70	2.85	0.34
		γ_2	0.51	0.05	0.15	—	—
^{72}Ga	14.3 hr	β^-, γ_1	2.51	0.26	2.83	13.47	1.6
		γ_2	2.21	0.33	3.34	—	—
		γ_3	1.87	0.08	0.71	—	—
		γ_4	1.59	0.05	0.39	—	—
		γ_5	1.2	<0.02	0.13	—	—
		γ_6	1.05	0.05	0.28	—	—
		γ_7	0.68	<0.02	0.08	—	—
		γ_8	0.63	0.24	0.88	—	—
^{72}Ge	2.9×10^{-7} sec	γ	0.84	1	4.83	0.356	0.042
^{73}As	76 days	K, γ_1	0.0539	1	0.356	—	—
		γ_2	0.0135	1	—	—	—
^{74}As	17.5 days	β^-, β^+, γ_1	0.635	0.245	0.93	5.33	0.63
		γ_2	0.596	0.445	1.60	—	—
		γ_3	0.51	0.94	2.80	—	—
^{86}Rb	19.5 days	β^-, γ	1.076	~0.20	1.25	1.25	0.15
^{127m}Te	115 days	γ	0.0885	1	0.41	0.41	0.049
^{131}I	8.14 days	β^-, γ_1	0.722	0.03	0.12	2.29	0.27
		γ_2	0.637	0.09	0.34	—	—
		γ_3	0.364	0.80	1.71	—	—
		γ_4	0.284	0.053	0.10	—	—
^{131m}Xe	12 days	γ_5	0.163	0.007	0.01	—	—
^{131}Xe	4.8×10^{-10} sec	γ_6	0.080	0.022	0.01	—	—
^{137}Cs	30 yr	β^-	—	—	—	—	—
^{137m}Ba	2.6 min	γ	0.6616	0.92	3.55	3.55	0.42
^{141}Ce	33.1 days	β^-, γ	0.145	0.67	0.47	0.47	0.056
^{186}Re	92.8 hr	β^-, γ_1	0.764	0.002	0.009	0.156	0.019
		γ_2	0.627	0.002	0.007	—	—
		γ_3	0.137	0.192	0.128	—	—
		γ_4	0.123	0.02	0.012	—	—

Table A.4 (continued)

Isotope	Half-life	Mode of transformation	Gamma-radiation energy (MeV)	Number of quanta of a given energy to a single disintegration	Gamma constant	Total gamma constant	Activity of one μCur of the gamma-radiation source (μg equivalent Ra)
197Pt	18 hr	β^-, γ_1	0.191	0.015	0.02	0.39	0.046
		γ_2	0.077	1.00	0.37	–	–
198Au	2.6 days	β^-, γ_1	1.09	0.0016	0.01	2.45	0.29
		γ_2	0.676	0.0082	0.04	–	–
		γ_3	0.411	0.998	2.40	–	–

* From STEHN [1960]
** Calculated by D. S. Parchevskaya

Table A.5

Radioactive waste in nuclear-powered ships (Committee on Oceanography [1959b])

Nuclides	Primary cooling circuit after 100 days in operation		Demineralizer after 50 days in use.
	Concentration	Total radioactivity	Total radioactivity in resin
	(Cur/l)	(Cur)	(Cur)

The nuclear-powered merchant vessel 'Savannah'

Corrosion products

Nuclides	Concentration (Cur/l)	Total radioactivity (Cur)	Total radioactivity in resin (Cur)
^{60}Co	1.2×10^{-5}	0.46	75.0
^{55}Fe	1.1×10^{-5}	0.43	75.0
^{59}Fe	5.5×10^{-7}	0.02	3.8
^{182}Ta	1.7×10^{-5}	0.67	103
^{51}Cr	4.0×10^{-5}	1.6	148

Fission products

Nuclides	Concentration (Cur/l)	Total radioactivity (Cur)	Total radioactivity in resin (Cur)
^{90}Sr ^{95}Zr ^{106}Ru ^{137}Cs ^{95}Nb ^{144}Ce	10^{-7}	0.006	–

The nuclear-powered submarine 'Nautilus'

Corrosion products

Nuclides	Concentration (Cur/l)	Total radioactivity (Cur)	Total radioactivity in resin (Cur)
^{60}Co	5.7×10^{-9}	–	10
^{58}Co	–	–	0.5
^{59}Fe	1.5×10^{-7}	–	0.5
^{51}Cr	1.0×10^{-8}	–	0.3
^{182}Ta	7.3×10^{-6}	–	–
^{54}Mn	–	–	0.2
^{64}Cu	1.5×10^{-8}	–	–
^{175}Hf	–	–	0.1

Fission products

Nuclides	Concentration (Cur/l)	Total radioactivity (Cur)	Total radioactivity in resin (Cur)
^{131}I ^{90}Sr ^{144}Ce ^{137}Cs	10^{-8}	–	–

APPENDIX II

REFERENCES

a. Preface to the Slavic titles

We have attempted to increase the usefulness of the Slavic language bibliography contained in the original edition by providing citations as complete as possible and by adding citations to English translations of those references for which we have been able to locate translations. This additional labor seemed justified in view of the very limited holdings of Slavic scientific literature in most American libraries and the consequent need for accurate, detailed citations in retrieving Slavic language articles and books. The author has generously contributed several additional references to update and supplement the bibliography and has supplied additional information on several incomplete entries that we were unable to complete with the resources readily available to us.

The style and format of the bibliography have been changed from the Russian edition to serve the needs of the non-Russian reader more appropriately. References to journal articles in the bibliography provide the following information: author, date, translated title, transliterated journal title, volume, number, and pages, followed by a citation to the English translation if one is known. Long journal titles are abbreviated and the abbreviations are explained in the accompanying table. Titles of the Russian journals in the table generally follow that in *Letopis' Periodicheskikh Izdaniy SSSR, 1955–1960 gg. Chast I. Zhurnaly, Trudy, Byulleteny* (List of Periodical Publications, USSR, 1955–1960. Part I. Journals, Periodicals, Bulletins). Moscow, Vsesoyuznaya Knizhnaya Palata, 1963. Journals are cited by transliterated title rather than by translated title because the former is more accurate and reliable. Transliteration is according to the system adopted by the Board on Geographic Names. Books, theses, documents, and similar materials are cited by author, date, transliterated title, translated title, place of publication, publisher, pagination, and translation, if any.

All references for which translations have been found are indicated by an asterisk. The publisher or source of the translation is indicated in parentheses after the citation. The two most frequently cited sources are the Atomic Energy Commission (AEC) and the Joint Publications Research Service (JPRS).

Several of the Russian journals cited in this bibliography are being (or were formerly) translated cover to cover into English on a continuing basis. Citations to articles contained in such cover-to-cover translations usually give the pagination since it often differs in the Russian and English versions. Further information on English translations of Russian scientific literature can be found in the booklet, *Providing U.S. Scientists with Soviet Scientific Information*, revised edition, 1962, by Boris I. Gorokhoff, available from the Publications Office of The National Science Foundation, Washington, D.C.

CHARLES F. LYTLE

Assistant Professor of Zoology

The Pennsylvania State University

b. Abbreviations and full titles of Slavic language journals

Atom. energiya
- Atomnaya energiya
- Atomic Energy
Biofizika
- Biofizika
- Biophysics
Biokhimiya
- Biokhimiya
- Biochemistry
Biol. zh.
- Biologicheskiy zhurnal
- Biological Journal (discontinued)
Bot. zh.
- Botanicheskiy zhurnal
- Botanical Journal
Byull. eksper. biol. med.
- Byulleten' eksperimental'noy biologii i meditsiny
- Bulletin of Experimental Biology and Medicine
Byull. Mosk. obshch. ispyt. prirody (biol.)
- Byulleten' Moskovskogo obshchestva ispytateley prirody. Otdel biologicheskiy
- Bulletin of the Moscow Society of Naturalists. Biology Section
Byull. Ural'skogo otd. Mosk. obshch. ispyt. prirody (biol.)
- Byulleten' Ural'skogo otdeleniya Moskovskogo obshchestva ispytateley prirody. Otdel biologicheskiy
- Bulletin of the Ural Section of the Moscow Society of Naturalists. Biology Section
Dokl. Akad. Nauk SSSR
- Doklady Akademii Nauk SSSR
- Reports of the Academy of Sciences USSR
Gidrobiol. zh.
- Gidrobiologicheskiy zhurnal
- Hydrobiological Journal (Kiev)
Hidrobiologia (Bucharest)*
- Hidrobiologia (Bucharest)
- Hydrobiology (Bucharest)

Izv. Akad. Nauk SSSR (biol.)
- Izvestiya Akademii Nauk SSSR. Seriya biologicheskaya
- Bulletin of the Academy of Sciences USSR. Biology Series
Izv. Akad. Nauk SSSR (geogr.)
- Izvestiya Akademii Nauk SSSR. Seriya geograficheskaya
- Bulletin of the Academy of Sciences USSR. Geography Series
Kommunist
- Kommunist
- Communist
Med. radiologiya
- Meditsinskaya radiologiya
- Medical Radiology
Nauchn. dokl. vyssh. shkoly (biol.)
- Nauchnye doklady vysshey shkoly. Biologicheskiye nauki
- Scientific Reports of the Institutions of Higher Education. Biological Sciences
Nauchni tr., nauchn.-izsled. inst. rib. i rib. prom., Varna (Bulgaria)
- Nauchni trudove, nauchno-izsledovatel-ski institut po ribarstvo i ribna promysh-lennost, Varna (Bulgaria)
- Proceedings of the Scientific Institute of Fisheries and the Fishing Industry, Varna (Bulgaria)
Nauk. zap. Odes. biol. stants.
- Naukovi zapiski Odeskoy biologichnoy stantsii. Kiev, Akademiya Nauk Ukr. SSR
- Scientific Notes of the Odessa Biological Station. Kiev, Academy of Sciences Ukr. SSR (in Ukrainian)
Novoye vremya
- Novoye vremya
- New Times
Okeanologiya
- Okeanologiya
- Oceanography

---

* Rumanian, summaries are in Russian.

Okhr. prirody zapoved. delo SSSR
- Okhrana prirody i zapovednoye delo v
  SSSR
- Nature conservation and national
  preserves in USSR
Pochvovedeniye
- Pochvovedeniye
- Soil Science
Pratsi Odes. rentg.-onkol. inst.
- Pratsi Odeskogo rentgeno-
  onkologichnogo institutu
- Proceedings of the Odessa Roentgeno-
  oncological Institute (in Ukrainian)
Priroda
- Priroda
- Nature (Moscow)
Radiobiologiya
- Radiobiologiya
- Radiobiology
Radiokhimiya
- Radiokhimiya
- Radiochemistry
Ryb. khozyaystvo
- Rybnoye khozyaystvo
- Fisheries
Sibirsk. arkhiv teor. klin. med.
- Sibirskiy arkhiv teoreticheskoy i
  klinicheskoy meditsiny
- Siberian Archives of Theoretical and
  Clinical Medicine
Tr. Azovo-Chernomorsk. nauchn.-issled.
inst. morsk. ryb. khoz. okeanogr.
- Trudy Azovo-Chernomorskogo nauchno-
  issledovatel'skogo instituta morskogo
  rybnogo khozyaystva i okeanografii
- Proceedings of the Azov-Black Sea
  Research Institute of Marine Fisheries
  and Oceanography
Tr. biol. stants. oz. Naroch'
- Trudy biologicheskoy stantsii na
  ozere Naroch' (Minsk, Belorusskiy
  Gosudarstvenniy Universitet)
- Proceedings of the Biological Station on
  Lake Naroch' (Minsk, Belorussian State
  University)
Tr. inst. biol., Akad. Nauk SSSR, Ural'skiy
filial

- Trudy instituta biologii, Sverdlovsk,
  Akademiya Nauk SSSR, Ural'skiy filial
- Proceedings of the Institute of Biology of
  the Ural Affiliate of the Academy of
  Sciences USSR, Sverdlovsk
Tr. inst. okeanol.
- Trudy instituta okeanologii, Akademiya
  Nauk SSSR
- Proceedings of the Institute of Oceano-
  logy, Academy of Sciences USSR
Tr. inst. morfol. zhivot.
- Trudy instituta morfologii zhivotnykh
  imeni A. N. Severtsova, Akademiya
  Nauk SSSR
- Proceedings of the Institute of Animal
  Morphology named for A. N. Severtsov,
  Academy of Sciences USSR
Tr. Odes. rentg.-onkol. inst.
- Trudy Odesskogo rentgeno-
  onkologicheskogo instituta
- Proceedings of the Odessa Roentgeno-
  oncological Institute
Tr. Saratov. otd. Kaspiyskogo filiala vses.
nauchn-issled. inst. morsk. ryb. khoz.
okeanogr.
- Trudy Saratovskogo otdeleniya
  Kaspiyskogo filiala vsesoyuznogo
  nauchno-issledovatel'skogo instituta
  morskogo rybnogo khozyaystva i
  okeanografii
- Proceedings of the Saratov Division of
  the Caspian Affiliate of the All-Union
  Research Institute of Marine Fisheries
  and Oceanography
Tr. Sevastopol. biol. stants.
- Trudy Sevastopol'skoy biologicheskoy
  stantsii, Akademiya Nauk SSSR
- Proceedings of the Sevastopol Biological
  Station, Academy of Sciences USSR
Tr. vses. gidrobiol. obshch.
- Trudy vsesoyuznogo
  gidrobiologicheskogo obshchestva,
  Akademiya Nauk SSSR
- Proceedings of the All-Union
  Hydrobiological Society, Academy of
  Sciences USSR
Tr. vses. nauchn.-issled. inst. morsk. ryb.

khoz. okeanogr.
- Trudy vsesoyuzniy nauchno-issledovatel'skiy institut morskogo rybnogo khozyaystva i okeanografii
- Proceedings of the All-Union Research Institute of Marine Fisheries and Oceanography

Uch. zap., Leningrad. gos. ped. inst., kaf. zool. darvinizma
- Uchenye zapiski, Leningradskiy gosudarstvenniy pedagogicheskiy institut, kafedra zoologii i darvinizma
- Scientific memoirs of the Leningrad State Pedagogical Institute, Chair of Zoology and Darwinism

Uch. zap., Mosk. gor. ped. inst.
- Uchenye zapiski, Moskovskiy gorodskoy pedagogicheskiy institut
- Scientific Memoirs of the Moscow City Pedagogical Institute

Usp. sovrem. biol.
- Uspekhi sovremennoy biologii
- Progress of Modern Biology

Vestn. Akad. Nauk SSSR
- Vestnik Akademii Nauk SSSR
- Bulletin of the Academy of Sciences USSR

Vestn. rentgenol. radiol.
- Vestnik rentgenologii i radiologii
- Bulletin of Roentgenology and Radiology

Vopr. ikhtiol.
- Voprosy ikhtiologii
- Problems of Ichthyology

Zh. eksp. biol.
- Zhurnal eksperimentalnoy biologii
- Journal of Experimental Biology (discontinued)

Zh. nauchn. prikl. fotogr. kinematogr.
- Zhurnal nauchnoy i prikladnoy fotografii i kinematografii
- Journal of Scientific and Applied Photography and Cinematography

Zh. obshch. biol.
- Zhurnal obshchey biologii
- Journal of General Biology

Zool. zh.
- Zoologicheskiy zhurnal
- Zoological Journal

c. Russian titles

Agranat, V. Z. 1958. Some data on accumulation of radioactive polonium (^{210}Po) by aquatic forms. Med. radiologiya 3(1): 65–69.

* Agre, A. L. and V. I. Korogodin. 1960. The distribution of radioactive contamination in a stagnant reservoir. Med. radiologiya 5(1): 67–73. (JPRS 5030).

All-Union Hydrobiological Society. 1965. *Voprosy Gidrobiologii.* (Problems of Hydrobiology). First Congress of the All-Union Hydrobiological Society, Moscow, 1–6 Feb. 1965, Abstracts of Reports. Akad. Nauk USSR, Otdeleniye obshchey biologii. Moscow, Nauka. 479 pp.

Anonymous. 1960. *Sanitarnye Pravila Raboty s Radioaktivnymi Veshchestvami i Istochnikami Ioniziruyushchikh Izlucheniy* (No. 333–60). [Sanitary Regulations for Work with Radioactive Substances and Sources of Ionizing Radiation (No. 333–60)]. Ministry of Health, Atomic Energy Commission USSR. Moscow, Gosatomizdat. 118 pp.

* Azhazha, E. G. and P. M. Chulkov. 1964. Strontium-90 in Atlantic surface waters in the first half of 1961. Okeanologiya 4(1): 68–73. (JPRS 24281).

* Balabukha, V. S. and G. Ye. Fradkin. 1958. *Nakopleniye Radioaktivnykh Elementov v Organizme i Ikh Vyvedeniye.* (Accumulation of Radioactive Elements in the Organism and Their Elimination). Moscow, Medgiz. 183 pp. (AEC-tr-5236).

Baranov, V. I. 1939. Assimilation of radioactive elements by plants. Dokl. Akad. Nauk SSR 24(9): 951–954.

Baranov, V. I. 1955. *Radiometriya.* (Radiometry). Moscow, Acad. Sci. USSR. 343 pp.

* Baranov, V. I. and L. M. Khitrov (eds.). 1964. *Radioaktivnaya Zagryaznennost' Morey i*

Okeanov. (Radioactive Contamination of Seas and Oceans). Moscow, Nauka. 224 pp. (JPRS 26002).

Baranov, V. I., V. N. Vdovenko, L. I. Gedeonov, L. M. Ivanova, A. G. Kolesnikov, S. A. Patin, N. I. Popov, L. M. Khitrov and P. M. Chulkov. 1964. Contamination of the oceans by long-lived radionuclides from Soviet research data. Document A/CONF 28/P384 of the 3rd International Conference on the Peaceful Uses of Atomic Energy. 19 pp.

Baranova, D. D. and G. G. Polikarpov. 1965. Sorption of strontium-90 and cesium-137 by Black Sea silts. Okeanologiya 5(4): 646–648.

Barinov, G. V. 1964. Radioisotopes and algae. Priroda (7): 82–83.

Barinov, G. V. 1965a. Exchange of ^{45}Ca, ^{137}Cs and ^{144}Ce between an alga and sea-water. Okeanologiya 5(1): 111–116.

Barinov, G. V. 1965b. Isotopic exchange in a hydrobiological system and its significance. Gidrobiol. zh. 1(2): 27–34.

Barinov, G. V. 1965c. The features of isotopic exchange in hydrobiological systems. In: *Voprosy Gidrobiologii*. (Problems of Hydrobiology). Moscow, Nauka. p. 27.

Barinov, G. V. 1966. Bioenergetical aspects of kinetics of ^{14}C and ^{32}P accumulation by algae. Gidrobiol. zh. 2(5) (in press).

Benevolenskiy, V. N., V. I. Korogodin and G. G. Polikarpov. 1957. Biophysical principles of the effect of ionizing radiation. In the series: *Itogi Nauki*. (Progress in Science). Biologicheskiye nauki. No. 1, Radiobiology. Kuzin, A. M. (ed.). Moscow, Acad. Sci. USSR. pp. 9–49.

Birukov, I. N., V. I. Korogodin and G. G. Polikarpov. 1958. Recent advances in the use of fluorescence microscopy to study the biological effect of ionizing radiation. Zh. nauchn. prikl. fotogr. kinematogr. 3(2): 128–130.

Blinov, V. A. 1956. The varying sensitivity of amphibian embryos to the effect of X-rays at various stages in their development. In: *Voprosy Radiobiologii*. (Problems of Radiobiology). Pobedinskiy, M. N. and P. N. Kiseleva (eds.). Leningrad, Medgiz. pp. 159–184.

Bogdanova, A. K. 1959. Water exchange through the Bosphorus and its role in mixing in the Black Sea. Tr. Sevastopol. biol. stants. 12: 410–420.

Bogorov, V. G. and Y. M. Kreps. 1958. Is it permissible to bury radioactive waste in deep water trenches? Priroda (9): 45–50.

Bogorov, V. G. and N. I. Popov. 1965. Problems of marine radioecology (review of book *Radioecology of Marine Organisms* by G. G. Polikarpov). Okeanologiya 5(3): 569–571.

* Bogorov, V. G. and B. A. Tareyev. 1960. Oceanic depths and the problem of burying radioactive wastes in them. Izv. Akad. Nauk SSSR (geogr.) (4): 3–10. (JPRS 4103).

* Bogoyavlenskaya, M. P. 1959. A study of calcium metabolism with the object of employing ^{45}Ca as a label for fishes. Vses. nauchn.-issled. inst. morskogo ryb. khoz. i okeanogr. Moscow. 55 pp. (Translation of conclusions, Fisheries Research Board of Canada transl. series No. 276).

Burovina, I. V., V. V. Glazunov, V. G. Leont'yev, V. P. Nesterov and I. A. Skul'skiy. 1964. Alkaline elements in evolution of marine organisms. Zh. obshch. biol. 25(2): 115–123.

Burovina, I. V., V. V. Glazunov, V. G. Leont'yev, V. P. Nesterov, I. A. Skul'skiy, D. G. Fleyshman and M. N. Shmitko. 1962. Lithium, sodium, potassium, rubidium and cesium content in the muscles of marine organisms in the Barents and Black Seas. Dokl. Akad. Nauk SSSR 149(2): 413–415.

Chasovnikov, N. 1928. The effect of X-rays on the fine structure of frog liver cells. Sibirsk.

arkhiv teor. klin. med. 3(1): 214–221.

Chulkov, P. M. and V. F. Gorbunov. 1963. The ^{137}Cs content of the surface waters of the Atlantic Ocean and the adjacent seas in 1961. Moscow, State Comm. of the USSR on the Utilization of Atomic Energy. 6 pp.

Dekhnik, T. V. 1961. Stages in embryonic development and the diurnal breeding rhythm of certain fishes in the Black Sea. Tr. Sevastopol. biol. stants. 14: 220–256.

Dimov, I. G. 1960. Zooplankton in the Black Sea off the Bulgarian coast in 1954, 1955 and 1956. Nauchni tr., nauchn.-izsled. inst. rib. i rib. prom., Varna (Bulgaria) 2: 85–146 (in Bulgarian).

1956. Tr. nauchn.-issled. inst. ryb. khoz. i prom., Varna (Bulgaria) 2: 85–146.

* Dolgopolskaya, M. A., L. A. Il'in, I. A. Puzanov and V. A. Tsenev. 1959. Use of radio-isotopes in the control of marine fouling. Atom. energiya 6(6): 674–676. (English translation 6: 508–510).

* Dubinin, N. P. 1961. *Problemy Radiatsionnoy Genetiki*. (Problems of Radiation Genetics). Moscow, Gosatomizdat. 468 pp. (AEC-tr-5376).

* Fedorov, A. F. 1960. Uptake of a mixture of long-lived beta-emitters by marine plankton. Med. radiologiya 6(6): 51–54. (JPRS 5649; pp. 122–132).

Fedorov, A. F. 1961. Concentration of certain radioisotopes in the plentiful forms of plankton organisms in the northern seas. Nauchn. dokl. vyssh. shkoly (biol.) 1(1): 95–98.

* Fedorov, A. F. 1962. Concise description of plankton samples for radiobiological analyses. Radiobiologiya 2(4): 620–625. (AEC-tr-5431; pp. 165–174).

Fedorov, A. F. and V. N. Podymakhin. 1962. The world ocean should be protected from radioactive contamination. Priroda (11): 47–50.

* Fedorov, A. F., V. N. Podymakhin, V. P. Kilezhenko, N. I. Buyanov and E. M. Golos-kina. 1964. The radiation situation in the North Atlantic fisheries (June–August 1961). Okeanologiya 4(3): 431–436. (JPRS 25966; pp. 42–52).

Fedorov, A. F., V. N. Podymakhin, N. T. Shchitenko and V. V. Chumachenko. 1964. The influence of low radioactive contamination of water on the development of *Pleuronectes platessa* L. Vopr. ikhtiol. 4[3(32)]: 579–585.

Fedorova, G. V. 1965. Contamination of fishes by radiocarbon at various stages in their development. In: *Voprosy Gidrobiologii*. (Problems of Hydrobiology). Moscow, Nauka. pp. 425–426.

Fritz-Niggli, H. 1961. Radiobiology: its basis and achievements. (Translation into Russian). Moscow, Gosatomizdat. 368 pp. Strahlenbiologie. Grundlagen und Ergebnisse. Georg Tieme Verlag, Stuttgart, 1959.

* Gedeonov, L. I. 1957. Radioactive contamination of the atmosphere. Atom. energiya 2(3): 260–271. (English translation 2: 313–325).

* Getsova, A. B. 1960. On the desorption of radioactive isotopes of certain representatives of aquatic invertebrates. Dokl. Akad. Nauk. SSSR 133(2): 459–461. (English translation pp. 509–514).

Getsova, A. B., N. A. Lyapunova, G. G. Polikarpov and Ye. A. Timofeyeva-Resovskaya. 1964. Concentration of chemical elements from aqueous solutions by freshwater organisms. Communication 6. Accumulation of the radioisotopes of eight different elements in the tissues of *Anodonta cellensis*. Nauchn. dokl. vyssh. shkoly (biol.) (4): 82–88.

* Getsova, A. B., Ye. A. Timofeyeva-Resovskaya and N. V. Timofeyev-Resovskiy. 1960. The effect of ethylenediaminetetracetic acid sodium salt on the uptake of various radioisotopes from an aqueous solution by leeches and midges. Dokl. Akad. Nauk

SSSR 130(2): 440–442. (English translation pp. 53–54).

* Gileva, E. A. 1960. Radioisotope concentration factors for freshwater algae. Dokl. Akad. Nauk SSSR 132(4): 948–949. (English translation pp. 343–344).

Gileva, E. A. 1965. Effect of the concentration and mode of occurrence of chemical elements in solution on their concentration by *Cladophora*. In: *Voprosy Gidrobiologii*. (Problems of Hydrobiology). Moscow, Nauka. pp. 90–91.

* Glazunov, V. V., V. P. Parchevskiy and D. G. Fleyshman. 1963. Variation of fission product content of *Cystoseira* in the Black Sea. Dokl. Akad. Nauk SSSR 152(5): 1222–1224. (JPRS 22714).

Golovinskaya, K. A. and D. D. Romashov. 1958. Effect of ionizing radiation on development and breeding of fishes. Vopr. ikhtiol. (11): 16–38.

* Gorodetskiy, A. A., B. R. Kirichinskiy and N. F. Lipkin. 1961. *Ocherki po Radiobiologii*. (Essays on Radiobiology). Kiev, Acad. Sci. Ukr. SSR. 222 pp. (AID No. S5853).

Grayevskiy, E. Ya. and N. I. Shapiro. 1957. *Sovremennyye Voprosy Radiobiologii*. (Modern Problems of Radiobiology). Nichipozovich, A. A. (ed.). Moscow, Acad. Sci. USSR. 95 pp.

Grechishkin, S. V. 1934. The biological effect of irradiation on *Elodea densa*, *Bacterium ponticum* and *Saccharomyces cerevisiae*. Bot. zh. 19(6): 527–533.

Gulyakin, I. V. and N. I. Seletkova. 1954. The uptake of radiothorium by plants. In: *Deystviye Ioniziruyushchikh Izlucheniy na Biologicheskiye Ob"ekty*. (The Effect of Ionizing Radiation on Biological Objects). Moscow, Acad. Sci. USSR. (Cited by Gulyakin and Yudintseva, 1962).

Gulyakin, I. V. and Ye. V. Yudintseva. 1956. Entering of the radioactive isotopes of strontium, cesium, ruthenium, zirconium and cerium into plants. Dokl. Akad. Nauk SSSR 111(1): 206–208.

* Gulyakin, I. V. and Ye. V. Yudintseva. 1958. Uptake of ^{60}Co by plants and its concentration in the harvest. Dokl. Akad. Nauk SSSR 123(2): 368–370.

Gulyakin, I. V. and Ye. V. Yudintseva. 1962. *Radioaktivnyye Produkty Deleniya v Pochve i Rasteniyakh*. (Radioactive Fission Products in Soil and Plants). Moscow, Gosatomizdat. 276 pp.

* Gusyev, N. G. 1961. *O Predel'no Dopustimykh Urovnyakh Ioniziruyushchikh Izlucheniy*. (Maximum Permissible Levels of Ionizing Radiation). Moscow, Medgiz. 199 pp. (JPRS 9706; pp. 8–15).

Isayev, B. M. 1961. Dosimetry and maximum permissible levels of ionizing radiation. In: *Kratkiy Spravochnik Inzhenera-Fizika. (Yadernaya Fizika, Atomnaya Fizika)*. [Brief Handbook of Engineering Physics (Nuclear Physics, Atomic Physics)]. Fedorov, N. D. (ed.). Moscow, Gosatomizdat. pp. 299–334.

* Ivanov, V. N. 1965a. Concentration of ^{90}Sr, ^{90}Y, ^{137}Cs, ^{144}Ce, ^{106}Ru, ^{59}Fe and ^{35}S by the eggs of the Black Sea haddock. Radiobiologiya 5(1): 57–60. (AEC-tr-6598; pp. 92–97).

Ivanov, V. N. 1965b. Radiosensitivity of the developing eggs of sea fishes. In: *Voprosy Gidrobiologii*. (Problems of Hydrobiology). Moscow, Nauka. pp. 181–182.

Ivanov, V. N. 1965c. Some features of radioecology of marine fishes in the early stages of ontogeny (an experimental study). Thesis for Degree of Candidate of Biological Sciences. Sevastopol. 197 pp.

* Ivanov, V. N. 1965d. Concentration of fission product radioisotopes by the eggs of Black Sea fishes. Radiobiologiya 5(2): 295–300. (AEC-tr-6599; pp. 201–209).

* Karzinkin, G. S. 1962. *Ispol'zovaniye Radioaktivnykh Izotopov v Rybnom Khozyaystve*. (The Use of Radioisotopes in the Fishing Industry). Moscow, Pishchepromizdat. 73 pp.

(JPRS 21, 109).

Kirpichnikov, V. S., A. N. Svetovidov and A. S. Troshin. 1956. The labeling of carp with the radioactive isotopes of phosphorus and calcium. Dokl. Akad. Nauk SSSR 111(1): 221–224.

* Klechkovskiy, V. M. (ed.). 1956. The behavior of radioactive fission products in soils, their uptake by plants and their concentration in the harvest. UN document A/AC.-82/G/R.41. (Hectograph editions). Moscow, Acad. Sci. USSR. 176 pp. (AEC-tr-2867).

* Klechkovskiy, V. M. and I. V. Gulyakin. 1958. The behavior of trace amounts of strontium, cesium, ruthenium and zirconium in soils and plants. Pochvovedeniye (3): 1–16.

Klechkovskiy, V. M., L. N. Sokolova and G. N. Tselishcheva. 1959. Absorption of trace amounts of strontium and cesium in soils. In: *Proc. 2nd UN Intern. Conf. on the Peaceful Uses of Atomic Energy* (Geneva, 1958) 18: 486–493.

Kodochigov, P. N. 1962. *O Prakticheskikh Voprosakh Dozimetrii Ioniziruyushchikh Izlucheniy.* (Practical Aspects of the Dosimetry of Ionizing Radiation). Moscow, Acad. Sci. USSR. 135 pp.

* Kolychev, B. S. 1961. Results of a conference on the dumping of radioactive wastes in seas and oceans (Vienna, January 1961). Atom. energiya 10(6): 634–639. (English translation pp. 630–632).

Kondrat'yeva, T. M. and Ye. V. Belogorskaya. 1961. Phytoplankton distribution in the Black Sea and its relationship to hydrological conditions. Tr. Sevastopol. biol. stants. 14: 44–63.

Korogodin, V. I. 1962. Genetics. In: *Vklad Radiologii v Razvitiye Medikobiologicheskikh Distsiplin.* (The Contribution of Radiology to the Growth of the Medico-biological Disciplines). Petrov, P. V., V. I. Korogodin, F. M. Lyass, A. A. Neyfakh, and Ye. F. Romantsev. Minsk, Ministry Special Technical and Vocational Training of the Belorussian SSR. pp. 45–100.

Korogodin, V. I. and G. G. Polikarpov. 1957. Primary processes in radiation injury. (On the problem of the mechanism of intensification of radiobiological effect). Usp. sovrem. biol. 44 [1(4)]: 93–102.

Koval', L. G. 1959. The zooplankton of coastal waters in the northwestern Black Sea between 1954 and 1957. Nauk. zap. Odes. biol. stants. (1): 34–51. (in Ukrainian).

Kovaleva, N. Ye. 1948. Issledovaniye deystviya rentgenovykh luchey na *Paramaecium caudatum.* (Investigation of the effect of X-rays on *Paramaecium caudatum*). [Thesis, Leningrad, 1947]. Uch. zap., Leningrad. gos. ped. inst., kaf. zool. darvinizma. Polyanskiy, Yu. U. (ed.). 70: 75–144.

Koval'skiy, V. V. 1958. Geochemical ecology. Priroda 47(9): 100–101.

Kreps, Ye. M. 1959. The problem of radioactive contamination of oceans and marine organisms. Izv. Akad. Nauk SSSR (biol.) (3): 321–334.

Kulikov, N. V. 1957. Studies in experimental biogeocoenology. II. The effect of soaking seeds in a mixture of beta-emitters on the biomass and structure of an experimental photocoenosis. Tr. inst. biol. Akad. Nauk SSSR, Ural'skiy filial (9). Collected works of the Biophysics Laboratory, Vol. 1: 252–291.

* Kulikova, G. M. 1960. The possibility of using filamentous algae to purify radioactively contaminated water. Dokl. Akad. Nauk SSSR 135(4): 978–980.

Kusmorskaya, A. P. 1955. Seasonal and annual variations in the zooplankton of the Black Sea. Tr. vses. gidrobiol. obshch. (6): 158–192.

Kuzin, A. M. 1959. *Chem Ugrozhayut Chelovechestvu Yadernye Vzryvy.* (The Threat of

Nuclear Explosions to Mankind). Moscow, Acad. Sci. USSR. 131 pp.

* Kuzin, A. M. 1960. The danger of the increasing ^{14}C concentration in the atmosphere. Vestn. Akad. Nauk SSSR (9): 48–51. (UN Document A/AC. 82/G/L.413) (AEC-tr-4532).

Kuzin, A. M. 1963. Radiobiology: its progress and tasks. Kommunist (1): 78–85.

Kuzin, A. M. 1964. Chap. 11, Radiation ecology. In: *Radiatsionnaya Biologiya* (Radiation Biology). Kuzin, A. M. and N. I. Shapiro (eds.). Moscow, Nauka. pp. 360–378.

Kuzin, A. M. and A. A. Peredel'skiy. 1956. Nature conservancy and some aspects of radioactive-ecological relations. Okhr. prirody zapoved. delo SSSR (1): 65–78.

* Lebedeva, G. D. 1959. Concentration of ^{144}Ce by yearling carp and *Elodea*. Med. radiologiya 4(4): 73–77. (JPRS 2592).

* Lebedeva, G. D. 1961. Concentration of strontium-90 by young *Leucaspius* and carp. In: *Raspredeleniye, Biologicheskoye Deystviye i Migratsiya Radioaktivnykh Isotopov.* (Distribution, Biological Effect and Migration of Radioisotopes). Lebedinskii, A. V. and Yu. I. Moskalev (eds.). Moscow, Medgiz. pp. 319–322. (JPRS 15067; pp. 51–56).

* Lebedeva, G. D. 1962. Principal points of entry of strontium-89 into benthos-eating fishes in freshwater. Radiobiologiya 2(1): 43–49. (AEC-tr-5428; pp. 62–71).

Lebedinskiy, A. V. 1957. The consequences of strontium-90 fallout. Med. radiologiya 2 (5): 22–33.

Leont'yev, V. G. and I. A. Skul'skiy. 1965. The interrelationship between the mineral and lipid compositions of the tissues of marine organisms. In: *Voprosy Gidrobiologii.* (Problems of Hydrobiology). Moscow, Nauka. p. 252.

* Leypunskiy, O. I. 1957. The radioactive hazard of pure hydrogen bomb and ordinary atomic bomb tests. Atom. energiya 3(12): 530–539. (English translation pp. 1413–1427).

* Leypunskiy, O. I. 1958. The radioactive hazard of continuous nuclear weapon tests. Atom. energiya 4(1): 63–70. (English translation pp. 71–81).

Lubyanov, I. P. 1965. The radioactivity of aquatic organisms in the waters of steppe zone and aspects of freshwater radioecology. In: *Voprosy Gidrobiologii.* (Problems of Hydrobiology). Moscow, Nauka. pp. 256–257.

Marey, A. N., M. M. Saurov and G. D. Lebedeva. 1958. A note on the transmission of radioactive strontium along a food chain from exposed water-bodies into the human organism. Med. radiologiya 3(1): 69–76.

Matsuda, H. and K. Hayashi. 1959. *Yadernoye Oruzhiye i Chelovek.* (Nuclear Weapons and Man). (Translation into Russian). Moscow, Foreign Lit. Press. 307 pp. (Original Japanese source published in Tokyo, 1956).

Medvedeva, G. A., M. N. Meysel' and Ya. L. Shekhtman. 1952. The use of irradiation of short duration and high energy to study the dynamics of the radiobiological effect. Zh. obshch. biol. 13(3): 243–245.

Moiseyev, P. A. 1957. The effect of an atomic explosion on fishing. Ryb. khozyaystvo (5): 81–84.

Moiseyev, P. A. 1958. New data on the effects of nuclear explosions on aquatic organisms. Ryb. khozyaystvo (7): 22–24.

Morozova-Vodyanitskaya, N. V. 1948. Black Sea phytoplankton. Part. 1. Tr. Sevastopol. biol. stants. 6: 39–172.

Morozova-Vodyanitskaya, N. V. 1954. Black Sea phytoplankton. Part 2. Tr. Sevastopol. biol. stants. 8: 11–99.

Morozova-Vodyanitskaya, N. V. 1957. Phytoplankton in the Black Sea and its quantitative development. Tr. Sevastopol. biol. stants. 9: 3–13.

Natali, V. F. 1940. The effect of X-rays on the sex gland of male *Lebistes reticulatus*. Uch. zap., Mosk. gor. ped. inst. (1): 5–29.

Natali, V. F. 1942a. Changes induced in the ovaries of fishes (Poecilidae) by X-rays in relation to the problem of change of sex. Byull. eksper. biol. med. (10): 57–61.

Natali, V. F. 1942b. The effect of X-rays on the sex gland of female *Lebistes reticulatus*. Byull. eksper. biol. med. [5(6)]: 56–59.

Nesterov, V. P. 1965. Sravnitel'noe kolichestvennoe izucheniye estestvennogo raspredeleniya shchelochnykh elementov (Li, Na, K, Rb i Cs) v myshechnoy tkani raznykh zhivotnykh. [Comparative quantitative investigation of natural distribution of alkaline elements (Li, Na, K, Rb, Cs) in muscle tissue of different animals]. Author's summary of thesis for Degree of Candidate of Biological Sciences. Leningrad. 15 pp.

Nesterov, V. P. and I. A. Skul'skiy. 1965. Lithium, sodium, potassium, rubidium and cesium content in the muscles of organisms inhabiting the Barents and Black Seas. In: *Voprosy Gidrobiologii*. (Problems of Hydrobiology). Moscow, Nauka. pp. 310–311.

Neyfakh, A. A. 1956a. Alteration of radiosensitivity in the process of fertilization in the loach. Dokl. Akad. Nauk SSSR 109(5): 943–946.

Neyfakh, A. A. 1956b. The effect of ionizing radiation on the sexual cells of *Misgurnus fossilis*. Dokl. Akad. Nauk SSSR 111(3): 585–588.

Neyfakh, A. A. 1959. The effect of ionizing radiation on the early development of fishes. Tr. inst. morfol. zhivot. (24): 135–159.

* Neyfakh, A. A. 1960. An investigation of the functions of the nuclei in the development of *Strongilocentrotus dröbachiensis* by the method of radiation inactivation. Dokl. Akad. Nauk SSSR 132(6): 1458–1461. (English translation pp. 376–379).

* Neyfakh, A. A. 1961a. Radiation studies of the morphogenetic functions of the nuclei in the early development of tailless amphibians. Dokl. Akad. Nauk SSSR 136(5): 1248–1251. (English translation pp. 41–44).

Neyfakh, A. A. 1961b. A comparative radiation study of the morphogenetic function of the nuclei in the development of animals. Zh. obshch. biol. 22(1): 42–57.

Neyfakh, A. A. 1962. Embryology. In: *Vklad Radiologii v Razvitiye Mediko-biologicheskikh Distsiplin*. (The Contribution of Radiology to the Growth of the Medico-biological Disciplines). Petrov, P. V., V. I. Korogodin, F. M. Lyass, A. A. Neyfakh, and Ye. F. Romantsev. Minsk, Ministry of Special Technical and Vocational Training of the Belorussian SSR. pp. 101–123.

Nikiforov, V. S. 1965. Algorithm for the description of the radiation environment. In: *Materialy Komi Respublikanskoy molodezhnoy nauchnoy konferentsii*. Tezisy dokladov. (Materials of a Scientific Conference of the Komi Republic Youth. Abstracts of reports). Syktyvkar. pp. 46–49.

Nikitin, S. A. 1932. Materials for a theory of the biological effect of X-rays. Tr. Odes. rentg.-onkol. inst. (1): 57–69.

Nikitin, S. A. 1938. The biological effects of X-rays (experimental data). Pratsi Odes. rentg.-onkol. inst. (3): 26–106. (in Ukrainian).

Nikitin, S. A. 1958. *Vvedeniye v Radiobiologiyu*. (Introduction to Radiobiology). Kiev, Gosmedizdat Ukrainian SSR. 184 pp.

Nikitin, V. N. 1945. Distribution of the plankton biomass in the Black Sea. Dokl. Akad. Nauk SSSR 47(7): 510–512. (in French).

* Parchevskiy, V. P. 1964a. The radioactivity of certain organisms from the Black Sea. In: *Radioaktivnaya Zagryaznennost' Morey i Okeanov*. (Radioactive Contamination of Seas

and Oceans). Baranov, V. I. and L. M. Khitrov (eds.). Moscow, Nauka. pp. 151–169 (JPRS 26002; pp. 87–109).

Parchevskiy, V. P. 1964b. Izucheniye radioaktivnosti nekotorykh gidrobiontov Chernogo Morya (radioekologicheskoye issledovaniye). [A study of the radioactivity of certain Black Sea aquatic organisms (a radio-ecological investigation)]. Author's summary of thesis for Degree of Candidate of Biological Sciences. Sevastopol-Dnepropetrovsk. 22 pp.

Parchevskiy, V. P. 1965a. The radionuclides of cerium, ruthenium and zirconium in Black Sea plants and animals. Okeanologiya 5(5): 860–866.

Parchevskiy, V. P. 1965b. Radioactivity of marine organisms. In: *Voprosy Gidrobiologii*. (Problems of Hydrobiology), Moscow, Nauka. p. 329.

* Parchevskiy, V. P., G. G. Polikarpov and I. S. Zaburunnova. 1965. Some features of yttrium and strontium concentration by marine organisms. Dokl. Akad. Nauk SSSR 164(4): 913–916.

Patin, S. A. 1965. On forms of existence and migration of artificial radioisotopes in sea water. Tr. inst. okeanol. 82 (in press).

Pechenko, B. F. 1922. The effect of radium rays on amoeba. Vestn. rentgenol. radiol. 2(1): 39–48.

* Peredel'skiy, A. A. 1957a. Ecological studies of the effect of ionizing radiation. In the series: *Itogi Nauki*. (Progress in Science), Biologicheskiye nauki. No. 1, Radiobiology. Kuzin, A. M. (ed.). Moscow, Acad. Sci. USSR. pp. 379–392. (JPRS 859D; OTS: 59–11, 758).

Peredel'skiy, A. A. 1957b. The principles and problems of radioecology. Zh. obshch. biol. 18(1): 17–30.

Peredel'skiy, A. A. 1958. Problems of radioecology. Priroda (8): 27–32.

Peredel'skiy, A. A. 1964. Penetrating radiation and radioecology. In: *Zemlya vo Vselennoy*. (The Earth in the Universe). Fedynskiy, V. V. (ed.). Moscow, Nauka. pp. 449–470.

* Peredel'skiy, A. A. and I. O. Bogatyrev. 1959. Radioactive contamination of the land by insects emerging from contaminated water-bodies. Izv. Akad. Nauk SSSR (biol.) (2): 186–192. (JPRS 17, 812).

Pertsov, L. A. 1964a. *Prirodnaya Radioaktivnost' Biosfery*. (The Natural Radioactivity of the Biosphere). Moscow, Atomizdat. 315 pp.

* Pertsov, L. A. 1964b. The content of natural fission product radioisotopes in the bio-sphere. Radiobiologiya 4(4): 619–623. (AEC-tr-6407; pp. 175–179).

Petipa, T. S., L. N. Sazhina and Ye. P. Delalo. 1963. Zooplankton distribution in the Black Sea 1951–1956. Okeanologiya 3(1): 110–122.

Petrov, R. V., V. I. Korogodin, F. M. Lyass, A. A. Neyfakh and Ye. F. Romantsev. 1962. *Vklad Radiologii v Razvitiye Mediko-Biologicheskikh Distsiplin*. (The Contribution of Radiology to the Growth of the Medico-biological Disciplines). Minsk, Ministry of Special Technical and Vocational Training of the Belorussian SSR. 147 pp.

Pitsyk, G. K. 1950. The quantitative development and horizontal distribution of phyto-plankton in the western half of the Black Sea (preliminary communication). Tr. Azovo-Chernomorsk. nauchn.-issled. inst. morsk. ryb. khoz. okeanogr. (14): 215–244.

Pitsyk, G. K. 1951. The phytoplankton of the Sea of Azov. Tr. Azovo-Chernomorsk. nauchn.-issled. inst. ryb. khoz. okeanogr. (15): 313–330.

Pitsyk, G. K. 1954. The quantity, composition and distribution of phytoplankton in the Black Sea. Tr. vses. nauchn.-issled. inst. morsk. ryb. khoz. okeanogr. 28: 224–239.

Pitsyk, G. K. 1955. The phytoplankton of the Sea of Azov following regulation of the discharge of the River Don. Tr. Azovo-Chernomorsk. nauchn.-issled. inst. morsk. ryb. khoz. okeanogr. (16): 279–310.

Pitsyk, G. K. 1960. The dynamics of phytoplankton and zooplankton in the Black Sea and some of its features. Hidrobiologia (Bucharest) 3: 243–255.

* Polikarpov, G. G. 1957a. On the problem of the development of the reactions of radiation after-effects. Biofizika 2(2): 174–177. (English translation, pp. 176–179).

Polikarpov, G. G. 1957b. Features of the reactions of radiation after effect in *Pelmatohydra oligactis*. In: *Biokhimicheskiye i Fiziko-Khimicheskiye Osnovy Biologicheskogo Deystviya Radiatsiy* (February 25–28, 1957). Tezisy dokladov. (Biochemical and Physico-chemical Basis of Radiation Effects. Abstracts of reports). Moscow, Moscow State Univ. pp. 34–35.

Polikarpov, G. G. 1958. Concentration of the radioisotope of cerium by freshwater molluscs. Priroda (5): 86–87.

* Polikarpov, G. G. 1960a. Radioisotopes and ionizing radiation in marine biology. Tr. Sevastopol. biol. stants. 13: 275–292. (JPRS 20, 630; pp. 1–25).

Polikarpov, G. G. 1960b. Absorption of radioactivity by marine organisms. Priroda (1): 104–105.

Polikarpov, G. G. 1960c. Concentration of fission product radioisotopes by marine organisms. I. Accumulation of strontium-90, yttrium-91 and cerium-144 by benthic plants and animals. Nauchn. dokl. vyssh. shkoly (biol.) (3): 97–105.

Polikarpov, G. G. 1960d. A note on the study of the phosphorus nutrition of *Ulva rigida* by the tagging method. Tr. Sevastopol. biol. stants. 13: 296–298.

Polikarpov, G. G. 1960e. The ability of the marine alga *Ulva rigida* to concentrate uranium-238 from an equilibrium mixture with thorium-234. Tr. Sevastopol. biol. stants. 13: 293–295.

Polikarpov, G. G. 1960f. The role of marine benthos in the migration of sulfates and sulphides. Nauchn. dokl. vyssh. shkoly (biol.) (4): 103–106.

* Polikarpov, G. G. 1961a. Data on the concentration factors of ^{32}P, ^{35}S, ^{90}Sr, ^{91}Y, ^{137}Cs and ^{144}Ce in marine organisms. Tr. Sevastopol. biol. stants. 14: 314–328. (JPRS 24, 227).

Polikarpov, G. G. 1961b. Concentration of fission product radioisotopes by marine organisms. II. Accumulation of germanium-71 and cesium-137 by algae, sea anemones and mussels, and of germanium-71, strontium-90, yttrium-91, cesium-137 and cerium-144 by flowering plants. Nauchn. dokl. vyssh. shkoly (biol.) (4): 92–98.

* Polikarpov, G. G. 1961c. The stability of concentration factors of strontium-90, yttrium-91 and cerium-144 in marine algae. Dokl. Akad. Nauk SSSR 140(5): 1192-1194. (English translation pp. 770–772).

Polikarpov, G. G. 1961d. Strontium-90 uptake by marine organisms. Priroda (2): 83.

* Polikarpov, G. G. 1961e. The role of detritus formation in the migration of strontium-90, cesium-137 and cerium-144. Experiments with the marine alga *Cystoseira barbata*. Dokl. Akad. Nauk SSSR 136(4): 921–923. (JPRS 18, 585; pp. 3–7).

Polikarpov, G. G. 1963. Problemy morskoy radioekologii. (Problems of marine radioecology). Doctorial thesis. Sevastopol. 480 pp. Author's summary of doctoral thesis (1964). Sevastopol-Kiev, Joint Academic Council for Biological Disciplines, Acad. Sci. Ukrainian SSR. 29 pp.

* Polikarpov, G. G. 1964. Some biological aspects of radioactive contamination of the seas and oceans. In: *Radioaktivnaya Zagryaznennost' Morey i Okeanov*. (Radioactive Contamination of the Seas and Oceans). Baranov, V. I. and L. M. Khitrov (eds.).

Moscow, Nauka. pp. 98–125. (JPRS 26, 002; pp. 20–55).

Polikarpov, G. G. 1965. The radioactivity of the hydrosphere and problems of the radio-ecology of aquatic organisms. In: *Voprosy Gidrobiologii*. (Problems of Hydrobiology). Moscow, Nauka. pp. 344–345.

Polikarpov, G. G. and A. D. Akamsin. 1960. An experimental study of yttrium concentration by marine algae, sea anemones and bottom sediments. Tr. Sevastopol. biol. stants. 13: 299–301.

Polikarpov, G. G. and V. N. Ivanov. 1961. The effect of ^{90}Sr-^{90}Y on developing anchovy eggs. Vopr. ikhtiol. 1 [3(20)]: 583–589.

* Polikarpov, G. G. and V. N. Ivanov. 1962a. The harmful effect of strontium-90-yttrium-90 in the early period of development of the red mullet, the green wrasse, the horse-mackerel and the Black Sea anchovy. Dokl. Akad. Nauk SSSR 144(1): 219–222.

* Polikarpov, G. G. and V. N. Ivanov. 1962b. Concentration of the radioisotopes of strontium and yttrium by the eggs of marine fishes. Radiobiologiya 2(2): 207–210. (AEC-tr-5429; pp. 50–56).

Polikarpov, G. G. and V. N. Ivanov. 1962c. The effect of strontium-90-yttrium-90 on the developing eggs of the Black Sea anchovy and the stone perch. Byull. Mosk. obshch. ispyt. prirody (biol.) 67(3): 153–154.

* Polikarpov, G. G. and L. A. Lanskaya. 1961. Proliferation of the mass unicellular alga *Prorocentrum micans* in the presence of sulphur-35. Tr. Sevastopol. biol. stants. 14: 329–333. (JPRS 20, 795; pp. 7–12).

* Polikarpov, G. G. and V. P. Parchevskiy. 1961. The radioactivity of algae in the Adriatic and the Black Seas. Okeanologiya 1(2): 338–339. (JPRS 18, 585; pp. 1–2).

Polikarpov, G. G. and V. S. Ten. 1961. Study of the kinetic features of UI and UX₁ by green, brown and red algae. Nauchn. dokl. vyssh. shkoly (biol.) (2): 116–119.

Polikarpov, G. G. and V. S. Ten. 1962. The kinetic features of the release of strontium-90 from *Cystoseira barbata* Good et Wood. Nauchn. dokl. vyssh. shkoly (biol.) (4): 89–97.

Polikarpov, G. G. and A. Ya. Zesenko. 1965. The potential application of the concept 'concentration factor' in marine radioecology. Okeanologiya 5(6): 1099–1107.

* Popov, N. I. 1964. Natural radioactivity of sea-water. Okeanologiya 4(2): 223–231. (JPRS 25068).

Popov, N. I., E. G. Azhazha, G. I. Kosourov and A. A. Yusefovich. 1962. Strontium-90 in Atlantic surface waters. Okeanologiya 2(5): 845–848.

* Popov, N. I., V. M. Orlov, S. A. Patin and N. P. Ushakova. 1964. Strontium-90 in the surface waters of the Indian Ocean in 1960–1961. Okeanologiya 4(3): 418–422. (UN Document A/AC.82/G/L.904) (JPRS 25,966; pp. 20–27).

Popov, N. I., V. M. Orlov and V. A. Pchelin. 1963. Strontium-90 in waters of the Pacific Ocean. Communication I. Western Pacific and adjacent seas, 1961. Okeanologiya 3(4): 666–668.

Popov, N. I., S. A. Patin, V. A. Pchelin and R. M. Polevoy. 1963. Strontsii-90 v vodakh Tikhogo okeana. (Strontium-90 in Pacific Waters). Moscow, State Committee for the Utilization of Atomic Energy, USSR. UN Document A/AC.82/G/L.907. 11 pp.

Popova, E. I. 1965. Concentration of radium by certain aquatic plants by increasing its content in natural waters. In: *Voprosy Gidrobiologii*. (Problems of Hydrobiology). Moscow, Nauka. pp. 349–350.

Popova, O. N. and R. P. Kodaneva. 1965. Regarding some modes of participation of radium and uranium in the mineral exchange of substances in plants. In: *Materialy*

Komi Respublikanskoy molodezhnoy nauchnoy konferentsii. Tezisy dokladov. (Materials of a Scientific Conference of the Komi Republic Youth. Abstracts of reports). Syktyvkar. pp. 44–46.

Povelyagina, Z. S. and M. M. Telitchenko. 1959. Concentration of the radioisotopes of phosphorus and strontium by various species of freshwater molluscs. Byull. Mosk. obshch. ispyt. prirody (biol.) 64(2): 79–83.

Roginskiy, S. Z. 1948. *Adsorbtsiya i Kataliz na Neodnorodnykh Poverkhnostyakh.* (Adsorption and Catalysis on Non-uniform Surfaces). Moscow, Acad. Sci. USSR. 643 pp.

Roginskiy, S. Z. 1956. *Teoreticheskiye Osnovy Izotopnykh Metodov Izucheniya Khimicheskikh Reaktsiy.* (Theoretical Principles of Isotope Methods in the Study of Chemical Reactions). Moscow, Acad. Sci. USSR. 611 pp.

Roginskiy, S. Z. and S. E. Schnol'. 1963. *Izotopy v Biokhimii.* (Isotopes in Biochemistry). Moscow, Acad. Sci. USSR. 379 pp.

Romashov, D. D. and K. A. Golovinskaya. 1960. Radiation biology and the genetics of fishes. In the series: *Itogi Nauki.* (Progress in Science). Biologicheskiye nauki. No. 3. Ionizing Radiation and Heredity. Dubinin, N. P. (ed.). Moscow, Acad. Sci. USSR. pp. 9–49.

Rudakov, N. P. 1958. Primeneniye radioizotopov dlya izucheniya zakonomernostey pogloshcheniya i vyvedeniya nekotorykh mineral'nykh ionov u ryb i ikh markirovki. (The use of radioisotopes for studying the uptake and release of certain mineral ions in fishes and their labeling). Author's summary of thesis for Degree of Candidate of Biological Sciences VNIORKh (All-Union Research Institute for Lake and River Fishing). 14 pp.

Sabinin, D. A. 1955. *Fiziologicheskiye Osnovy Pitaniya Rasteniy.* (The Physiological Principles of Plant Nutrition). Moscow, Acad. Sci. USSR. 512 pp.

Samokhvalova, G. V. 1935. The effect of X-rays on the sexual gland and secondary sexual characteristics of *Lebistes reticulatus.* In the series: *Trudy po Dinamike Razvitiya.* (Works on Growth Dynamics). Vol. 10, Moscow, Acad. Sci. USSR. pp. 213–229.

Samokhvalova, G. V. 1938. The effect of X-rays on fishes (*Lebistes,* swordtailed minnows and crucian carp). Biol. zh. 7: 1023–1034.

* Saurov, M. M. 1957. The radioactive contamination of a fish living in water containing radioactive strontium. In: *Trudy Vsesoyuz. Konf. po Med. Radiol. Voprosy Gigieny i Dozimetrii.* [Proceedings of the All-Union Conference on Medical Radiology (problems of hygiene and dosimetry)]. Moscow, Medgiz. pp. 66–73. (AEC-tr-3746).

* Schmidt, N. K. 1946. The axial gradients of injury due to the effect of X-rays and other agents in *Dendrocoelum lacteum.* Dokl. Akad. Nauk SSSR 53(8): 761–764. (in English).

Schmidt, N. K. 1949. Vliyaniye rentgenovykh luchey na organizm v svyazi s ego polyarnoy differentsirovkoy. (The effect of X-rays on the organism in relation to its polar differential). Author's summary of thesis for Degree of Candidate of Biological Sciences. Leningrad, Leningrad State Univ. 15 pp.

Semenov, D. I. and I. P. Tregubenko. 1957. The effect of complexing agents on the behavior of metals and emitters *in vivo.* 2. Ethylenediaminetetracetic acid sodium salt. Tr. inst. biol., Akad. Nauk SSSR, Ural'skiy filial (9). Collected papers of the Biophysics Laboratory, Vol. 1: 20–56.

* Sereda, G. A. 1962. Contamination of the seas and oceans by artificial radioactive substances. In: *Voprosy Yadernoy Meteorologii.* (Problems of Nuclear Meteorology). Karol', I. L. and S. G. Malakhov (eds.). Moscow, Gosatomizdat. pp. 259–271. (AEC-tr-6128; pp. 312–328).

Shakhidzhanyan, L. G., D. G. Fleyshman, V. V. Glazunov, V. G. Leont'yev and V. P Nesterov. 1960. A method of measuring beta-activity in biological subjects employing scintillating jelly. Med. radiologiya 5(3): 72–74.

Shekhanova, I. A. 1959. *Izucheniye Fosfornogo Obmena u Molodi Karpovykh i Osetrovykh Ryb s Primeneniyem Radioaktivnogo Fosfora.* The Study of Phosphorus Metabolism in Carp and Sturgeon Fry Using Radioactive Phosphorus. Moscow, Ryb. Khozyaystvo Press. 78 pp.

Shekhtman, Ya. L. and V. A. Klyupfel'. 1930. A roentgenometric study of the mechanism of the biological effect of X-rays. Communication 1. The immediate of long-term effect of X-rays. Zh. eksp. biol. 6(4): 280–282.

Sherstnev, A. I. and O. P. Vasil'yev. 1965. Composition and radioactivity level of the biomass in the equatorial area of the Atlantic Ocean and on the northwest African shelf. In: *Voprosy Gidrobiologiya.* (Problems of Hydrobiology). Moscow, Nauka. pp. 455–456.

* Shvedov, V. P. (ed.). 1959. *Opredeleniye Zagryazneniy Biosfery Produktami Yadernykh Ispytaniy: Sbornik Statey.* (Evaluation of the contamination of the biosphere by products of nuclear tests: Collected Papers). Moscow, Akad. Nauk SSSR. UN Document A/AC.-82/G/L.323. 253 pp. (AEC-tr-4599).

* Shvedov, V. P., V. A. Blinov, L. I. Gedeonov and Ye. P. Ankudinov. 1958. Radioactive fallout in the Leningrad area. Atom. energiya 5(5): 577–582. (English translation pp. 1500–1507).

* Shvedov, V. P., L. M. Ivanova, A. M. Maksimova and A. V. Stepanov. 1962. Content and distribution of radioactive substances in sea-water. In: *Radioaktivnye Zagryazneniya Vneshney Sredy.* (Radioactive Contamination of the External Environment). Shvedov, V. P. and S. I. Shirokov (eds.). Moscow, Gosatomizdat. pp. 236–242. (AEC-tr-6049; pp. 250–257).

* Shvedov, V. P. and S. I. Shirokov (eds.). 1962. *Radioaktivnye Zagryazneniya Vneshney Sredy.* (Radioactive Contamination of the External Environment). Moscow, Gosatomizdat. 275 pp. (AEC-tr-6049).

* Shvedov, V. P., G. V. Yakovleva and M. I. Zhilkina. 1960. The external gamma-radiation dose of radioactive fallout in 1959. Atom. energiya 9(4): 323–324. (English translation pp. 868–869).

Shvedov, V. P., A. A. Yuzefovich, L. I. Belyayev, N. I. Popov, L. I. Gedeonov, L. M. Ivanova, N. P. Ushakova and S. A. Patin. 1961. Issledovaniye radioaktivnykh zagryazneniy v Rayone Chernogo Morya v 1959g. (Investigation of radioactive pollution in the Black Sea area in 1959). UN Document A/AC.82/G/L.703. 28 pp.

Shvedov, V. P., A. A. Yuzefovich, L. I. Gedeonov, A. M. Maksimova and S. A. Patin. 1963. *Opredeleniye strontsiya-90 v vodakh Atlanticheskogo Okeana v 1961g.* (Strontium-90 determination in Atlantic waters in 1961). Moscow, State Commission on the Utilization of Atomic Energy, USSR. UN Document A/AC.82/G/L.899. 41 pp.

* Shvedov, V. P., A. A. Yuzefovich, V. A. Yeroshev-Shak, S. A. Patin, L. M. Ivanova, A. V. Stepanova and A. M. Maksimova. 1964. Determination of strontium-90 content in the Black Sea. In: *Radioaktivnaya Zagryaznennost' Morey i Okeanov.* (Radioactive Contamination of Seas and Oceans). Baranov, V. I. and S. M. Khitrov (eds.). Moscow, Nauka. pp. 76–80. (JPRS 26002).

* Shvedov, V. P., M. I. Zhilkina, V. K. Zinov'cyva, L. M. Ivanova, T. P. Makarova and N. A. Pavlova. 1959. ^{90}Sr fallout from the atmosphere in the Leningrad area. Atom.

energiya 7(5): 479. (English translation pp. 948–949).

* Sinochkin, Yu. D., D. A. Perumov, S. A. Patin and E. G. Azhazha. 1962. Zirconium-based ionites. Communication 2. The use of zirconium oxalate to determine ^{90}Sr from the daughter isotope ^{90}Y in sea-water samples. Radiokhimiya 4(2): 198–205. (English translation pp. 177–182).

Skopintsev, B. A. 1959. Features of salt content and distribution in the Black Sea. Priroda (12): 87–90.

Spitsyn, V. I. 1960. Comment in Panel Discussion No. 6, Effects of radiation on marine organisms. In: *Disposal of Radioactive Wastes*, Vol. II. Vienna, Intern. Atomic Energy Agency. p. 315.

* Spitsyn, V. I. and B. S. Kolychev. 1960. Results of the international conference on the treatment and burial of radioactive waste, Monaco. Atom. energiya 9(1): 58–62. (English translation pp. 560–567).

Starik, I. Ye. 1960. Osnovy Radiokhimii. (Principles of Radiochemistry). Moscow, Leningrad, Acad. Sci. USSR. 459 pp.

Strakhov, N. M., N. G. Brodskaya, L. M. Knyazeva, A. N. Razzhivina, M. A. Rateyev, D. S. Sapozhnikov and Ye. S. Shishova. 1954. *Obrazovaniye Osadkov v Sovremennykh Vodoyemakh.* (Formation of Sediments in Present-day Water-bodies). Moscow, Acad. Sci. USSR. 791 pp.

Strelin, G. S. 1934. The effects of X-rays on frog corneal epithelium. Vestn. rengtenol. i radiol. 13(1): 98–113.

Sukachev, V. N. 1947. Principles of Biogeocoenological Theory. In: *Yubileyniy Sbornik, Posvyashchenniy 30-Letiyu Velikoy Oktyabr'skoy Sotsialisticheskoy Revolyutsii.* (Collected Papers on the Occasion of the 30th Anniversary of the Great October Socialist Revolution). Vol. 2. Moscow, Acad. Sci. USSR. pp. 283–305.

Tarusov, B. N. 1954. *Osnovy Biologicheskogo Deystviya Radioaktivnykh Izlucheniy.* (Principles of the Biological Effect of Radioactive Radiations). Moscow, Medgiz. 140 pp.

Telitchenko, M. M. 1958. Chronic effects of small doses of ^{238}U, ^{232}Th, and $^{89+90}Sr$ on a number of generations of *Daphnia magna* Straus. Nauchn. dokl. vyssh. shkoly (biol.) (1): 114–118.

* Telitchenko, M. M. 1961. Concentration of the radioisotopes of strontium-(89+90)-yttrium-90 by mirror carp. Ryb. khozyaystvo (5): 40–43. (JPRS 21, 989; pp. 5–10).

Telitchenko, M. M. 1962. The role of metabolism in the accumulation of radioisotopes by fishes. Nauchn. dokl. vyssh. shkoly (biol.) (3): 90–93.

Timofeyev-Resovskiy, N. V. 1957. Use of radiations and radiation sources in experimental biogeocoenology. Bot. zh. 42(2): 161–194.

Timofeyev-Resovskiy, N. V. and N. V. Luchnik. 1960. The cytological principles of radio-stimulation of plants. Tr. inst. biol., Akad. Nauk SSSR, Ural'skiy filial (13). Collected papers of the Biophysics Laboratory, Vol. 3: 5–17.

Timofeyev-Resovskiy, N. V. and Ye. A. Timofeyeva-Resovskaya. 1959. Distribution of emitters in water-bodies. In: *Soveshchaniye po Voprosam Ekspluatatsii Kamskogo Vodokhranilishcha.* (A Conference on Aspects of the Exploitation of the Kama Reservoir). Perm', Ural'sk. fil. Akad. Nauk SSSR. pp. 1–21.

* Timofeyev-Resovskiy, N. V., Ye. A. Timofeyeva-Resovskaya, G. A. Milyutina and A. B. Getsova. 1960. Concentration factors of the radioisotopes of 16 different elements in freshwater organisms and the effect of EDTA on some of them. Dokl. Akad. Nauk SSSR 132(5): 1191–1194. (English translation pp. 369–372) (French translation CEA-tr-

R-1243).

Timofeyeva, N. A. 1965. Features of radiostrontium concentration by freshwater organisms. In: *Voprosy Gidrobiologii*. (Problems of Hydrobiology). Moscow, Nauka. pp. 416–417.

Timofeyeva-Resovskaya, Ye. A. 1956. Some experiments on the decontamination of water. In: *Vsesoyuznaya Konferentsiya po Meditsinkoy Radiologii*. Tezisy dokladov. (Gigiyenicheskaya Sektsiya). [All-Union Conference on Medical Radiology. Abstracts of proceedings. (Hygiene Commission)]. Moscow, Medgiz. pp. 24–25.

Timofeyeva-Resovskaya, Ye. A. 1957. Soil biological decontamination of water in sedimentation tanks. Byull. Mosk. obshch. ispyt. prirody (biol.) 62(1): 37–41.

Timofeyeva-Resovskaya, Ye. A. 1958a. The effect of ethylenediaminetetracetic acid sodium salt on strontium, ruthenium, cerium, cobalt, zinc and cesium concentration factors. In: *Kompleksony (sintez, svoystva, primeneniye v biologii i meditsine)*. [Complexing agents (their synthesis, properties and uses in biology and medicine)]. Sverdlovsk. pp. 159–163.

Timofeyeva-Resovskaya, Ye. A. 1958b. The rate of subwater fouling (formation of periphyton) in the presence of weak concentrations of emitters. Byull. Ural'skogo otd. Mosk. obshch. ispyt. prirody (biol.) (1): 87–96.

* Timofeyeva-Resovskaya, Ye. A. 1963. Distribution of radioisotopes in the main components of freshwater-bodies. (Monograph). Tr. inst. biol., Akad. Nauk SSSR, Ural'skiy filial (30): 1–78. (JPRS 21, 816).

Timofeyeva-Resovskaya, Ye. A., B. M. Agafonov and N. V. Timofeyev-Resovskiy. 1960. Soil-biological decontamination of water. Tr. inst. biol., Akad. Nauk SSSR, Ural'skiy filial (13). Collected Papers of the Biophysics Laboratory, vol. 3: 35–48.

Timofeyeva-Resovskaya, Ye. A., E. I. Popova and G. G. Polikarpov. 1958. Concentration of chemical elements from aqueous solutions by freshwater organisms. 1. Concentration of radioisotopes of phosphorus, zinc, strontium, ruthenium, cesium and cerium by various species of freshwater mollusc. Byull. Mosk. obshch. ispyt. prirody (biol.) 63(3): 65–78.

Timofeyeva-Resovskaya, Ye. A. and N. V. Timofeyev-Resovskiy. 1958. Uptake of chemical elements from the water by freshwater organisms. 2. Rate of uptake of different radioisotopes by the pond snail (*L. stagnalis*). Byull. Mosk. obshch. ispyt. prirody (biol.) 63(5): 123–131.

Timofeyeva-Resovskaya, Ye .A. and N. V. Timofeyev-Resovskiy. 1960. Distribution of trace elements in the components of water-bodies. 2. Soil. biological deactivation of water in sedimentation tanks. Tr. inst. biol., Akad. Nauk SSSR, Ural'skiy filial (12). Collected Papers of the Biophysics Laboratory, vol. 3: 194–223.

Timofeyeva-Resovskaya, Ye. A., N. A. Timofeyeva and N. V. Timofeyev-Resovskiy. 1959. Concentration of chemical elements from aqueous solutions by freshwater organisms. 3. The concentration factors of various radioisotopes by three species of water plants. Byull. Mosk. obshch. ispyt. prirody (biol.) 64(5): 117–131.

* Titlyanova, A. A. and V. I. Ivanov. 1960. Concentration of cesium by three species of freshwater plants from solutions of different concentrations. Dokl. Akad. Nauk SSSR 136(3): 721–722.

Troshin, A. S. 1956. *Problema Kletornoy Pronitsayemosti*. (The problem of Cell Permeability). Moscow, Leningrad, Acad. Sci. USSR. 474 pp.

Vasil'yev, O. P. and A. I. Sherstnev. 1965. Effect of strontium-90 on the formed elements of fish blood. In: *Voprosy Gidrobiologii*. (Problems of Hydrobiology). Moscow, Nauka.

pp. 56–57.

* Vavilov, P. P., I. N. Verkhovskaya, R. P. Kodaneva and O. N. Popova. 1963. The growth and development of *Vicia faba* L. under conditions of an increased soil content of U and Ra. Radiobiologiya 3(1): 132–138. (AEC-tr-5434; pp. 186–196).

Vel'tishcheva, I. F. 1951. Some features of metabolism in sturgeon and sevryuga fry reared under various conditions. Tr. Saratov. otd. Kaspiyskogo filiala vses. nauchn.-issled inst. morsk. ryb. khoz. okeanogr. 1: 96–112.

* Vende, G. V. and V. P. Parchevskiy. 1964a. Radioactive contamination of organisms in the Black Sea. Results and methodological peculiarities in the determination of ^{90}Sr in *Cystoseira*. In: *Radioaktivnaya Zagryaznennost' Morey i Okeanov*. (Radioactive Contamination of Seas and Oceans). Baranov, F. I. and L. M. Khitrov (eds.). Moscow, Nauka. pp. 143–150. (JPRS 26, 002; pp. 77–86).

* Vende, G. V. and V. P. Parchevskiy. 1964b. A 4π scintillation solution counter for absolute measurements of beta-active samples in films and preparation procedure. Radiobiologiya 4(3): 465–466. (AEC-tr-6406; pp. 202–205).

Verkhovskaya, I. N. 1962. *Brom v Zhivotnom Organizme i Mekhanizm Ego Deystviya*. (Bromine in the Animal Organism and the Mechanism of Its Action). Moscow, Acad. Sci. USSR. 307 pp.

Verkhovskaya, I. N., N. A. Gabelova, Ye. G. Zinov'yeva, V. M. Klechkovskiy, A. M. Kuzin, Ya. V. Mamul', Ye. G. Plyshevskaya, G. M. Frank and Ya. L. Shekhtman. 1955. *Metod Mechenykh Atomov v Biologii*. (The Method of Tagged Atoms in Biology). Kuzin, A. M. (ed.). Moscow, Moscow Univ. 452 pp.

Vernadskiy, V. I. 1929. Concentration of radium by living organisms. Dokl. Akad. Nauk SSSR ser. A, 2: 33–34.

Vernadskiy, V. I. 1940. *Biogeokhimicheskiye Ocherki*. 1922–1932. (Biogeochemical Outlines. 1922–1932). Moscow, Acad. Sci. USSR. 249 pp.

Vinberg, G. G. and V. I. Gaponenko. 1958. The development of various aquatic organisms in cultures at various levels of ^{32}P activity. Tr., biol. stants. oz. Naroch' (1): 193–196.

Vinogradov, A. P. 1956. The biological role of the radioisotope ^{40}K. Dokl. Akad. Nauk SSSR 110(3): 375–378.

* Vinogradov, A. P. 1957a. The isotope ^{40}K and its biological role. Biokhimiya 22(1–2): 14–20.

* Vinogradov, A. P. 1957b. *Geokhimiya Redkikh i Rasseyannykh Khimicheskikh Elementov v Pochvakh*. (The Geochemistry of Rare and Trace Elements in Soils). 2nd. ed., Moscow, Acad. Sci. USSR. 238 pp. (English translation Consultants Bureau, New York, 1959. 209 pp.)

Vinogradova, Z. A. 1965. The role of marine plankton in chemical element migration. In: *Voprosy Gidrobiologii*. (Problems of Hydrobiology). Moscow, Nauka. pp. 75–76.

* Vodyanitskiy, V. A. 1958. Is it permissible to dispose of the wastes of atomic works in the Black Sea? Priroda (2): 46–52. (AEC-tr-3296).

Voynar, A. O. 1960. *Biologicheskaya Rol' Mikroelementov v Organizme Zhivotnykh i Cheloveka*. (The Biological Role of Trace Elements in Animals and Man). Moscow, Vysshaya shkola. 544 pp.

Yasvoin, G. V. 1926. The sequence of development of morphological changes in the cell under the influence of X-rays and radium. Vestn. rentgenol. i radiol. 4(5–6): 305.

Yemel'yanov, V. S. (ed.). 1958. Radioecology of agricultural plants and animals. In: *Kratkaya Entsiklopediya "Atomnaya Energiya"*. (Short Encyclopedia of Atomic Energy).

Moscow, Bol'skaya sovetskaya entsiklopedia. p. 363.

Yemel'yanov, V. S. 1962. A further note on the problem of radioactive waste. Novoye vremya (39): 16–18.

Yemel'yanov, V. S. 1964. *Atcm i Mir.* (Atoms and Peace). Moscow, Atomizdat. 275 pp.

Zakutinskiy, D. I. 1959. *Voprosy Toksikologii Radioaktivnykh Veshchestv.* (Problems of the Toxicology of Radioactive Substances). Moscow, Medgiz. 150 pp.

Zaytsev, Yu. P. 1960. Existence of the neuston biocoenosis in the marine pelagial. Nauchn. zap. Odes. biol. stants. (2): 37–42. (in Ukrainian).

Zaytsev, Yu. P. 1961a. An uninvestigated pelagic biocoenosis of the Black Sea. *Nauchnaya Konferentsiya, Posvyashchennaya 40-Letney Deyatel'nosti Novorossiyskoy Biologicheskoy Stantsii (Mart 1961).* Tezisy dokladov. [Scientific Conference Dedicated to the 40th Anniversary of the Activity of the Novorossiysk Biological Station (March 1961). Abstracts of Reports]. Novorossiysk. pp. 41–44.

Zaytsev, Yu. P. 1961b. Indication of the spawning of the anchovy and the horse-mackerel in the northwestern Black Sea by the use of a new method. Nauk. zap. Odes. biol. stants. (3): 45–59. (in Ukrainian).

Zaytsev, Yu. P. 1961c. The near surface pelagic biocoenosis of the Black Sea. Zool. zh. 40(6): 818–825.

Zaytsev, Yu P. 1962. Some features of development of the hyponeuston in the northwestern Black Sea. Nauk. zap. Odes. biol. stants. (4): 19–31 (in Ukrainian).

Zaytsev, Yu. P. 1963. On the boundary of two oceans. Priroda (11): 27–31.

Zaytsev, Yu. P. 1965. Some means of adaptation of the hyponeuston organisms to the life conditions of the near-surface layer of a sea. In: *Voprosy Gidrobiologii.* (Problems of Hydrobiology). Moscow, Nauka. pp. 160–161.

* Zaytsev, Yu. P. and G. G. Polikarpov. 1964. Problems of the radioecology of the hyponeuston. Okeanologiya 4(3): 423–430. (JPRS 25,966).

Zaytsev, Yu. P. and G. G. Polikarpov. 1965. The hyponeuston and aspects of its radioecology. In: *Voprosy Gidrobiologii.* (Problems of Hydrobiology). Moscow, Nauka. p. 161.

Zenkevich, L. A. 1951. *Morya SSSR, ikh fauna i flora.* (The Seas of the USSR, Their Fauna and Flora). Moscow, Uchpedgiz. 168 pp.

* Zenkevich, L. A. 1960. Adsorption and biocirculation in oceans. In: *Disposal of Radioactive Wastes.* Vol. II. Vienna, Intern. Atomic Energy Agency. pp. 99–100.

* Zenkevich, L. A. 1963. *Biologiya Morey SSSR.* (Biology of the Seas of the USSR). Moscow, Acad. Sci. USSR, 739 pp. (Trans. avail. from Interscience, New York).

Zenkevich, L. A. and D. I. Shcherbakov. 1960. Progress in contemporary oceanology. Priroda (4): 56–64.

Zernov, S. A. 1949. *Obshchaya Gidrobiologiya.* (General Hydrobiology). Moscow and Leningrad, Acad. Sci. USSR. 503 pp.

Zesenko, A. Ya. 1965. The distribution of radionuclides within marine animals. In: *Voprosy Gidrobiologiya.* (Problems of Hydrobiology). Moscow, Nauka. pp. 174–175.

Zesenko, A. Ya. and V. N. Ivanov. 1965. Accumulation of phosphorus-32 by the developing eggs of sea fishes. Vopr. ikhtiol. 6(2) (in press).

* Zesenko, A. Ya. and G. G. Polikarpov. 1965. Ruthenium-106 concentration and distribution factors in the organs and tissues of marine molluscs. Radiobiologiya 5(1): 320–322. (AEC-tr-6598).

Zhadin, V. I. 1958. Use of radioisotopes in hydrobiology and pisciculture. Priroda (6): 58–62.

Zhadin, V. I. 1960. Biological purification of water basins. Vestn. Akad. Nauk SSSR 9: 61–64.

Zhadin, V. I., N. B. Il'inskaya, A. N. Svetovidov and A. S. Troshin. 1955. Problems and methods of tagging insects and fishes with radioactive isotopes. In: *Trudy nauchnoy sessii posvyashchennoy dostizheniyam i Zadacham Sovetskoy biofiziki v sel'skom khozyaystve.* (Proceedings of the Conference on the Achievements and Tasks of Soviet Biophysics in Agriculture). Kuzin, A. M. (ed.). Moscow, Akad. Nauk SSSR. pp. 276–284.

Zhadin, V. I., S. I. Kuznetsov and N. V. Timofeyev-Resovskiy. 1959. Radioactive isotopes in solving the problems of hydrobiology. In: *Trudy Vtoroy Mezhdunarodnoy Konferentsii po Mirnomu Ispol'zovaniyu Atomnoy Energii* (Zheneva, 1958). Papers by Soviet scientists, Vol. 6, Moscow, Atomizdat. pp. 335–346. [*Proc. 2nd UN Intern. Conf. on the Peaceful Uses of Atomic Energy* 27: 200–207. (Geneva, 1958)].

Zharova, T. V. 1965. Filamentous bacteria such as *Sphaerotilus* in water-bodies and their role in the accumulation of radioisotopes. In: *Voprosy Gidrobiologii.* (Problems of Hydrobiology). Moscow, Nauka. pp. 152–153.

* Zhogova, V. M. 1962. The role of aquatic flora and fauna in modifying the sanitary condition of bodies of water. Bacterial flora. In: *Radiatsionnaya Gigiyena.* (Radiation Hygiene). Vol. 2, Marey, A. N. (ed.), Moscow, Medgiz. pp. 50–59. (JPRS 21799; pp. 27–39).

d. Non-Russian titles

Adams, C. E., N. H. Farlow and W. R. Schell. 1960. The compositions, structures and origins of radioactive fallout particles. Geochim. Cosmochim. Acta 18 (1/2): 42–56.

Agnedal, P-O, N-E. Barring, J. S. Lindhe and J. W. Smith. 1958. Biological investigations in the water recipient at Studsvik, the research establishment of the Swedish Atomic Energy Company. In: *Proc. 2nd UN Intern. Conf. on the Peaceful Uses of Atomic Energy.* 18: 400–403.

Aliverti, G. 1960. Panel discussion. In: *Disposal of Radioactive Wastes,* Vol. II. Intern. Atomic Energy Agency. Vienna. pp. 249–252.

Allen, A. L. and L. M. Mulkay. 1960. X-ray effects on embryos of the paradise fish, with notes on normal stages. Growth 24(2): 131–168.

Allen, B. M., O. A. Schjeide, M. J. Millard and R. Piccirillo. 1952. Relation of X-irradiation dosage and elapsed time to the destruction of hematopoietic cells of tadpoles. Univ. of California at Los Angeles, U.S. AEC report UCLA-232. 21 pp.

Allen, B. M., O. A. Schjeide and R. Piccirillo. 1954. Influence of anoxia upon hematopoietic cells of tadpoles exposed to X-irradiation and colchicine. J. Cell. Comp. Physiol. 44(2): 318–322.

Anderson, J. B., E. C. Tsivoglou and S. D. Shearer. 1963. Effects of uranium mill wastes on biological fauna of the Animas River (Colorado-New Mexico). In: *Radioecology.* pp. 373–383. (see: Schultz, V. and A. W. Klement, Jr., 1963).

Anghileri, L. J. 1957. Decontamination and potability of the waters of the Rio de la Plata after a nuclear explosion. Publs. Commission Nacional de Energia Atomica, Buenos Aires, ser. quim. 1(12): 141–188. (in Spanish).

Anghileri, L. J. 1959. Estudio de la contaminacíon del *Prochilodus platensis* (Sábalo) con productos de fisión. Comision Nacional de Energia Atomica, Buenos Aires, Informe No. 15. 19 pp.

Anghileri, L. J. 1960a. Study of the contamination and absorption of Sr^{90} and Cs^{137} by *Prochilodus platensis* (shad). Comision Nacional de Energia Atomica, Buenos Aires,

Informe No. 31. 14 pp. (in Spanish) (English translation: Hanford Labs., U.S. AEC. report HW-tr-41).

Anghileri, L. J. 1960b. Estudio de la adsorcion de productos de fision por tierra de Ezeiza. Comision Nacional de Energia Atomica, Buenos Aires, Informe No. 35. 13 pp.

Anonymous. 1960a. Disposal of radioactive waste. Nature 185(4705): 1–2.

Anonymous. 1960b. The biological effects of atomic radiation. Summary reports. National Academy of Sciences – National Research Council, Washington, D.C. xiii, 90 pp.

Anonymous. 1960c. Rapport van de wetenschappelijke conferentie over de verwijdering van radioactieve afvalstoffen. Gehouden 16–21 november 1959 te Monaco. Nederl. Akad. Wet. Rapp. 12(1): 34–42.

Anthenisse, L. J. and J. Lever. 1956. I^{131}-accumulation in some invertebrates. Koninkl. Nederl. Akad. Wet. Proc. 59(4): 562–565.

Aubert, M., R. Chesselet and D. Nordemann. 1962. Mesures de radioactivité (α, β, γ) effectuées sur les échantillons de plancton et d'algues prélevés en Méditerranée, dans la zone côtière de la Ville de Nice. Cahiers C.E.R.B.O.M., (2): 23–41.

Auerbach, S. I. and J. S. Olson. 1963. Biological and environmental behavior of ruthenium and rhodium. In: Radioecology. pp. 509–519. (see: Schultz, V. and A. W. Klement, Jr., 1963).

Bachmann, R. W. and E. P. Odum. 1960. Uptake of Zn^{65} and primary productivity in marine benthic algae. Limnol. Oceanogr. 5(4): 349–355.

Back, A. 1939. Sur un type de lésions produites chez Paramecium caudatum par les rayons X. Compt. rend. soc. biol. 131(19): 1103–1106.

Bacq, Z. M. and P. Alexander. 1955. Fundamentals of Radiobiology. Butterworths Sci. Publ., London. xii, 389 pp.

Bacq, Z. M., J. Damblon and A. Herve. 1955. Radiorésistance d'une algue Acetabulatia mediterranea. Compt. rend. soc. biol. 149(13/14): 1512–1515.

Berner, L., Jr., R. Bieri, E. D. Goldberg, D. Martin and R. L. Wisner. 1962. Field studies of uptake of fission products by marine organisms. In: Oceanographic Studies During Operation 'Wigwam'. Limnol. Oceanogr. Suppl. to Vol. 7, pp. lxxxii–xci.

Bevelander, G. 1952. Calcification in mollusks. III. Intake and deposition of ^{45}Ca and ^{32}P in relation to shell formation. Biol. Bull. 102(1): 9–15.

Björnerstedt, R. 1960. Health hazards from fission products and fallout. II. Gamma radiation from nuclear weapons fallout. Arkiv för Fysik 16(4): 293–313.

Black, W. A. P. and R. L. Mitchell. 1952. Trace elements in the common brown algae and in sea-water. J. Marine Biol. Assoc. U.K. 30(3): 575–583.

Blinks, L. R. 1952. Effects of radiation on marine algae. J. Cell. Comp. Physiol. 39 (Suppl. 2): 11–18.

Blinks, L. R. and J. P. Nielsen. 1940. The cell sap of Hydrodictyon. J. Gen. Physiol. 23(5): 551–559.

Bolter, E., K. K. Turekian and D. F. Shutz. 1964. The distribution of rubidium, cesium and barium in the oceans. J. Geochem. Soc. 28: 1459–1466.

Bonham, K. 1955. Sensitivity to X-rays of the early cleavage stages of the snail Helisoma subcrenatum. Growth 19(1): 9–18. (also: Univ. of Washington, U.S. AEC report UWFL-39).

Bonham, K., L. R. Donaldson, R. F. Foster, A. D. Welander and A. H. Seymour. 1948. The effects of X-ray on mortality, weight, length, and counts of erythrocytes and hematopoietic cells in fingerling chinook salmon, Oncorhyncus tschawytscha Walbaum. Growth

12(2): 107–121. (also: Univ. of Washington, U.S. AEC report UWFL-3a; MDDC-1416).

Bonham, K. and R. F. Palumbo. 1951. Effects of X-rays on snails, crustacea and algae. Growth 15(3): 155–188. (also: Univ. of Washington, U.S. AEC report UWFL-26).

Bonham, K., A. H. Seymour, L. R. Donaldson and A. D. Welander. 1947. Lethal effects of X-rays on marine microplankton organisms. Science 106(2750): 245–246.

Bonham, K. and A. D. Welander. 1963. Increase in radioresistance of fish to lethal doses with advancing embryonic development. In: *Radioecology*. pp. 353–358. (see: Schultz, V. and A. W. Klement, Jr., 1963).

Boroughs, H., W. A. Chipman and T. R. Rice. 1957. Laboratory experiments on the up-take, accumulation and loss of radionuclides by marine organisms. In: *The Effects of Atomic Radiation on Oceanography and Fisheries*. National Academy of Sciences – National Research Council, Washington, D.C. Publ. 551. pp. 80–87.

Boroughs, H. and D. F. Reid. 1958. The role of the blood in the transportation of strontium-90-yttrium-90 in teleost fish. Biol. Bull. 115(1): 64–73.

Boroughs, H., S. J. Townsley and W. Ego. 1958. The accumulation of Y^{90} from equilibrium of Sr^{90}-Y^{90} by *Artemia salina* L. Limnol. Oceanogr. 3(4): 413–417.

Boroughs, H., S. J. Townsley and R. W. Hiatt. 1956a. The metabolism of radionuclides by marine organisms. I. The uptake, accumulation and loss of strontium-89 by fishes. Biol. Bull. 111(3): 336–351.

Boroughs, H., S. J. Townsley and R. W. Hiatt. 1956b. The metabolism of radionuclides by marine organisms. II. The uptake, accumulation and loss of yttrium-91 by marine fish and the importance of short-lived radionuclides in the sea. Biol. Bull. 111(3): 352–357.

Boroughs, H., S. J. Townsley and R. W. Hiatt. 1956c. Method for predicting amount of strontium-89 in marine fishes by external monitoring. Science 124(3230): 1027–1028.

Boroughs, H., S. J. Townsley and R. W. Hiatt. 1957. The metabolism of radionuclides by marine organisms. III. The uptake of calcium-45 in solution by marine fish. Limnol. Oceanogr. 2(1): 28–32.

Bowen, H. J. M. 1956. Strontium and barium in sea water and marine organisms. J. Marine Biol. Assoc. U.K. 35(3): 451–460.

Bowen, V. 1959. Unpublished material presented at meeting of Subcommittee on Radio-activity in the Ocean. Special Committee on Oceanic Research, at Intern. Oceanogr. Congr., New York, 2 Sept. 1959. (Cited from Anonymous, 1960b, page 88, table 3).

Bowen, V. T. and T. T. Sugihara. 1960. Strontium-90 in the mixed layer of the Atlantic Ocean. Nature 186(4718): 71–72.

Bowen, V. T. and T. T. Sugihara. 1963. Cycling and levels of strontium-90, cerium-144, and promethium-147 in the Atlantic Ocean. In: *Radioecology*. pp. 135–139. (see: Schultz, V. and A. W. Klement, Jr., 1963).

Bray, H. G. and K. White. 1957. *Kinetics and Thermodynamics in Biochemical Processes*. Academic Press, New York. xii, 343 pp.

Bridgman, J. and R. F. Kimball. 1954. The effects of X-rays on division rate and survival of *Tillina magna* and *Colpoda* sp. with an account of delayed death. J. Cell. Comp. Physiol. 44(3): 431–445.

Brown, F. A., Jr., M. F. Bennett and C. L. Ralph. 1955. Apparent reversible influence of cosmic-ray-induced showers upon biological systems. Proc. Soc. Exptl. Biol. Med. 89(3): 332–337.

Brown, F. A., Jr., H. M. Webb and M. F. Bennett. 1958. Comparisons of some fluctuations in cosmic radiation and in organismic activity during 1954, 1955 and 1956. Am. J. Physiol. 195(1): 237–243.

Brown, V. M. 1962. The accumulation of strontium-90 and yttrium-90 from a continuously flowing natural water by eggs and alevins of the Atlantic salmon and the sea trout. United Kingdom Atomic Energy Authority, report PG-288. 16 pp.

Brown, V. M. and W. L. Templeton. 1964. Resistance of fish embryos to chronic irradiation. Nature 203(4951): 1257–1259.

Brues, A. M. and G. A. Sacher. 1952. Analysis of mammalian radiation injury and lethality. In: *Symposium on Radiobiology* (The Basic Aspects of the Radiation Effects on living Systems). John Wiley and Sons, New York. pp. 441–465.

Bryan, G. W. 1961. The accumulation of radioactive caesium in crabs. J. Marine Biol. Assoc. U.K. 41(3): 551–575.

Bryan, G. W. 1963. The accumulation of ^{137}Cs by brackish water invertebrates and its relation to the regulation of potassium and sodium. J. Marine Biol. Assoc. U.K. 43(2): 541–565.

Bryan, G. W. and E. Ward. 1962. Potassium metabolism and the accumulation of ^{137}Cs by decapod crustacea. J. Marine Biol. Assoc. U.K. 42(2): 199–241.

Bryan, G. W. and E. Ward. 1965. The absorption and loss of radioactive and non-radioactive manganese by the lobster, *Homarus vulgaris*. J. Marine Biol. Assoc. U.K. 45(1): 65–95.

Buchsbaum, R. 1958. Chap. 5. Species response to radiation; radioecology. In: *Radiation Biology and Medicine*, (W. D. Claus, ed.), Addison-Wesley Publ. Co., Reading, Mass. pp. 124–141.

Buzzati-Traverso, A. A. 1961. The biological aspects of sea water contamination through radioactivity. In: Intern. Symp. on Legal and Administrative Problems of Protection in the Peaceful Uses of Atomic Energy, Bruxelles, 1960. EURATOM, Brussels, pp. 611–623.

Cerrai, E., L. Pelati, B. Schreiber and C. Triulzi. 1962. Misure di radioattività di campioni di zooplancton del Mare Adriatico e del Mar Ligure pescati fra il maggio 1961 ed il gennaio 1962. Energia nucleare 9(3): 173–175.

Chakravarti, D. and E. E. Held. 1960. Potassium and cesium-137 in *Birgus latro* (coconut crab) muscle collected at Rongelap Atoll. J. Marine Biol. Assoc. (India) 2(1): 75–81. (also: Univ. of Washington, U.S. AEC report UWFL-64).

Cheek, C. H. and V. J. Linnenbom. 1960. Calculation of absorbed doses. Power Apparatus and Systems 51: 1004–1016.

Cherry, R. D. 1964. Alpha-radioactivity of plankton. Nature 203(4941): 139–143.

Chesselet, R. and C. Lalou. 1964. Concentrations en radionuclides émitteurs gamma présentées par des holothuries prélevées dans la zone côtière d'Antibes au cap Ferrat en août 1962. Bull. Inst. Oceanogr. 63(1305); 16 pp.

Chipman, W. A. 1958a. Accumulation of radioactive materials by fishery organisms. Proc. Gulf and Caribbean Fish. Inst., 11th Ann. Session. pp. 97–110.

Chipman, W. A. 1958b. Biological accumulation of radioactive materials. In: Proc. First Texas Conf. on the Utilization of Atomic Energy. Special Publ. Agr. Mech. Coll. Texas, College Station, Texas. pp. 36–41.

Chipman, W. A. 1959a. Accumulation of radioactive pollutants by marine organisms and its relation to fisheries. In: Trans. 2nd Seminar on Biological Problems in Water Pollution, April 20–24, 1959, Cincinnati, Ohio, U.S. PHS, R. A. Taft Sanitary Engineering Center, Cincinnati, Technical report W60–3. pp. 8–14.

Chipman, W. A. 1959b. Disposal of radioactive materials and its relation to fisheries. Proc.

Natl. Shellfisheries Assoc. 49: 5–12.

Chipman, W. A. 1960. Biological aspects of disposal of radioactive wastes in marine environments. In: *Disposal of Radioactive Wastes*, Vol. II. Intern. Atomic Energy Agency, Vienna. pp. 3–15.

Chipman, W. A. 1961. General history of the radiobiology program of the biological laboratory of the U.S. Bureau of Commercial Fisheries at Beaufort, North Carolina. U.S. Bur. Commerical Fisheries, U.S. Fish and Wildlife Service. mimeo 9 pp.

Chipman, W. A., T. R. Rice and T. J. Price. 1958. Uptake and accumulation of radioactive zinc by marine plankton, fish, and shellfish. U.S. Fish and Wildlife Service, Fish. Bull. 135, pp. 279–292: (in: Fishery Bull. of the U.S. Fish and Wildlife Service, Vol. 58).

Coffin, C. C., F. R. Hayes, L. H. Jodrey and S. G. Whiteway. 1949. Exchange of materials in a lake as studied by radioactive phosphorus. Nature 163(4155): 963–964.

Cohen, P. and C. Gailledreau. 1960. Adsorption des produits de fission sur des vases de Méditerranée. Centre d'Études Nucléaires, Saclay, France, rapport CEA-1704. 13 pp.

Collins, J. C. (ed.). 1960. *Radioactive Wastes, Their Treatment and Disposal*. E. and F. N. Spon Ltd., London. xxi, 239 pp.

Committee on Oceanography. 1959a. Oceanography 1960–1970: 5 – Artificial radioactivity in the marine environment. National Academy of Sciences – National Research Council, Washington, D.C. 31 pp.

Committee on Oceanography. 1959b. Radioactive waste disposal into the Atlantic and Gulf coastal waters. National Academy of Sciences – National Research Council, Washington, D.C. Publ. 655. viii, 37 pp.

Committee on the Effects of Atomic Radiation on Oceanography and Fisheries. 1959. Considerations on the disposal of radioactive wastes from nuclear-powered ships into the marine environment. National Academy of Sciences – National Research Council, Washington, D.C. Publ. 658. xviii, 52 pp.

Copenhaver, W. M., R. H. van Dyke and R. Rugh. 1960. Effects of X-irradiation on embryos at critical stages of heart development. Yale J. Biol. Med. 32(6): 421–430.

Corcoran, E. F. and J. F. Kimball, Jr. 1963. The uptake, accumulation and exchange of strontium-90 by open sea phytoplankton. In: *Radioecology*. pp. 187–191. (see: Schultz, V. and A. W. Klement, Jr., 1963).

Cousteau, J. Y. (with J. Dugan). 1963. *The Living Sea*. Harper and Row, New York. 325 pp.

Crossley Jr., D. A. and H. F. Howden. 1961. Insect-vegetation relationships in an area contaminated by radioactive wastes. Ecology 42(2): 302–317.

Crowther, J. A. 1926. The action of X-rays on *Colpidium colpoda*. Proc. Roy. Soc. London 100 (B704): 390–404.

Danckwerte, P. V. 1956. Fission product disposal from Windscale. Nuclear Engineering 1: 25–27.

Davis, J. J. 1963. Cesium and its relationships to potassium in ecology. In: *Radioecology*. pp. 539–556. (see: Schultz, V. and A. W. Klement, Jr., 1963).

Davis, J. J. and R. F. Foster. 1958. Bioaccumulation of radioisotopes through aquatic food chains. Ecology 39(3): 530–535.

Davis, J. J., R. W. Perkins, R. F. Palmer, W. C. Hanson and J. F. Cline. 1958. Radioactive materials in aquatic and terrestrial organisms exposed to reactor effluent water. In: *Proc. 2nd UN Intern. Conf. on the Peaceful Uses of Atomic Energy* 18: 423–428. (also: Congress of the United States, Joint Committee on Industrial Radioactive Waste Disposal 2: 1103–1123).

Davis, J. J., D. G. Watson and C. C. Palmiter. 1956. Radiobiological studies of the Columbia River through December 1955. Hanford Atomic Products Operation, General Electric Co., U.S. AEC report HW-36074 (Del). 177 pp.

Donaldson, L. R. and R. F. Foster. 1957. Effects of radiation on aquatic organisms. In: *The Effects of Atomic Radiation on Oceanography and Fisheries*. National Academy of Sciences – National Research Council, Washington, D.C. Publ. 551. pp. 96–102.

Dunster, H. J. 1956. The discharge of radioactive waste products into the Irish Sea. Part 2. The preliminary estimate of the safe daily discharge of radioactive effluent. In: *Proc. Intern. Conf. on the Peaceful Uses of Atomic Energy* 9: 712–715.

Dunster, H. J. 1958. The disposal of radioactive liquid wastes into coastal waters. In: *Proc. 2nd UN Intern. Conf. on the Peaceful Uses of Atomic Energy* 18: 390–399.

DuShane, G. 1959. Nuclear ships. Science 127(3310): 1315.

Ellinger, F. 1939. Note of action of X-rays on goldfish (*Carassius auratus*). Proc. Soc. Exptl. Biol. Med. 41(2): 527–529.

Evans, T. C. 1952. The influence of quantity and quality of radiation on the biological effect. In: *Symposium on Radiobiology* (The Basic Aspects of Radiation Effects on Living Systems). John Wiley and Sons, New York. pp. 393–410.

Fage, L. 1960. Panel discussion. In: *Disposal of Radioactive Wastes*, Vol. II. Intern. Atomic Energy Agency, Vienna. pp. 251–252.

Fair, D. R. R. and A. S. McLean. 1956. The disposal of waste products into the sea. Part 3. The experimental discharge of radioactive effluents. In: *Proc. Intern. Conf. on the Peaceful Uses of Atomic Energy* 9: 716–717.

Federov, A. F. 1964. Mathematical formulas for concentration coefficient study of radioactive material to sea biota. Bull. Inst. Oceanogr. 63(1304): 11 pp.

Fitzgerald, B. W., J. S. Rankin and D. M. Skauen. 1962. Zinc-65 levels in oysters in the Thames River (Connecticut). Science 135(3507): 926.

Folsom, T. R. and J. H. Harley. 1957. Comparison of some natural radiations received by selected organisms. In: *The Effects of Atomic Radiation on Oceanography and Fisheries*. National Academy of Sciences – National Research Council, Washington, D.C. Publ. 551. pp. 28–33.

Folsom, T. R. and G. J. Mohanrao. 1961. 1960–1961 cesium-137 in the Pacific: a preliminary summary. Unpublished paper presented at 10th Pacific Sci. Congr., Honolulu, Hawaii. 6 pp.

Folsom, T. R., G. J. Mohanrao and P. Winchell. 1960. Fall-out caesium in surface sea water of the California coast (1959–1960) by gamma-ray measurements. Nature 187 (4736): 480–482.

Folsom, T. R., D. R. Young, J. N. Johnson and K. C. Pillai. 1963. Manganese-54 and zinc-65 in coastal organisms of California. Nature 200(4904): 327–329.

Fontaine, M. 1956. Les océans et les dangers résultants de l'utilisation de l'énergie atomique. J. Conseil. Intern. Explor. Mer 21(3): 241–249.

Fontaine, M. 1959. Océanographie et radioéléments. Journées des 24 et 25 Févr. 1958. Centre belge d'Océanographie et de récherches sousmarines, F.N.R.S., Liège. pp. 91–107.

Fontaine, Y. 1960. La contamination radioactive des milieux et des organismes aquatiques. Centre d'Études Nucléaires, Commissariat à l'Énergie, Saclay, France, rapport CEA-N-1588. 155 pp. (also: U.S. AEC translation AEC-tr-5358).

Fontaine, Y.-A. and A. Aeberhardt. 1961. Étude de la contamination d'une biocénose artificielle d'eau douce par le cérium radioactif. Compt. Rend. 252(20): 3151–3153.

Foreman, E. E. and W. L. Templeton. 1955. The uptake of zirconium-95 and niobium-95 by *Porphyra* sp. UKAEA, Ind. Group, Windscale Works, Shellafield, Cumb., England report RDB (w)/TN-187. 15 pp.

Foster, R. F. 1963. Environmental behaviour of chromium and neptunium. In: *Radio-ecology*. pp. 569–576. (see: Schultz, V. and A. W. Klement, Jr., 1963.)

Foster, R. F. and J. J. Davis. 1956. The accumulation of radioactive substances in aquatic organisms. In: *Proc. Intern. Conf. on the Peaceful Uses of Atomic Energy* 13: 364–367. [also: Intern. J. Appl. Radiation and Isotopes 1(1–2): 131–132].

Foster, R. F., L. R. Donaldson, A. D. Welander, K. Bonham and A. N. Seymour. 1949. The effect on embryos and young of rainbow trout from exposing the parent fish to X-rays. Growth 13(2): 119–142. (also: Univ. of Washington, U.S. AEC report UWFL-12).

Freiling, E. C. and N. E. Ballou. 1962. Nature of nuclear debris in sea water. Nature 195(4848): 1283–1287.

Fretter, V. 1953. Experiments with radioactive strontium (^{90}Sr) on certain mollusks and polychaetes. J. Marine Biol. Assoc. U.K. 32(2): 367–384.

Fretter, V. 1955. Uptake of radioactive sodium (^{24}Na) by *Nereis diversicolor* Mueller and *Perinereis cultrifera* Grude. J. Marine Biol. Assoc. U.K. 34(1): 151–160.

Frye, A. 1962. *The Hazards of Atomic Wastes*: perspectives and proposals on oceanic disposal. Public Affairs Press, Washington, D.C. iv, 45 pp.

Fukai, R. 1964. A note on the strontium-90 content of marine organisms as an index for variations of strontium-90 in sea-water. Bull. Inst. Oceanogr. 63(1307); 16 pp.

Fukai, R. and W. W. Meinke. 1962. Activation analyses of vanadium, arsenic, molybdenum, tungsten, rhenium, and gold in marine organisms. Limnol. Oceanogr. 7(2): 186–200.

Fukai, R. and N. Yamagata. 1962. Estimation of the levels of caesium-137 in sea-water by the analysis of marine organisms. Nature 194(4827): 466.

Gajewskaja, N. 1923. Einfluss der Roentgenstrahlen auf *Artemia salina*. Verhandl. Intern. Ver. theor. u. angew. Limnol. Grund-Vers. Z. Kiel. pp. 359–362.

Gilet, R. and P. Ozenda. 1961. Analyse de l'action des rayons X sur la multiplication d'un organisme chlorophyllien unicellulaire: la chlorophycée *Scenedesmus crassus* Chod. Compt. rend. 252(24): 3867–3869.

Glaser, R. 1960. Zur Frage der Anreicherung von Radiojod (J^{131}) in der Muschel *Dreissensia polymorpha* Pall. Naturwissenschaften 47(12): 284.

Glaser, R. 1961a. Über die Aufnahme von Radiojod (J^{131}) aus wässriger Lösung durch die Muschel *Dreissensia polymorpha* Pall. Zool. Jb. Physiol. 69(3): 339–378.

Glaser, R. 1961b. Über die Radioactivität des Planktons des Stechlin- und Nehmitzsees bei Rheinsberg (Mark Brandenburg). Kernenergie 4(5): 398–399.

* Glaser, R. 1962a. Die Stellung der Wasserorganismen im Rahmen des Strahlenschutz-problems der Umgebung kerntechnischer Anlagen. Kernenergie 5(7): 515–533. (also: U.S. AEC translation AEC-tr-5754).

Glaser, R. 1962b. Die Inkorporation und Dekorporation von Radiocerium (^{144}Ce) durch die Muschel *Dreissensia polymorpha* Pall. Zool. Jb. Physiol. 70(1): 91–110.

Glaser, R. 1962c. Ein Beitrag zur Frage der Trägerabhängigkeit bei der Anreicherung radioaktiver Isotope durch Wasserorganismen aus wässriger Lösung. Biol. Zentralblatt 81(5): 539–547.

Glaser, R. and E. Spode. 1960. Über die Aufnahme von Radioyttrium Y^{91} in einzelligen Algen. Physikalische Gesellschaft in der DDR. In: Probleme und Ergebnisse aus Biophysik und Strahlenbiologie. II. Bericht über die Vierte Arbeitstagung Biophysik der Physikalischen Gesellschaft Deutschen Demokratischen Republik vom 13–15 Okt. 1958 in Oberhof. Pfenningsdorf, G. and W. Eckart. Akademie-Verlag, Berlin, German Democratic Rep. pp. 54–59.

Glocker, R., H. Langendorff and M. Langendorff. 1933. Gibs es eine obere Dosis-Grenze für die biologische Wirkung der Röntgenstrahlen. Naturwissenschaften 21(17): 316.

Glocker, R., H. Langendorff and A. Reuss. 1933. Über die Wirkung von Röntgenstrahlen verschiedener Wellenlange auf biologische Objekte. III. Strahlentherapie 46(2): 517–528.

Godward, M. B. E. 1960. Resistance of algae to radiation. Nature 185(4714): 706.

Goldberg, E. D. 1957. Biogeochemistry of trace metals. In: *Treatise on Marine Ecology and Paleontology*. Hedgpeth, J. W. (ed.) Memoir Geological Soc. Am. 67(1): 345–357.

Gong, J. K., W. H. Shipman, H. V. Weiss and S. H. Cohn. 1957. Uptake of fission products and neutron induced radionuclides by the clam. Proc. Soc. Exptl. Biol. Med. 95(3): 451–454.

Gorbman, A. and M. S. James. 1963. An exploratory study of radiation damage in the thyroids of coral reef fishes from the Eniwetok Atoll. In: *Radioecology*. pp. 385–399. (see: Schultz, V. and A. W. Klement, Jr., 1963).

Goreau, T. F. and V. T. Bowen. 1955. Calcium uptake by a coral. Science 122(3181): 1188–1189.

Graham, J. W. 1959. Metabolically induced precipitation of trace elements from sea water. Science 129(3360): 1428–1429.

Greendale, A. E. and N. E. Ballou. 1954. Physical state of fission product elements following their vaporization in distilled water and sea water. U.S. Naval Radiological Defense Lab., San Francisco, report USNRDL 436. 28 pp.

Gromadska, M. 1961. Radioekologia – nowa gałaź ekologii. Ekologia Polska, ser. B 7(3): 175–182.

Grosch, D. S. and H. Smith. 1957. X-ray experiments with *Molgula manhattensis*: adult sensitivity and induced zygotic lethality. Biol. Bull. 112(2): 171–179.

Hance, R. T. and A. Clarce. 1926. Studies on X-ray effect. The effect of X-rays on the division rate of paramecium. J. Exptl. Med. 43(1): 61–70.

Hanson, W. C. and H. A. Kornberg. 1956. Radioactivity in terrestrial animals near an atomic site. In: *Proc. Intern. Conf. on the Peaceful Uses of Atomic Energy* 13: 385–388.

Harley, J. H. (ed.). 1956. Operation Troll. N.Y. Operations Office, U.S. AEC report NYO-4656. v, 37 pp.

Harriss, E. B., L. F. Lamerton, M. J. Ord and J. F. Danielli. 1952. Site of action of mutagenic reagents. Nature 170(4335): 921–922.

Harvey, R. S. 1964. Uptake of radionuclides by fresh water algae and fish. Health Physics 10(4): 243–248.

Healy, J. W., B. V. Anderson, H. V. Clukey and J. K. Soldat. 1958. Radiation exposure to people in the environs of a major production atomic energy plant. In: *Proc. 2nd UN Intern. Conf. on the Peaceful Uses of Atomic Energy* 18: 309–318.

Hela, I., 1963. Alternative ways of expressing the concentration factors for radioactive substances in aquatic organisms. Bull. Inst. Oceanogr. 61 (1280); 8 pp.

Henshaw, P. S. 1932. Studies of the effect of roentgen rays on the time of the first cleavage some marine invertebrate eggs. I. Recovery from roentgen-ray effect in *Arbacia* eggs. Am. J. Roentgenol. Radium Therapy 27(6): 890–898.

Henshaw, P. S. 1940. Further studies on the action of roentgen rays on the gametes of

Arbacia punctata. IV. Changes in radiosensitivity during the first cleavage cycle. Am. J. Roentgenol. Radium Therapy 43: 917–920.

Hess, D. W. 1959. Radiation ecology of the Front Range, report on background gamma radiation of sites in the Rocky Mountain Front Range west of Boulder, Colorado. Inst. of Arctic and Alpine Research, Univ. of Colorado, Boulder, Colorado. 26 pp.

Hibiya, T. and T. Yagi. 1956. Effects of fission materials upon development of aquatic animals. In: *Research in the Effects and Influences of the Nuclear Bomb Test Explosions*, Vol. II. Japan Society for the Promotion of Science, Ueno, Tokyo. pp. 1219–1224.

Higano, R. 1959. Radiochemical analysis of the equatorial Pacific surface water. Intern. Oceanogr. Congr.: Preprints. American Association for the Advancement of Sciences, Washington, D.C. pp. 815–816.

Hiyama, Y. (comp.) 1960. Biogeochemical studies with strontium-90, cesium-137, and others. Nippon Gakujutsu Shinkokai pp. 79–93. (also: U.S. AEC translation AEC-tr-4245).

Hiyama, Y. 1962. Studies in uptake of radioisotopes by edible marine fishes. Material presented at the Intern. Atomic Energy Agency Panel on coordination of research projects on radioactivity in the marine environment, Vienna, Nov. 1962. (cited from Wallauschek and Lützen, 1964).

Hiyama, Y. and R. A. Ichikawa. 1961. A measure on level of strontium-90 concentration in sea water around Japan, at the end of 1956. Rec. Oceanogr. Works Japan 4(1): 49–54.

Hiyama, Y. and J. M. Khan. 1964. On the concentration factors of radioactive I, Co, Fe and Ru in marine organisms. Rec. Oceanogr. Works Japan 7(2): 79–106.

Hiyama, Y. and M. Shimizu. 1964. On the concentration factors of radioactive Cs, Sr, Cd, Zn and Ce in marine organisms. Rec. Oceanogr. Works Japan 7(2): 43–77.

Hoagland, D. P. and T. C. Broyer. 1936. General nature of the process of salt accumulation by roots with description of experimental methods. Plant Physiol. 11(3): 471–507.

Hollaender, A. (ed.). 1954. *Radiation Biology*. Vol. I, Parts 1 and 2. McGraw-Hill Book Co., N.Y. ix, 1265 pp.

Hsiao, S. C. and H. Boroughs. 1958. The uptake of radioactive calcium by sea urchin eggs. I. Entrance of Ca^{45} into unfertilised egg cytoplasm. Biol. Bull. 114(2): 196–204.

Hsiao, S. C. and P. C. Daniel. 1960. Studies on the effect of ionising radiation upon developing sea urchin. II. (abstr.) Anat. Rec. 137(3): 365–366.

Hunter, H. F. and N. E. Ballou. 1951. Fission-product decay rates. Nucleonics 9(5): C2-C7.

Ichikawa, R. 1961. On the concentration factors of some important radionuclides in the marine food organisms. Bull. Japanese Soc. Sci. Fisheries 27(1): 66–74.

International Atomic Energy Agency. 1960. *Disposal of Radioactive Wastes*. Conf. Proc., Monaco, 16–21 November 1959. Vienna, Austria. Vol. 2, 584 pp.

International Atomic Energy Agency. 1962. Disposal of radioactive wastes into marine and fresh waters. Vienna, Austria. Bibliographical Series No. 5. 368 pp.

Jacobson, B. S. 1957. Evidence for recovery from X-ray damage in *Chlamydomonas*. Radiation Res. 7(4): 394–406.

Jodrey, L. H. 1953. Studies on shell formation. III. Measurement of calcium deposition in shell and calcium turnover in mantle tissue using the mantle-shell preparation and Ca^{45}. Biol. Bull. 104(3): 398–407.

Joint Committee on Atomic Energy, Congress of the U.S. 1959a. Industrial Radioactive Waste Disposal. Hearing before the Special Subcommittee on Radiation. (5 Vols.). U.S.

Government Printing Office, Washington, D.C. 3142 pp.

Joint Committee on Atomic Energy, Congress of the U.S. 1959b. Fallout from Nuclear Weapons Tests, Hearings before the Special Subcommittee on Radiation, May 5–8, 1959. U.S. Government Printing Office, Washington, D.C. 4 Vols., 2693 pp.

Jones, R. F. 1960. The accumulation of nitrosil ruthenium by fine particles and marine organisms. Limnol. Oceanogr. 5(3): 312–325.

Joseph, W. and G. Prowazek. 1902. Versuche über die Einwirkung von Röntgen-Strahlen auf einige Organismen, besonders auf deren Plasmatätigkeit. Z. allgem. Physiol. 1(2): 142–153.

Joyner, T. and R. Eisler. 1961. Retention and translocation of radioactive zinc by salmon fingerlings. Growth 25(2): 151–156.

Katcoff, S. 1960. Fission-product yields from neutron-induced fission. Nucleonics 18(11): 201–208.

Ketchum, B. H. and V. T. Bowen. 1958. Biological factors determining the distribution of radioisotopes in the sea. In: Proc. 2nd UN Intern. Conf. on the Peaceful Uses of Atomic Energy 18: 429–433.

Kimball, R. F. and N. Gaither. 1952. The role of externally produced hydrogen peroxide in damage to Paramecium aurelia by X-rays. Proc. Soc. Exptl. Biol. Med. 80(3): 525–529.

Klement, Jr., A. W. and V. Schultz. 1962. Terrestrial and freshwater radioecology: a selected bibliography. U.S. AEC report TID-3910. ii, 79 pp.

Klement, Jr., A. W. and V. Schultz. 1963. Terrestrial and freshwater radioecology: a selected bibliography. U.S. AEC report TID-3910 (Suppl. 1). ii, 95 pp.

Klement, Jr., A. W. and V. Schultz. 1964. Terrestrial and freshwater radioecology: a selected bibliography. U.S. AEC report TID-3910 (Suppl. 2). ii, 123 pp.

Klement, Jr., A. W. and V. Schultz. 1965. Terrestrial and freshwater radioecology: a selected bibliography. U.S. AEC report TID-3910 (Suppl. 3). iii, 115 pp.

Knauss, H. J. and J. W. Porter. 1954. The absorption of inorganic ions by Chlorella pyrenoidosa. Plant Physiol. 29(3): 229–234.

Kobayashi, S. and H. Hirata, 1957. Effects of X-irradiation upon rainbow trout (Salmo irideus) I. Influence on the feeding activity in rainbow trout fry. Bull. Fac. Fish. Hokkaido Univ. 8(1): 23–35.

Krauskopff, K. B. 1956. Factors controlling the concentrations of thirteen rare metals in the sea water. Geochim. Cosmochim. Acta 9: 1–32.

Krough, A. 1939. Osmotic regulation in aquatic animals. Cambridge Univ. Press, London. 242 pp. (cited from Karzinkin, 1962).

Krumholz, L. A. 1956. Observations on the fish population of a lake contaminated by radioactive wastes. Bull. Am. Mus. Nat. Hist. 110(4): 281–367.

Krumholz, L. A. and R .F. Foster. 1957. Accumulation and retention of radioactivity from fission products and other radiomaterials by fresh-water organisms. In: The Effects of Atomic Radiation on Oceanography and Fisheries. National Academy of Sciences – National Research Council, Washington, D.C. Publ. 551 .pp. 88–95.

Krumholz, L. A., E. D. Goldberg and H. A. Boroughs. 1957. Ecological factors involved in the uptake, accumulation, and loss of radionuclides by aquatic organisms. In: The Effects of Atomic Radiation on Oceanography and Fisheries. National Academy of Sciences-National Research Council, Washington, D.C. Publ. 551. pp. 69–79.

Lea, D. E. 1955. Actions of radiations on living cells. rev. ed. Cambridge Univ. Press, London. 416 pp.

Levi, H. W. 1960. Die Beseitigung radioaktiver Abfälle – ein Lagebericht. Atomwirtschaft 5(2): 57–59.

Lewin, R. A. and T. J. Chow. 1961. La enpreno de strontio en kokolitoforoj. Plant and Cell. Physiol. 2(2): 203–208.

Lowman, F. G. 1958. Radionuclides in plankton near the Marshall Islands, 1956. Univ. of Washington, U.S. AEC report UWFL-54. v, 31 pp.

Lowman, F. G. 1963a. Iron and cobalt in ecology. In: Radioecology. pp. 561–567. (see: Schultz, V. and A. W. Klement, Jr., 1963).

Lowman, F. G. 1963b. Radionuclides in plankton and tuna from the Central Pacific. In: Radioecology. pp. 145–149. (see: Schultz, V. and A. W. Klement, Jr., 1963).

Lynch, W. F. 1958. The effect of X-rays, irradiated sea water and oxidizing agents on the rate of attachment of Bugula larvae. Biol. Bull. 114(2): 215–225.

Marcovich, M. H. 1956. The problem of the biological action of low doses of ionizing radiation. In: Proc. Intern. Conf. on the Peaceful Uses of Atomic Energy 11: 244–247.

Marshall, J. S. 1963. The effects of continuous, sub-lethal gamma radiation on the intrinsic rate of natural increase and other population attributes of Daphnia pulex. In: Radioecology. pp. 363–366. (see: Schultz, V. and A. W. Klement, Jr., 1963).

Martin, D., Jr. 1957. The uptake of radioactive wastes by benthic organisms. In: Proc. 9th Pacific Sci. Congr., Bangkok, Thailand. pp. 167–169.

Matsue, Y. and R. Hirano. 1956. Accumulation of radioactivity in plankton cultured or kept in water infusions of the Bikini ashes. In: Research in the Effects and Influences of the Nuclear Bomb Test Explosions, Vol. II. Japan Society for the Promotion of Science. Ueno, Tokyo. pp. 1099–1104.

Mauchline, J. 1961. A review of the biological significance of certain neutron induced radioisotopes in the marine environment. United Kingdom Atomic Energy Authority, Production and Engineering Groups, Risley report PG-248(W). 17 pp.

Mauchline, J. 1963. The biological and geographical distribution in the Irish Sea of radioactive effluent from Windscale Works 1959 to 1960. United Kingdom Atomic Energy Authority, Production Group, Windscale, report AHSB(RP)-R-27. 74 pp.

Mauchline, J. and W. L. Templeton. 1964. Artificial and natural radioisotopes in the marine environment. In: Oceanogr. Mar. Biol. Ann. Rev., H. Barnes (ed.). London 2: 229–279. Hafner Publ. Co., New York.

Merejkowsky, M. C. 1879–1880. Sur une anomalie chez les Hydroméduses et sur leur mode du nutrition au moyen de l'ectoderme. Arch. Zool. Experim. B8 (cited from Karzinkin, 1962).

Mikami, Y., H. Watanabe and K. Takano. 1956. The influence of radioactive rain-water on the growth and differentiation of a tropical fish Zebra danio. In: Research in the Effects and Influences of the Nuclear Bomb Test Explosions, Vol. II. Japan Society for the Promotion of Science, Ueno, Tokyo. pp. 1225–1229.

Miyake, Y. and K. Saruhashi. 1960. Vertical and horizontal mixing rates of radioactive material in the ocean. In: Disposal of Radioactive Wastes, Vol. II. Intern. Atomic Energy Agency, Vienna. pp. 167–173.

Miyake, Y., K. Saruhashi, Y. Katsuragi and T. Kanazawa. 1961. Penetration of artificial radioactivity in deep waters of the Pacific and vertical diffusion of sea water. Unpublished paper presented at 10th Pacific Sci. Congr., Honolulu, Hawaii. [see: 1962. Penetration of ^{90}Sr and ^{137}Cs in deep layers of the Pacific and vertical diffusion rate of deep water. J. Radiation Res. (Tokyo) 3(3): 141–147].

Morgan, A. and D. G. Stanbury. 1961. The contamination of rivers with fission products from fallout. Health Physics 5(3/4): 101–107.

Mori, T. and M. Saiki. 1956. Studies on the distribution of administered radioactive zinc in the tissues of fish. In: *Research in the Effects and Influences of the Nuclear Bomb Test Explosions*, Vol. II. Japan Society for the Promotion of Science, Ueno, Tokyo. pp. 1205–1208.

Nakai, Z., R. Fukai, H. Tozawa, S. Hattori, K. Okubo and T. Kidachi. (1960/1961). Radioactivity of marine organisms and sediments in the Tokyo bay and its southern neighbourhood. Collected Reprints of Tokai Regional Fish. Res. Lab., Chuo-ku, Tokyo. Contr. NB-333, pp. 18–38.

National Committee on Radiation Protection. 1959. Maximum permissible body burdens and maximum permissible concentrations of radionuclides in air and in water. Natl. Bur. Stds. Handbook 69, Washington, D.C. viii, 95 pp. (Handbook 69 supersedes Handbook 52).

Nelson, D. J. 1963. The strontium and calcium relationships in Clinch and Tennessee River mollusks. In: *Radioecology*. pp. 203–211. (see: Schultz, V. and A .W. Klement, Jr., 1963).

Norris, W. and W. Kisieleski. 1948. Comparative metabolism of radium, strontium, and calcium. *Cold Spring Harbor Symposia on Quantitative Biology* 13: 164–172.

O'Brien, R. D. and L. S. Wolfe. 1964. *Radiation, Radioactivity, and Insects*. Academic Press, New York. xv, 211 pp.

Odum, E. P. 1956. Consideration of the total environment in power reactor waste disposal. In: *Proc. Intern. Conf. on the Peaceful Uses Atomic Energy* 13: 350–353.

Odum, E. P. 1959. Chap. 14, Radiation ecology. In: *Fundamentals of Ecology* (2nd. ed.). W. B. Saunders Co., Philadelphia. pp. 452–486.

Okada, I., I. Osakabe, T. Kikuchi and K. Konno. 1956. On the influence of γ-ray radiation on the aquatic animals. On the influence in the early development of gold-fish (*Carassius auratus* L.). In: *Research on the Effects and Influences of the Nuclear Bomb Test Explosions*, Vol. II. Japan Society for the Promotion of Science, Ueno, Tokyo. pp. 1211–1218.

Ophel, I. L. 1963. The fate of radiostrontium in a freshwater community. In: *Radioecology*. pp. 213–216. (see: Schultz, V. and A. W. Klement, Jr., 1963).

Oppermann, K. 1913a. Die Entwicklung von Forelleneiern nach Befruchtung mit radium-bestrahlten Samenfaden. II. Das Verhalten des Radiumchromatins während des ersten Teilungsstadien. Arch. f. microsc. Anat. Z. Abt. 83(4): 307–323.

Oppermann, K. 1913b. Die Entwicklung von Forelleneiern nach Befruchtung mit radium-bestrahlten Samenfaden. Arch. f. microsc. Anat. Z. Abt. 83(2): 141–189, pls. 5–7.

Osterberg, C., W. G. Pearcy, and H. Curl, Jr. 1964. Radioactivity and its relationship to food chains. J. Marine Res. 22(1): 2–12.

Owens, M., N. S. Thom, and G. E. Eden. 1961. The uptake and release of radioactive strontium by freshwater plants. Proc. Soc. Water Treatment Exam. 10(1): 53–65.

Palumbo, R. F. 1957. Uptake of iodine-131 by the red alga *Asparagopsis taxiformis*. Univ. of Washington, U.S. AEC report UWFL-44. 8 pp.

Palumbo, R. F. 1963. Factors controlling the distribution of the rare earths in the environment and its living organisms. In: *Radioecology*. pp. 533–537. (see: Schultz, V. and A. W. Klement, Jr., 1963).

Parker, J. 1956. Translocation of P^{32} and due behavior in two species of marine algae. Naturwissenschaften 43(19): 452.

Pendleton, R. C. and W. C. Hanson. 1958. Absorption of caesium-137 by components of an aquatic community. In: *Proc. 2nd UN Intern. Conf. on the Peaceful Uses of Atomic Energy* 18: 419–422.

Perkins, R. W., J. M. Nielsen, W. C. Roesch and R. C. McCall. 1960. Zinc-65 and chromium-51 in foods and people. Science 132(3443): 1895–1897.

Peters, T. 1960. Über die Wirkung von Röntgenstrahlen und befruchtete Eizellen von *Triton alpestris* unter besonderer Berücksichtigung der Wirkung kleiner Strahlendosen und geringer Schaden. Strahlentherapie 112(4): 525–542.

Piccotti, M. 1961. Expansion des déchets radioactifs et dynamique des eaux de la Méditerranée. Rapport et procès-verbaux des réunions (Commission Internationale pour l'Exploration Scientifique de la Méditerranée, Monaco), 16(3): 569–574.

Pickering, D. C. and J. W. Lucas. 1962. Uptake of radiostrontium by an alga, and the influence of calcium ion in the water. Nature 193(4820): 1046–1047.

Pillai, K. C. and A. K. Ganguly. 1961. Evaluation of maximum permissible concentration of radioisotopes in sea water of Bombay. Indian Atomic Energy Establishment Trombay, Bombay, report AEET/HP/R-11. 37 pp.

Platt, R. B. 1959. Studies in radiation ecology (abstr.). Proc. 9th Intern. Bot. Congr. 2(A): 28.

Polikarpov, G. G. 1961f. Ability of some Black Sea organisms to accumulate fission products. Science 133(3459): 1127–1128.

Pomeroy, L. R. and H. H. Haskin. 1954. The uptake and utilisation of phosphate ions from sea water by the American oyster *Crassostrea virginica* (Gmel.). Biol. Bull. 107(1): 123–129.

Pora, E. A., J. Oros, D. Rusdea, F. Stoicovici and C. Wittenberger. 1961. Fixation et élimination du P^{32} par quelques organismes de la Mer Noire. J. physiol. (France) 53(2): 449–450.

Pora, E. A., I. Oros, D. Rusdea, C. Wittenberger and F. Stoicovici. 1962. Inglolarea si eliminarea P^{32} la cîteva organisme din Marea Neagra. Acad. rep. populare Romine Studii si cerc. biol. fil. Cluj. 12(2): 293–326.

Pritchard, D. W. 1960. The application of existing oceanographic knowledge to the problem of radioactive waste disposal into the sea. In: *Disposal of Radioactive Wastes*, Vol. II. Intern. Atomic Energy Agency, Vienna. pp. 229–249.

Pritchard, D. W. 1961. Disposal of radioactive wastes in the ocean. Health Physics 6(3/4): 103–109.

Prosser, C. L., W. Pervinsek, J. Arnold, G. Svihla and P. C. Tompkins. 1945. Accumulation and distribution of radioactive strontium, barium-lanthanum, fission mixture and sodium in goldfish. Univ. of Chicago, U.S. AEC report MDDC-496; CH-3233. 42 pp.

Pütter, A. 1907. Der Stoffhaushalt des Meeres. Z. allgem. Physiol. 7(2/3): 321–368.

Ralston, H. J. 1939. The immediate and delayed action of X-rays upon the protozoan *Dunaliella salina* (abstr.). Am. J. Cancer 37(4): 609.

Revelle, R. R., T. R. Folsom, E. D. Goldberg and J. D. Isaacs. 1956. Nuclear science and oceanography. In: *Proc. Intern. Conf. on the Peaceful Uses of Atomic Energy* 13: 371–380.

Revelle, R. and M. Schaefer. 1957. General considerations concerning the oceans as a receptacle for artificially radioactive materials. In: *The Effects of Atomic Radiation on Oceanography and Fisheries*. National Academy of Sciences – National Research Council, Washington, D.C. Publ. 551. pp. 1–25.

Rice, T. R. 1956. The accumulation and exchange of strontium by marine planktonic algae. Limnol. Oceanogr. 1(2): 123–138.

Rice, T. R. 1963a. Accumulation of radionuclides by aquatic organisms. In: Studies of the Fate of Certain Radionuclides in Estuarine and Other Aquatic Environments. (J. J. Sabo and P. H. Bedrosian, eds.). Proc. Symposium held in Savannah, Georgia. U.S. Dept of Health, Education, and Welfare. Environmental Health Series. Public Health Service Publ. No. 999-R-3, xiii, 73 pp.

Rice, T. R. 1963b. Review of zinc in ecology. In: *Radioecology*. pp. 619–631. (see: Schultz, V. and A. W. Klement, Jr., 1963).

Rice, T. R. 1963c. The role of phytoplankton in the cycling of radionuclides in the marine environment. In: *Radioecology*. pp. 179–185. (see: Schultz, V. and A. W. Klement, Jr., 1963).

Rice, T. R. and V. M. Willis. 1959. Uptake, accumulation, and loss of radioactive cerium-144 by marine planktonic algae. Limnol. Oceanogr. 4(3): 277–290.

Robeck, G. G., C. Henderson and R. C. Palange. 1954. Water quality studies on the Columbia River. U.S. PHS, R. A. Taft Sanitary Engineering Center, Cincinnati, 293 pp.

Roche, J., S. André and I. Covelli. 1960. Sur la fixation de l'iode par la moule (*Mytilus galloprovincialis* L.) et la nature des combinaisons iodées elaborées. Compt. Rend. Soc. Biol. 154(12): 2201–2206.

Roche, J. and Y. Yagi. 1952. Sur la fixation de l'iode radioactif par les algues et sur les constituants iodés des Laminaires. Compt. Rend. Soc. Biol. 146(4): 642–645.

Ronkin, R. R. 1950. The uptake of radioactive phosphate by the excited gill of the mussel *Mytilus edulis*. J. Cell. Comp. Physiol. 35(2): 241–250.

Rosenthal, H. L. 1957. Uptake of calcium-45 and strontium-90 from water by freshwater fishes. Science 126(3276): 699–700.

Rosenthal, H. L. 1960. Accumulation of strontium-90 and calcium-45 by freshwater fishes. Proc. Soc. Exptl. Biol. Med. 104(1): 88–91.

Rugh, R. 1960. General biology: gametes, the developing embryo, and cellular differentiations. In: *Mechanisms in Radiobiology*, Vol. II. (M. Errera and A. Forssberg, (eds.).) Academic Press, N.Y. pp. 1–94.

Rugh, R. and H. Clugston. 1955a. Effects of various levels of X-irradiation on the gametes and early embryos of *Fundulus heteroclitus*. Biol. Bull. 108(3): 318–325.

Rugh, R. and H. Clugston. 1955b. Hydratation and radiosensitivity. Proc. Soc. Exptl. Biol. Med. 88(3): 467–472.

Russell, L. B. and W. L. Russell. 1954. An analysis of the changing radiation response of the developing mouse embryo. J. Cell. Comp. Physiol. 43 (Suppl. 1): 103–149.

Sabo, J. J. and P. H. Bedrosian (eds.). 1963. Studies of the fate of certain radionuclides in estuarine and other aquatic environments. Proc. Symposium held in Savannah, Georgia. U.S. Dept. of Health, Education, and Welfare. Environmental Health Series. Public Health Service Publ. No. 999-R-3, xiii, 73 pp.

Saddington, K. and W. L. Templeton. 1958. *Disposal of Radioactive Waste*. George Newnes Ltd., London. 112 pp.

Sándi, E. 1962. Radioactivity in snail shells due to fallout. Nature 193 (4812): 290.

Schaefer, H. J. 1955. Biological significance of natural background of ionizing radiation. Observations of sea level and at extreme altitude. J. Aviat. Med. 26(6): 453–462.

Schaefer, M. B. 1961. Some fundamental aspects of marine ecology in relation to radioactive wastes. Health Physics 6(3/4): 97–102.

Schreiber, B. 1960. Ecology of acantharia and strontium circulation in the sea. In: *Disposal of Radioactive Wastes*, Vol. II. Intern. Atomic Energy Agency, Vienna. pp. 25–38.

Schroeder, B. W. and R. D. Cherry. 1962. Caesium-137 in the seas off the Cape of Good Hope. Nature 194(4829): 669.

Schultz, V. and A. W. Klement, Jr. (eds.). 1963. *Radioecology*. Proc. First National Symposium on Radioecology, Held at Colorado State Univ., Ft. Collins, Colorado, Sept. 10–15, 1961. Reinhold Publ. Corp., N.Y. xvii, 746 pp.

Scott, C. M. 1937. Some quantitative aspects of the biological action of X- and γ-rays. Medical Research Council Special Report Series No. 223, Her Majesty's Stationary Office, London. 99 pp.

Scott, R. 1954. A study of cesium accumulation by marine algae. In: *Radioisotope Conference, 1954*. Vol. I. Medical and Physiological Applications. Proc. 2nd Conf., Oxford., 19–23 July. pp. 373–380. Academic Press Inc., New York.

Seymour, A. H. 1963. Radioactivity of marine organisms from Guam, Palau and the Gulf of Siam, 1958–1959. In: *Radioecology*. pp. 151–157. (see: Schultz, V. and A. W. Klement, Jr., 1963).

Seymour, A. H., E. E. Held, F. G. Lowman, J. R. Donaldson and D. J. South. 1956. Survey of the radioactivity in the sea and in pelagic marine life west of the Marshall Islands, Sept. 1–20, 1956. Univ. of Washington, U.S. AEC report UWFL-47. vi, 57 pp.

Slater, J. V. 1961. Comparative accumulation of radioactive zinc in young rainbow, cutthroat and brook trout. Copeia 2: 158–161.

Smales, A. A. and L. Salmon. 1955. Determination by radioactivation of small amounts of rubidium and caesium in sea-water and related materials of geochemical interest. The Analyst 80(946): 37–50.

Smith, J. M., Jr. 1959. Disposal of radioactive liquids from nuclear powered ships. Sewage and Industrial Wastes 31(11): 1323–1326.

Solberg, A. N. 1938. The susceptibility of *Fundulus heteroclitus* embryos to X-radiation. J. Exptl. Zool. 78(40): 441–469.

Spooner, G. M. 1949. Observations of the absorption of radioactive strontium and yttrium by marine algae. J. Marine Biol. Assoc. U.K. 28(3): 587–625.

Stehn, J. R. 1960. Table of radioactive nuclides. Nucleonics 18(11): 186–195.

Stoklasa, J. and J. Pĕnkava. 1932. *Biologie des Radiums und der Radioactiven Elemente*. Vol. 1, Biologie des Radiums and Uraniums. Verlag von Paul Parey, Berlin. xiv, 958 pp.

Sugihara, T. T., H. I. James, E. J. Troianello and V. T. Bowen. 1959. Radiochemical separation of fission products from large volumes of sea water. Anal. Chem. 31: 44–49.

Summers, D. L. and M. C. Gaske. 1961. Maximum permissible activity (MPA) for fission-products in air and water. Health Physics 4(3/4): 289–292.

Suyehiro, Y., S. Yoshino, Y. Tsukamoto, M. Akamatsu, K. Takahashi and T. Mori. 1956. Transmission and metabolism of strontium-90 in aquatic animals. In: *Research in the Effects and Influences of the Nuclear Bomb Test Explosions*, Vol. II. Japan Society for the Promotion of Science, Ueno, Tokyo. pp. 1135–1142.

Swan, E. F. 1956. The meaning of strontium-calcium ratios. Deep-sea Research 4(1): 71.

Swift, E. and W. R. Taylor. 1960. Uptake and release of calcium-45 by *Fucus vesiculosus* (abstr.). Biol. Bull. 119(2): 342.

Taya, N. 1956. On the capacity of water bacteria for adsorption of fission products. In: *Research in the Effects and Influences of the Nuclear Bomb Test Explosions*, Vol. II. Japan Society for the Promotion of Science, Ueno, Tokyo. pp. 1245–1250.

Taylor, W. R. and E. P. Odum. 1960. Uptake of iron-59 by marine benthic algae (abstr.). Biol. Bull. 119(2): 343.

Templeton, W. L. 1959. Fission products and aquatic organisms. In: *The Effects of Pollution on Living Material*. Inst. of Biology, London, Sept. 1958. No. 8. pp. 125–140.

Templeton, W. L. 1962. The transfer of radionuclides from the environment through aquatic food products to man. In: *Proc. Seminar on Agricultural and Public Health Aspects of Radioactive Contamination in Normal and Emergency Situations*. December 11–15, 1961, Scheveningen, Holland. FAO, Rome. pp. 49–72.

Templeton, W. L. 1965a. Ecological aspects of the disposal of radioactive wastes to the sea. In: *Ecology and the Industrial Society*. Fifth Symposium of the British Ecological Society. Blackwell Sci. Publ., Oxford, England. pp. 65–97.

Templeton, W. L. 1965b. Personal communication.

Templeton, W. L. and V. M. Brown. 1963. Accumulation of calcium and strontium by brown trout from waters in the United Kingdom. Nature 198(4876): 198–200.

Teresi, J. D. and C. L. Newcombe. 1961. Calculations of maximum permissible concentrations of radioactive fallout in water and air based upon military exposure criteria. Health Physics 4(3/4): 275–288.

Thomas, I. M. 1956. The accumulation of radioactive iodine by *Amphioxus*. J. Marine Biol. Assoc. U.K. 35(1): 203–210.

Timofeeff-Ressovsky, N. W. and K. G. Zimmer. 1947. *Biophysik*. Vol. 1. Das Trefferprinzip in der Biologie. Hirzel Verlag, Leipzig. 329 pp.

Tomiyama, T., S. Ishio and K. Kobayashi. 1956a. Absorption of dissolved ^{45}Ca by marine fishes. In: *Research in the Effects and Influences of the Nuclear Bomb Test Explosions*, Vol. II. Japan Society for the Promotion of Science, Ueno, Tokyo. pp. 1163–1167.

Tomiyama, T., S. Ishio and K. Kobayashi. 1956b. Absorption of dissolved ^{45}Ca by *Carassius auratus*. In: *Research in the Effects and Influences of the Nuclear Bomb Test Explosions*, Vol. II. Japan Society for the Promotion of Science, Ueno, Tokyo. pp. 1151–1156.

Tomiyama, T., S. Ishio and K. Kobayashi. 1956c. Absorption by *Carassius auratus* of ^{45}Ca contained in *Rhizodrilus limasus*. In: *Research in the Effects and Influences of the Nuclear Bomb Test Explosions*, Vol. II. Japan Society for the Promotion of Science, Ueno, Tokyo. pp. 1157–1162.

Tomiyama, T., K. Kobayashi and S. Ishio. 1956a. Excretion of absorbed ^{45}Ca by goldfish. In: *Research in the Effects and Influences of the Nuclear Bomb Test Explosions*, Vol. II. Japan Society for the Promotion of Science, Ueno, Tokyo. pp. 1169–1172.

Tomiyama, T., K. Kobayashi and S. Ishio. 1956b. Absorption of dissolved ^{45}Ca by *Rhizodrilus limasus*. In: *Research in the Effects and Influences of the Nuclear Bomb Test Explosions*, Vol. II. Japan Society for the Promotion of Science, Ueno, Tokyo. pp. 1173–1176.

Tomiyama, T., K. Kobayashi and S. Ishio. 1956c. Excretion of absorbed ^{45}Ca by *Rhizodrilus limasus*. In: *Research in the Effects and Influences of the Nuclear Bomb Test Explosions*, Vol. II. Japan Society for the Promotion of Science, Ueno, Tokyo. pp. 1177–1180.

Tomiyama, T., K. Kobayashi and S. Ishio. 1956d. Absorption of ^{90}Sr (^{90}Y) by carp. In: *Research in the Effects and Influences of the Nuclear Bomb Test Explosions*, Vol. II. Japan Society for the Promotion of Science, Ueno, Tokyo. pp. 1181–1187.

Tomiyama, T., K. Kobayashi and S. Ishio. 1956e. Distribution and excretion of intramuscularly administered ^{90}Sr (^{90}Y) in carp. In: *Research in the Effects and Influences of the Nuclear Bomb Test Explosions*, Vol. II. Japan Society for the Promotion of Science,

Ueno, Tokyo. pp. 1189–1193.

Tomiyama, T., K. Kobayashi and S. Ishio. 1956f. Absorption of $^{32}PO_4$ ion by carp. In: *Research in the Effects and Influences of the Nuclear Bomb Test Explosions*, Vol. II. Japan Society for the Promotion of Science, Ueno, Tokyo. pp. 1195–1200.

Tomiyama, T., K. Kobayashi and S. Ishio. 1956g. Distribution and excretion of intramuscularly administered $^{32}PO_4$ by carp. In: *Research in the Effects and Influences of the Nuclear Bomb Test Explosions*, Vol. II. Japan Society for the Promotion of Science, Ueno, Tokyo. pp. 1201–1203.

Townsley, S. J., D. F. Reid and W. T. Ego. 1960. Uptake of radioisotopes and their transfer through food chains by marine organisms. Annual report (1959–1960), Univ. of Hawaii contract AT(04-3)-56 with the U.S. AEC. 40 pp.

Townsley, S. J., D. F. Reid and W. T. Ego. 1961. The accumulation of radioactive isotopes by tropical marine organisms. Annual Report (1960–1961), Univ. of Hawaii contract AT(04-3)-235 with the U.S. AEC. 35 pp.

Tozawa, H., Y. Tokue and K. Amano. 1957. Studies on the radioactivity in certain pelagic fish. V. Confirmation and determination of radioactive elements in contaminated fishes caught in 1956 (Pt. 1). Bull. Japan Soc. Sci. Fish. 23: 335–340.

Trapnell, B. M. W. 1955. *Chemisorption*. Academic Press, New York. 265 pp.

Turner, R. C., J. M. Radley and W. V. Mayneord. 1958. The naturally occurring α-ray activity of foods. Health Physics 1(3): 268–275.

United Nations. 1962. Report of the United Nations Scientific Committee on the Effects of Atomic Radiation. General Assembly, Seventeenth Session, Supplement No. 16 (A/5216). New York. iv, 442 pp.

Vichney, N. 1960. Un problème non résolu: l'élimination des résidus radioactifs. La Nature 88(3297): 28–29.

Vinogradov, A. P. 1953. *The Elementary Chemical Composition of Marine Organisms*. Mem. Sears Found. Mar. Res. 2: 647 pp.

Vintemberger, P. 1931. Cited in 'The biological effects of short radiation' by C. Packard. Quart. Rev. Biol. 6: 253–280.

Wallauschek, E. and J. Lützen. 1964. Study of problems relating to radioactive waste disposal into the North Sea. II. General survey on radioactivity in sea water and marine organisms. Organisation for Economic Co-operation and Development, European Nuclear Energy Agency, Paris. 96 pp.

Wangersky, P. J. 1963. Manganese in ecology. In: *Radioecology*. pp. 499–508. (see: Schultz, V. and A. W. Klement, Jr. ,1963).

Way, K. and E. P. Wigner. 1948. The rate of decay of fission products. Phys. Rev. 73(11): 1318–1330.

Webb, D. A. 1937. Studies on the ultimate composition of biological material. II. Spectrographic analyses of marine invertebrates, with special reference to chemical composition of their environment. Roy. Dublin Soc. Sci. Proc. 21(45): 505–539.

Welander, A. D. 1954. Some effects of X-irradiation of different embryonic stages of the trout (*Salmo gairdnerii*). Growth 18(4): 227–255. (also: Univ. of Washington, U.S. AEC report UWFL-38).

Welander, A. D., L. R. Donaldson, R. F. Foster, K. Bonham and A. H. Seymour. 1948. The effects of roentgen rays on the embryos and larvae of the chinook salmon. Growth 12(3): 203–242. (also: Univ. of Washington, U.S. AEC report UWFL-8; MDDC-1689).

Welander, A. D., L. R. Donaldson, R. F. Foster, K. Bonham, A. H. Seymour and F. G.

Lowman. 1949. The effects of roentgen rays on adult rainbow trout. Univ. of Washington, U.S. AEC report UWFL-17; AECU-188. 17 pp.

Whittaker, R. H. 1953. Removal of radiphosphorus contaminant from the water in an aquarium community. Biology Research – Annual Report 1952. Hanford Atomic Products Operation, General Electric Co., U.S. AEC report HW-28636. pp. 14–19.

Wichterman, R. and F. H. J. Figge. 1954. Lethality and the biological effects of X-rays in *Paramecium*: radiation resistance and its variability. Biol. Bull. 106(2): 253–263.

Wilke-Dörfurt, E. 1928. The iodine content of mussel shells. A chemical contribution to the goiter problem. II. Biochem. Z. 192: 73–82.

Willard, W. K. 1960. Avian uptake of fission products from an area contaminated by low-level atomic wastes. Science 132 (3420): 148–150.

Williams, L. G. 1960. Uptake of caesium-136 by cells and detritus of *Euglena* and *Chlorella*. Limnol. Oceanogr. 5(3): 301–311.

Williams, L. G. and Q. Pickering. 1961. Direct and foodchain uptake of cesium-137 and strontium-85 in bluegill fingerlings. Ecology 42(1): 205–206.

Williams, L. G. and H. D. Swanson. 1958. Concentration of cesium-136 by algae. Science 127(3291): 187–188.

Wilson, S. H. and M. Fieldes. 1941. Studies in spectrographic analyses. II. Minor elements in a sea-weed (*Macrocystis pyrifera*). New Zealand J. Sci. Technol. 23(2B): 47–48.

Wiseman, J. D. H. 1955. Marine organisms and biochemistry. Nature 176 (4487): 818–819.

Wong, A.-C. and C.-Y. Wang. 1960. The effect of embryonic development of goldfish from exposing the parent fish to X-rays. Acta zool. sinica 12(1): 127–130.

Wyker, H. 1961. Some remarks on the maximum permissible concentration of unidentified radionuclides in water and related disposal formulae. Health Physics 4(3/4): 309–311.

Yamagata, N. 1959. Concentration of cesium-137 in the coastal waters of Japan (1959). Nature 184 (Suppl. 23): 1813–1814.

Yamagata, N. and S. Matsuda. 1959a. Cesium-137 in the coastal waters of Japan. Bull. Chem. Soc. Japan 32(5): 497–502.

Yamagata, N. and S. Matsuda. 1959b. Unpublished material presented at meetings of Subcommittee on Radioactivity in the Ocean, Special Committee on Oceanic Research, at Intern. Oceanogr. Congr., New York, 2 Sept. 1959. (cited from Anonymous, 1960b, page 88, table 3).

Yoshii, G., N. Watabe and Y. Okada. 1956. Biological decontamination of fission products. Science 124(3216): 320–321.

Young, E. G. and W. H. Langille. 1958. The occurrence of inorganic elements in marine algae of the Atlantic Provinces of Canada. Can. J. Bot. 36(3): 301–310.

Zimmer, K. G. 1961. *Studies on Quantitative Radiation Biology*. (translated into English by H. D. Griffith). Hafner Publ. Co. 124 pp.